HUANGPU RIVER
SUZHOU CREEK
THE WATERFRONT SPACE & ARCHITECTURE IN SHANGHAI

一江一河
上海城市滨水空间与建筑

上海市规划和自然资源局 编著

世纪出版集团 | 上海文化出版社

编委会
Editorial Committee

主　　任：	徐毅松
副 主 任：	许　健
主　　编：	胡晓忠　赵宝静　高文艳　魏珏欣
特约顾问：	郑时龄　汪孝安　唐玉恩　沈　迪　支文军
	章　明　吴　蔚　柳亦春　张　斌　李孔三
	张俊杰　汤朔宁　潘　琳　张　亮　朱祥明
	钱少华　沈果毅　刘　伟　汤明华　王　静
	刘　安　牟　娟　程　军　周建非　叶可央
	金卫红　黄立勋　彭　波　郭　海　王　滨
	刘夏夏　伍攀峰
综述执笔：	赵宝静　李　锴　沈　璐　奚东帆　沈　璇
	邹钧文　石　慧　徐晨炜　过甦茜　曹　韵
编辑统筹：	胡晓忠　高文艳　魏珏欣　周晓文　沈　瑾
	张颖倩　刘　芳　余诗阳　姚元佳　金　山
	王欣蕊　张懿文
翻　　译：	郑　超

Director of Editorial Board: Xu Yisong
Vice Director of Editorial Board: Xu Jian
Chief Editors:
Hu Xiaozhong, Zhao Baojing, Gao Wenyan, Wei Juexin
Advisors:
Zheng Shiling, Wang Xiaoan, Tang Yuen, Shen Di, Zhi Wenjun, Zhang Ming, Wu Wei, Liu Yichun, Zhang Bin, Li Kongsan, Zhang Junjie, Tang Shuoning, Pan Lin, Zhang Liang, Zhu Xiangming, Qian Shaohua, Shen Guoyi, Liu Wei, Tang Minghua, Wang Jing, Liu An, Mou Juan, Cheng Jun, Zhou Jianfei, Ye Keyang, Jin Weihong, Huang Lixun, Peng Bo, Guo Hai, Wang Bin, Liu Xiaxia, Wu Panfeng
Writers of Overview:
Zhao Baojing, Li Kai, Shen Lu, Xi Dongfan, Shen Xuan, Zou Junwen, Shi Hui, Xu Chenwei, Guo Suqian, Cao Yun
Editorial coordination:
Hu Xiaozhong, Gao Wenyan, Wei Juexin, Zhou Xiaowen, Shen Jin, Zhang Yingqian, Liu Fang, Yu Shiyang, Yao Yuanjia, Jin Shan, Wang Xinrui, Zhang Yiwen
Translator: Zheng Chao

序一
PREFACE 1

城市之光

郑时龄

中国科学院院士、法国建筑科学院院士
同济大学建筑与城市规划学院教授，博士生导师

　　城市是社会和文化的物质形态，是生活方式的反映。进入后工业社会后，我们正在重新认识城市，重新反思复杂的现代生活方式，重新改造城市的产业结构，重新思考城市的未来，重组城市的空间结构。习近平总书记在2019年视察杨浦滨江时作出"人民城市人民建，人民城市为人民"的指示，成为城市规划建设的根本指导思想。

　　今天的人民城市关注宜居，关注生态，关注自然，关注公共空间，关注公共艺术。自20世纪末以来，上海将视野投向黄浦江和苏州河的滨水空间，还之于民，让滨水空间成为人文内涵丰富的城市会客厅，让人们在这里生活、游憩、交往、观赏、冥想和创造。滨水空间带来浦东的发展，也带来中国2010年上海世博会，滨水空间正在焕发出无限的活力与魅力。

　　黄浦江和苏州河一江一河的再开发显示了上海作为国际文化大都市的理想和价值取向，也体现了城市的决心、经济实力和组织协调能力。一江一河的再开发需要全面协调规划、交通、市政设施、工程技术、环境保护、建筑、景观、公共艺术、建设、历史文化保护和精细化管理等领域，坚持"创新、协调、绿色、开放、共享"的发展理念，遵循"开放为本、生态为先、活力为主"的基本原则，着力将黄浦江和苏州河两岸打造成为服务于人民的世界级滨水公共开放空间，推进两岸城市更新及用地转型，建设生态廊道，贯通滨水空间，营造水陆景观，提升生态质量。

　　2017年上海实现黄浦江两岸45公里全线贯通，同时建设一系列公共文化设施，如文化公园、博物馆、美术馆、展览馆、文化中心、剧院、艺术中心、公共服务中心等。杨浦滨江段是上海滨水空间的"东大门"，通过发掘沿岸百年工业传承的文化特质，彰显世界一流滨水空间的魅力，杨浦滨江生活秀带在2020年被选为国家级文物保护利用示范区。

　　黄浦江滨水空间的定位是国际大都市发展能级的集中展示区和核心功能的空间载体、人文内涵丰富的城市公共客厅，以及具有宏观尺度价值的生态廊道。黄浦江的滨水岸线在市区蜿蜒流转60公里，连着广袤的腹地，哺育了沿岸数百平方公里的城区。千百年来黄浦江的岸线沿着老城厢、外滩向下游和上游拓展，孕育了中国的民族工业，发展了商贸和金融业，又塑造了陆家嘴、杨树浦、北外滩、南外滩、世博园、西岸、后滩和前滩等一个又一个独特的地区，高潮迭起，从最初的原生态、农耕时代迈向工业化，迈向世界商港，跃居全球城市，这一切都源自黄浦江。

　　苏州河经历了悠久的发展历程，在历史上不同时期承担起航运、产业和居住等功能，在这里孕育了中国最早的一批纺织、面粉、火柴、化工等民族工业。由于历史的原因，苏州河作为航道，象征着工业化进程，沿岸遍布工厂、仓库，但也成为"下水道"，变成城市的污染源。自1996年起，上海全面启动苏州河环境综合整治，实施苏州河环境综合整治工程，苏州河干流全部消除黑臭。2020年又实现苏州河21公里中心城区段滨水空间全线贯通，长风生态园、梦清园、苏河湾等地区以及一系列优秀历史建筑和工业遗存彰显了上海的城市之光。

　　上海在历史上位于水网地带，水是上海城市生命的源泉和文化资源。上海的滨水空间独树一帜，一江一河见证了城市的历史，上海控江踞海的地理位置，优越的港口条件使上海不断发展繁荣一跃而成全球城市。上海因水而更形卓越，因水而让人民更感幸福，一江一河已经成为有温度、有生活的人民共享空间。

两千多年前，孟子曾经高度赞颂水，将水看作是人的隐喻："源泉混混，不舍昼夜，盈科而后进，放乎四海"（《孟子·离娄下》），描述了有源头的泉水滚滚奔流，日夜不停，注满了洼地又向前奔腾，一直流向大海。水是现有的人类资源中历史最悠久的，河流给城市带来活力，提供交通，灌溉土地，滋养生命。河流自然流淌，沉淀杂质，荡涤万物，澄清自我。水在中国文化中，自古以来就是生命和智慧的象征，是知识的源泉，是理解人类行为的准则，水培育了城市文化和城市精神。水是维系人类生命不可或缺的物质，人类与城市伴随着水，伴随着对未来城市的理想一起成长。

自20世纪60年代起，为了重振城市发展的活力，世界上许多城市都将滨水空间作为城市后工业时代的发展核心，将工业化时代形成的滨水地区的港口、工厂、仓库、船坞、码头、堆场、站场、构筑物等工业遗存加以改造和再利用，将工业化时代为追随利润而塑造的工业污染环境转换为优美的公共空间。这既是挑战，也是机遇。每座城市都以自己的方式开发滨水空间，创造了一系列优秀的案例，一些名不见经传的城市也一跃进入世界城市的行列，一些案例成为城市设计的典范。

由于滨水空间的规模，以及所处城市中优越的地理位置，滨水空间从原先作为城市的边缘转而成为城市空间的核心，滨水空间的再开发不仅是城市繁荣的驱动力，也重塑了城市的形象，让水和水岸焕发出新时代的魅力和城市发展的无限潜力。

黄浦江、苏州河是上海建设国际文化大都市的代表性空间和标志性载体，规划站在新时代，面向2035，展望2050，以更高站位谋求规划升级，以更高标准优化建设管理，着力推动高质量发展，创造高品质生活。本书汇集了一江一河开发建设中的80多个优秀案例，介绍了人民城市的建设实践，集中展示了"发展为要、人民为本、生态为基、文化为魂"的指导思想，凝聚了城市管理人员、规划师、建筑师、景观建筑师、工程师、园艺师和建筑工人等长年累月的智慧和辛勤劳动，将一江一河打造成为具有全球影响力的世界级滨水区，未来的黄浦江沿岸将成为全球城市发展能级的集中展示区，苏州河沿岸将建成城市宜居生活的典型示范区，城市之光在这里闪耀。

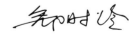

THE GLORY OF A CITY

Zheng Shiling

Academician, Chinese Academy of Sciences
Member of the French Academy of Architectural Sciences
Professor, Doctoral Supervisor
College of Architectureand Urban Planning, Tongji University

A city is an embodiment of society's culture, reflecting the lives of its people. In a post-industrial society, we perceive our city in a novel way – reflecting the complexity of modern life, reshaping its industrial structure, planning on its future, and reorganizing its spatial structure. On President Xi Jinping's visit to the Yangpu section of the Huangpu River waterway in 2019, he proposed that Shanghai should be a "people's city by the people and for the people"; this is the guiding principle of Shanghai's urban planning and construction.

Today's a "people's city" should focus on topics such as livability, ecology, nature, public space and public art. Since the end of the 20th century, Shanghai has been prioritizing utmost attention to the waterfront along the Huangpu River and the Suzhou Creek. A revitalization campaign was unveiled to provide waterfront access to all. The aim is to make the waterfront the living room of the city for all its citizens. Everyone can enter, play, and meditate here; everyone can savor its views and seek inspirations for respective endeavors. The waterfront has fueled the development of Pudong and made possible the 2010 Shanghai World Expo. The waterfront is brimming with infinite charm and vitality.

The redevelopment of the Huangpu River and Suzhou Creek demonstrates Shanghai's commitment to achieving an international metropolis status through economic tenacity and organizational capability. The redevelopment of the River and Creek requires the intricate coordination amongst planning, transportation, infrastructure, engineering technology, environmental protection, architecture, landscape, public art, construction, historical and cultural preservation, and refined management. By adhering to core values of "innovation, coordination, green, openness and sharing" and following the principle of "openness, ecology and vitality," significant effort is contributed to promote urban regeneration and land transformation on both banks of the Huangpu River and the Suzhou Creek. Such efforts include building ecological corridors, linking up separate waterfront sections, creating waterscape and landscape; and improving ecological quality.

In 2017, Shanghai made the banks on both sides of the Huangpu River open to the public. There is a wide array of cultural facilities, such as cultural theme parks, museums, art galleries, exhibition halls, cultural centers, theaters, art centers, public service centers and other facilities. The Yangpu section of the Huangpu waterfront is the "eastern gate" of Shanghai's waterfront, exploring the cultural qualities of the centuries-old industrial heritage along the riverfront and highlighting the charm of the world-class waterfront. The Yangpu Waterfront Lifestyle Showcase was nominated to be a national heritage protection and utilization zone in 2020.

The Huangpu River waterfront is planned to be a major showcase of both Shanghai's development capabilities and its core functions. It is a cultural treasure trove and an ecological corridor that testifies the city's values. The waterfront of the Huangpu River winds through the city in a 60-kilometer-course, connecting various parts of the city and supporting an expansive hinterland. For hundreds of years, the Huangpu River's waterfront has been extending downstream and upstream from the old Shanghai town and the Bund. It has nurtured China's native industrial, commercial and financial sectors. It has built the reputation of such localities as Lujiazui, Yangshupu, North Bund, South Bund, Expo Park, West Bund, Houtan and Qiantan; smoothly transitioning its development from the agrarian era to an industrial one. All these success can be attributed to the Huangpu River.

The Suzhou Creek has also undergone a long development process. It played a major role as a navigation channel, and a base for industrial and residential establishments. It is a cradle of some of China's earliest native industrial enterprises, such as textile and flour mills, match factories and chemical works. For historical reasons, the Suzhou Creek had waterway properties thereby symbolizing industrialization and was flanked with factories and warehouses. It also became a drain and a source of pollution. In 1996, Shanghai launched a comprehensive environmental improvement project to decontaminate it. Its water is no longer black or odorous. The revitalization campaign of its 21 km waterfront in the city center was completed in 2020, and a range of historical buildings and industrial heritage sites, such as the Changfeng Ecological Park, Mengqing Garden, Suhe Bay, represent the city's glory.

Historically, Shanghai had an intricate waterway network: water provided the source of the city's vibrancy and shaped the city's culture. Shanghai's waterfront is unique: The River and Creek have witnessed the city's evolution. Due to Shanghai's advantageous geographical location at the mouth of the Yangtze River, this has forged its global position and presence. Rivers have brought prosperity to Shanghai and happiness to its people. Shanghai's waterfront has become cherished civic spaces.

More than 2,000 years ago, Mencius once highly praised river by comparing it to an aspiring person: "It thrusts ahead day and night, filling up all the lying land. It never loses its momentum until it meets the sea." Rivers bring vitality to cities; they can facilitate transportation, irrigate land, and nourish life. In a river's natural course, it can precipitate impurities hence cleansing everything and itself. Water has been a symbol of life and wisdom in Chinese culture since ancient times, a source of knowledge that helped understand human behavior. It has nurtured urban culture and spirit. Water is an indispensable substance that sustains human life. Humans and cities advance concurrently with the support of water, advancing towards the ideal future city.

Since the 1960s, many cities around the world have made waterfront the core space for urban development in the post-industrial era for the purpose of urban development. They have converted and reused waterfront industrial heritage sites such as harbors, factories, warehouses, wharves, yards that took shape during the industrialization era, converting the pollution-inducing yet profit-making facilities in the industrial era into scenic spots. This is both a challenge and an opportunity. Each city has developed its waterfront in its own way, and many successful cases have emerged. Some lesser-known cities have risen to fame, while others have become classic examples of urban design.

Due to the scale of waterfronts and their advantageous locations in their respective cities, waterfront spaces are no longer absent of public attention; they have become much sought-after urban spaces. Waterfront redevelopment programs are not only a driving force for urban prosperity, but have also reshaped the images of cities, providing charm to the waterways and waterfront in a new era.

The Huangpu River and Suzhou Creek are iconic landmarks of Shanghai as an international metropolis. To envision its future in 2035 and 2050, we have high aims to upgrade our planning, apply construction management at higher standards, carry forward quality development and create high-quality life. This book brings together more than 80 cases; they have witnessed the wisdom and hard work of urban administrators, planners, architects, landscape architects, engineers, horticulturists and construction workers, etc. Their commitments have contributed to the new imagery of the River and the Creek and the glory of this city.

序二
PREFACE 2

让上海因水而更加卓越

徐毅松
上海市规划和自然资源局局长

黄浦江、苏州河（以下统称"一江一河"）是上海重要的发展轴和标志性空间，其沿岸变迁是上海城市发展历程的缩影。立足"两个一百年"历史交汇的新发展阶段，一江一河沿岸作为上海未来发展不可或缺的宝贵资源，不仅是强化城市核心功能、提升城市软实力的重要空间载体，更是践行"人民城市"重要理念的示范区域。

一江一河沿岸地区以建设具有全球影响力的世界级滨水区为目标，贯彻落实创新、协调、绿色、开放、共享的新发展理念，坚持"以发展为要、以人民为本、以生态为基、以文化为魂"，明确将一江一河沿岸地区打造为上海全球城市建设重要的发展纽带、高品质中央活动区功能的承载区、标志性的城市空间、重要的生态廊道，着力推进三个方面：

一是坚持着眼长远强化规划引领，对标全球城市定位，实现以创新为动力的能级提升。提升产业能级和集聚度，植入激发区域创新活力的新兴功能，促进多元复合、错位互补。推动公共功能进一步从滨水向腹地延伸，从重点地区向一般地区辐射，带动区域整体发展。深入挖掘人文内涵，打造具有国际竞争力的文化、旅游功能。二是坚持以人为本强化空间品质，对标世界级滨水区要求，打造城市客厅和景观标志。构建滨水贯通并向腹地延伸的公共游憩网络，完善公共服务配套设施。塑造具有韵律感、层次感的滨江天际线，打造具有标识性、辨识度的节点空间。探索历史风貌资源多元化利用方式，展现历史底蕴和文化魅力。三是坚持生态优先强化绿色发展，对标先进生态理念，体现人与自然和谐共生。以生态安全为底线，重视一江一河重大生态廊道的定位，构建水绿系统交融、更加稳固、安全的生态格局。

随着《黄浦江沿岸地区建设规划（2018—2035年）》《苏州河沿岸地区建设规划（2018—2035年）》和《关于提升黄浦江、苏州河沿岸地区规划建设工作的指导意见》的出台实施，一江一河滨水岸线实现基本贯通，沿岸开发建设成效显著，市民的获得感、幸福感不断增强。杨浦滨江、虹口滨江、黄浦九子公园等滨水空间得到重塑，杨浦绿之丘、浦东上海国际时尚中心、静安四行仓库等历史建筑得到激活，徐汇西岸美术馆、上海市少年儿童图书馆新馆等多元建筑不断创新，实现了将最好的滨水岸线资源留给市民，保护和活化利用历史文化资源，打造滨水公共空间的标志性节点，一江一河沿岸地区成为集聚人气和活力的"生活秀带""发展秀带"。这些阶段性成果精雕细琢、寸寸用心，凝聚着广大城市规划管理者、建设者、设计师、市民和社会等方方面面的智慧和力量，全面开启了上海建设具有全球影响力的世界级滨水区的美好篇章。

上海因水而生，因水而兴。我们将按照市委、市政府工作部署，继续坚持高起点规划、高标准建设、高水平管理，以更高标准、更宽视野、更大格局推进一江一河沿岸规划建设，令上海的母亲河散发出更加耀眼的光芒。未来，期待上海因水而更加卓越！

徐毅松

CREATE A BETTER SHANGHAI WITH ITS WATERWAYS
Xu Yisong
Director of Shanghai Municipal Bureau of Planning and Natural Resources

The Huangpu River and Suzhou Creek are important development axes of Shanghai, and their changes provided a cross-section of Shanghai's urban evolution. In the new development stage, the waterfront of the River and the Creek is indispensable resource for Shanghai's future development. It is not only an important space for strengthening the core functions of the city and enhancing its soft power, but also exemplified the concept of "People's City."

With the goal of building a world-class waterfront area with global influence, implementing the new development concept of innovation, coordination, green, openness and sharing, and insisting on the fundamental importance of "development, people, ecology and culture". The waterfront of the River and the Creek will be built into a powerhouse to make Shanghai a global city, a bearing area for the functions of a high-quality central activity district, an iconic urban space, and an important ecological corridor.

The revitalization campaign focuses on three aspects. First, it should be led by a long-term plan, to benchmark global cities, and realize capacity enhancement driven by innovation. We will pay attention to enhance the industrial capacity and industry concentration, assign new functions to stimulate regional innovation and vitality, and promote diversified elements that are mutually complementary. We will also extend public functions from the waterfront to the rest urban areas and radiate them from key areas to general areas, thus driving the overall development of the region. Cultural resources will be explored to create internationally competitive cultural and tourism functions. Second, we will build high quality of spaces on the basis of people-oriented-ness. We will pursue high standard for world-class waterfront to create a civic space. We will build a recreation space grid that extends from the waterfront to the rest urban areas and introduce public service facilities. Our objective is to make a waterfront skyline with a sense of rhythm and hierarchy, and create recognizable facilities. We will also explore the diversified use of historical landscape resources to celebrate historical and cultural heritage. Third, we will adhere to the policy to give top priority to ecology to realize green development. This requires us to adopt advanced ecological concepts so as to reflect the harmonious coexistence of man and nature. With ecological security as the bottom line of our effort, we will make the waterfront a major ecological corridor, building a solid and secure ecological system.

Since the announcement of "The Waterfront Development Plan for the Huangpu River and Suzhou Creek (2018-2035)" and "Guidelines for the Planning and Construction of the Waterfront of the Huangpu River and Suzhou Creek", the waterfront of both the River and the Creek has been opened to the public, and the public's sense of benefit and happiness is constantly enhanced. Many spaces, e.g., the Yangpu and Hongkou sections of the Huangpu River waterfront and the Jiuzi Park in the Huangpu Park, have been reshaped. Historic buildings, such as the Green Hill, Pudong Shanghai International Fashion Center, and the Joint Warehouse have been renovated. The Westbund Museum in Xuhui and the new Shanghai Children's Library are all hailed as fine examples of architectural innovation. These projects have realized the goal to reserve the best waterfront resources to the public, to protect and use historical and cultural resources, to create iconic facilities on the waterfront, and to make the waterfront a showcase of lifestyle. These milestones are meticulously achieved with the wisdom of urban planning managers, builders, designers, citizens and the society at large, opening a new chapter in the construction of Shanghai's world-class waterfront.

Shanghai is a city that thrives on the water. We will continue to adhere to the plan of the municipal government, and pursue high standard for the planning and redevelopment, so that Shanghai's mother rivers will shine more brightly. And better yet, we expect Shanghai can be brilliant because of its waterways!

目录 CONTENTS

02 序一
PREFACE 1

06 序二
PREFACE 2

12 综述
OVERVIEW

34 **黄浦江篇**
HUANGPU RIVER WATERFRONT

37 **滨水空间重塑**
REVITALIZATION OF WATERFRONT

涤岸之兴，向水而生
LIVE AND THRIVE BY THE WATER — 设计·师说 MASTERS' WORDS

42 杨浦滨江公共空间
YANGPU SECTION OF HUANGPU WATERFRONT

60 虹口滨江公共空间
HONGKOU SECTION OF HUANGPU WATERFRONT

62 黄浦滨江公共空间
HUANGPU SECTION OF HUANGPU WATERFRONT

68 徐汇滨江公共空间
XUHUI SECTION OF HUANGPU WATERFRONT

74 浦东滨江公共空间
PUDONG SECTION OF HUANGPU WATERFRONT

92 宝山滨江公共空间
BAOSHAN YANGTZE WATERFRONT

98 闵行滨江公共空间
MINHANG SECTION OF HUANGPU WATERFRONT

101 **历史建筑激活**
RENOVATION OF HISTORICAL BUILDINGS

黄浦江畔 工业历史的当代风景
HUANGPU WATERFRONT: A LIVING
INCARNATION OF SHANGHAI'S INDUSTRIAL PAST — 设计·师说 MASTERS' WORDS

106 上海国际时尚中心
SHANGHAI INTERNATIONAL FASHION CENTER

110 灰仓美术馆
ASH GALLERY

114 绿之丘
GREEN HILL

118 世界技能博物馆
WORLD SKILLS MUSEUM

120 明华糖厂
MING HUA SUGAR REFINERY

124 世界会客厅
THE GRAND HALLS

128 外滩公共服务中心
BUND PUBLIC SERVICE CENTER

130 上海世博会城市最佳实践区
URBAN BEST PRACTICES AREA OF SHANGHAI WORLD EXPO

134 上海当代艺术博物馆
POWER STATION OF ART

138	余德耀美术馆 Yuz Museum	176	外滩 SOHO Bund SOHO
142	星美术馆 Start Museum	180	外滩金融中心 The Bund Finance Center
144	龙美术馆西岸馆 Long Museum West Bund	184	董家渡金融商业中心 Dongjiadu Financial Center
148	油罐艺术公园 Tank Shanghai	188	世博会博物馆 World Expo Museum
152	民生码头八万吨筒仓 80,000-ton Silos at Minsheng Wharf	192	西岸美术馆 West Bund Museum
156	上海浦东船厂 1862 Shanghai Pudong Shipyard 1862	196	西岸智慧谷 West Bund AI Valley
160	上海艺仓美术馆及长廊 Shanghai Modern Art Museum and Its Walkways	202	西岸传媒港 West Bund Media Center
164	上海宋城·世博大舞台 Shanghai Songcheng · World Expo Grand Stage	208	浦东美术馆 Museum of Art Pudong

人类与水的共存
Regeneration of Waterfront Built Environment *设计·师说 Masters' Words*

212	上海中心大厦 Shanghai Tower
216	上海 JW 万豪侯爵酒店 Shanghai JW Marriott Marquis Hotel
220	新开发银行总部大楼 New Development Bank Headquarters
224	上海大歌剧院 Shanghai Grand Opera House
228	上海久事国际马术中心 Shanghai Juss International Equestrian Center
232	前滩国际商务区 New Bund International Business District

169 多元融合创新
INNOVATION OF NEW BUILDINGS

新建筑开发赋能
Forging a New Trail In Architectural Development *设计·师说 Masters' Words*

172 上海白玉兰广场
Sinar Mas Plaza

250 上海长滩
SHANGHAI LONG BEACH

传承 延续
INHERITANCE AND CONTINUATION
设计·师说 MASTERS' WORDS

257 活力水岸连接
DYNAMIC WATERFRONT CONNECTION

为公共性赋形：
景观基础设施与浦江两岸贯通
BUILT FOR THE PUBLIC: THE LANDSCAPE
INFRASTRUCTURE AND THE HUANGPU RIVER WATERFRONT
REVITALIZATION
设计·师说 MASTERS' WORDS

262 日晖港步行桥
FOOTBRIDGE ON RIHUI CREEK

264 游客集散中心
TOURIST DISTRIBUTION CENTER

268 龙华港桥
LONGHUA CREEK BRIDGE

272 洋泾港步行桥
YANGJING CREEK PEDESTRIAN BRIDGE

276 民生轮渡站
MINSHENG FERRY STATION

280 东昌栈桥
DONGCHANG ELEVATED PASSAGE

282 白莲泾 M2 游船码头
BAILIANJING M2 TOURIST TERMINAL

286 倪家浜桥
NIJIABANG BRIDGE

舒适人居，趣味城市
MAKING CITIES FUN, GIVING PEOPLE COMFORT
设计·师说 MASTERS' WORDS

292 苏州河篇
SUZHOU CREEK

295 滨水空间重塑
REVITALIZATION OF WATERFRONT

296 虹口区北苏州路滨河空间
NORTH SUZHOU ROAD WATERFRONT

300 介亭
PAVILLION INBETWEEN

304 中石化一号加油站
SINOPEC NO. 1 GAS STATION

306 飞鸟亭
FLYING BIRD PAVILION

310 九子公园
JIUZI PARK

314 蝴蝶湾花园
BUTTERFLY BEACH PARK

316 华东政法大学长宁校区
CHANGNING CAMPUS OF EAST CHINA UNIVERSITY
OF POLITICAL SCIENCE AND LAW

320 天原河滨公园
TIANYUAN RIVERSIDE PARK

322 临空滑板公园
AIRPORT SKATEPARK

325 历史建筑激活
RENOVATION OF HISTORICAL BUILDINGS

苏州河的新生
THE REBIRTH OF THE SUZHOU CREEK 　　*设计·师说 MASTERS' WORDS*

328 外滩源 33#
BUND 33

330 洛克·外滩源
ROCKBUND

332 华侨城苏河湾
OCT SUHE CREEK

336 四行仓库
JOINT TRUST WAREHOUSE

341 多元融合创新
INNOVATION OF NEW BUILDINGS

342 海鸥饭店
SEAGULL HOTEL

344 中美信托金融大厦
SINO-AMERICAN T. & F. TOWER

348 上海市少年儿童图书馆新馆
NEW SITE OF SHANGHAI CHILDREN'S LIBRARY

352 江森自控亚太总部大楼
JOHNSON CONTROLS ASIA-PACIFIC HEADQUARTERS BUILDING

357 活力水岸连接
DYNAMIC WATERFRONT CONNECTION

358 昌平路桥
CHANGPING ROAD BRIDGE

360 苏州河桥下空间
SPACE UNDER SUZHOU CREEK BRIDGE

从标志性到公共性与日常性：
上海城市空间转型的新维度 　　*设计·师说 MASTERS' WORDS*
FROM ICONIC LANDMARKS TO PUBLIC SPACE
FOR EVERYDAY USE: A NEW DIMENSION OF SHANGHAI'S URBAN SPACE
TRANSFORMATION

376 附录
APPENDIXES

376 设计·师说作者介绍
THE AUTHORS OF MASTERS' WORDS

377 参编单位
INSTITUTIONS INVOLVED

377 摄影师
PHOTOGRAPHERS

378 后记
POSTSCRIPT

综述
OVERVIEW

黄浦江与苏州河宛如两条蜿蜒的玉带，分别呈南北与东西向流经上海市域境内。一江一河的变迁是上海这座城市发展历程的缩影，与上海的发展紧密相连，是上海建设"卓越全球城市"的代表性空间和标志性载体。

2019年11月，习近平总书记在视察杨浦滨江的时候，提出"人民城市人民建、人民城市为人民"的重要理念。一江一河的发展正是深入贯彻"人民城市"理念和"上海2035"总体规划，体现创新、协调、绿色、开放、共享，从而建设具有全球影响力的世界级滨水区，成为城市魅力的名片、城市活力的项链和城市实力的塔尖。

Like two winding ribbons, the Huangpu River and the Suzhou Creek, flow north-south and east-west respectively across Shanghai. Their sweeping changes highlight Shanghai's evolution and are closely linked to the city's development. The waterfront proves to be a representative space for making Shanghai a "global city of excellence".

In November 2019, during his inspection of the Yangpu section of the Huangpu River waterfront, Chinese President Xi Jinping proposed to make Shanghai "a people's city built by the people and for the people." The development of the River and the Creek reflects the effort to implement this grand mission and the Shanghai Master Plan 2035. Embracing innovation, coordination, green, openness and sharing, the revitalization will build a superb, world-class waterfront with global influence.

一江一河全景图
The estuary of the Suzhou Creek

历史演变
Intriguing History

上海因水而生，依水而兴。

19世纪中叶前，上海就凭借江海交汇的优越地理位置，挟黄浦江水运之利和江南富庶之地，逐步形成发达的航运业，实现了从农耕社会的邑城向商业繁荣的港口城市的转变。当时，黄浦江已开辟有沿海、长江、内河等多条航线，根据《上海县志》记载："闽、广、辽、沈之货，鳞萃羽集，远及西洋暹罗之舟，岁亦间至。"上海县城厢沿江一带舟楫林立，商肆鼎沸，正如清嘉庆时诗人施润所描绘的"一城烟火半东南"的繁荣景象，赫然已是东南大港和著名商邑，享有"江南通津，东南都会"的声誉。

近代上海的起点

1843年上海开埠，成为中国主要的对外口岸之一，拉开了近代上海城市发展的帷幕。开埠后的上海进出口贸易遽增，外资银行纷纷入驻外滩一带，外滩成了最早的"洋场"，各国商行争先在这里兴建楼房。1865年，上海对外贸易总额，占全国贸易总额的53%，居国内首位。而洋泾浜（今延安路）到虹口港的滨江之地，先后有英国、美国、德国、日本、俄国的领事馆建驻。同时，伴随着19世纪西方工业文明引入，造船、纺织、发电厂、水厂等近代工业在黄浦江畔相继兴起，逐渐形成杨树浦和沪南两大滨水工业区，成为中国近代工业的发源地。至1937年，杨树浦工业区民族工业301家，外商工厂57家，成为中国近代重要的工业基地。据统计，中国近代工业上海约占70%，而位于黄浦江畔的杨浦约占上海企业比重的三分之一。贸易和工商业的繁荣进一步促进了黄浦江航运功能的快速发展，江中千帆竞发，江畔实业兴盛，带动了城市的发展，奠定了中心城区的空间格局。至20世纪30年代，上海已成为闻名遐迩的远东大都会，而黄浦江沿线可以说是这个大都会的核心空间。

开埠后的苏州河，水路运输便捷，承担了重要的航运功能，沿岸密布知名民族企业，涉及纺织、粮食加工、机械制造等诸多领域，苏州河也因此被称为近代上海民族工业的发祥地。工业的繁荣带来人口集聚、沿河两岸往来交通需求激增，如今苏州河上的外白渡桥、西藏路桥、恒丰路桥等正是建于这一时期。20世纪20至30年代，苏州河自东向西的发展序列，也体现了上海空间发展的脉络和那个时期的城市形态：从外白渡桥到河南路桥是高级商住和公建区，而河南路桥至恒丰路桥沿河主要聚集一些码头仓库以及石库门住宅。

Shanghai is a city that thrives on the water.

Before the mid-19th century, with its advantageous geographical location at the estuary of the Yangtze River, Shanghai, saw a flourishing shipping industry and realized the transformation from a small, agricultural town to a port city with thriving commerce. At that time, a number of coastal and hinterland-bound navigation routes were already available starting from the Huangpu River. *A History of Shanghai County* extols: "Shanghai is the gathering place of commodities from Fujian, Guangdong, and Liaoning provinces, and the port is crowded with ships. West-bound and Thailand-bound ships can also be found a few times each year." The waterfront of the Shanghai County was thronged with masts and shops. Shi Run, a poet in the early 18th century described that the bustling scene of Shanghai is almost unrivaled in Southeast China, and it was the foremost commercial hub in the region.

The cradle of modern Shanghai

In 1843, Shanghai was opened as one of the major Treaty Ports of China, and that marks the beginning of the development of Shanghai in that era. In 1865, Shanghai's foreign trade volume accounted for 53% of the country's total trade, ranking first in China. The waterfront between the Yangjingbang Creek (today's East Yan'an Road) and the Hongkou Creek was also home to British, American, German, Japanese and Russian consulates. At the same time, along with the introduction of Western industrial civilization in the 19th century, industries such as shipbuilding, textiles, power plants and waterworks sprang up along the Huangpu River, gradually forming two waterfront industrial zones, one in Yangshupu and the other in the south of the city, which became the birthplace of modern Chinese industry. By 1937, there were 301 native industrial enterprises and 57 foreign ones in the Yangshupu Industrial Zone. The latter became an important industrial base of post-Opium War China. According to statistics, Shanghai accounted for about 70% of China's modern industrial enterprises. Yangpu District on the northern bank of the Huangpu River is home to about one-third of industrial enterprises in Shanghai. The thriving trade and industry and commerce led to the rapid development of the navigation function of the Huangpu River. That fueled the development of the city and defined the spatial pattern of the central city area. By the 1930s, Shanghai had become a well-known metropolis in the Far East, and the Huangpu River was the core space of this metropolis.

After the opening of Shanghai as a Treaty Port, the Suzhou Creek became an important navigation channel due to its convenient traffic condition. Many famous native-run industrial enterprises were located along the Creek, involving many sectors such as textile, grain processing, and machinery manufacturing. The Suzhou Creek is dubbed the birthplace of Chinese native industrial sector in modern Shanghai. The flowering industrial sector led to a surging population and a rise in demand for traffic infrastructure along the Creek. The present-day Waibaidu Bridge (Garden Bridge), Xizang Road Bridge and Hengfeng Road Bridge across the Suzhou Creek were all built during this period. The waterfront from the Waibaidu Bridge to the Henan Road Bridge, was lined with upscale commercial, residential and public buildings, while the section between the Henan Road Bridge and the Hengfeng Road Bridge were mainly occupied by docks, warehouses and shikumen residences ("stone-framed door", a dominant type of lane complex in the early 20th century in Shanghai).

1922年上海租界地图
Map of Shanghai's International Settlement and French Concession dating from 1922

原件藏于澳大利亚图书馆，由上海工部局制作。地图详细标示了上海各街道、码头、铁路、商业公司和重要建筑物的名字和位置，不同的颜色表示公共租界四个区域和法租界的区域。地图也展示了20世纪20年代上海沿一江一河发展的格局。
The original copy of the map is kept in an Australian library. Produced by the Shanghai Municipal Council, the map shows the names and locations of Shanghai's streets, docks, railways, commercial companies and important buildings. It uses different colors to indicate the four sections of the International Settlement and the area of the French Concession. The map also shows Shanghai's development in the 1920s was mostly driven by facilities along the River and the Creek.

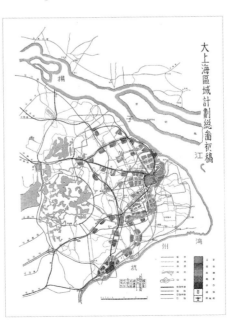

1947—1949年大上海都市计划
1947-1949, The Greater Shanghai Metropolitan Plan

1947-1949年，大上海都市计划指出，上海为一港埠城市，亦将成为全国最大工商业中心之一，并预测至1996年，上海人口将增至1500万人。城市的发展主要沿黄浦江，呈现南北向延伸的特点
From 1947 to 1949, The Greater Shanghai Metropolitan Plan pointed out that Shanghai is a port city and will become one of the largest industrial and commercial centers in China. It is predicted that by 1996, the population of Shanghai will increase to 15 million. The development of the city is mainly along the Huangpu River, showing the characteristics of extending from north to south.

社会主义建设的窗口

1949年以后，社会主义建设开始启动，随着中央的工业战略布局调整，上海在建设新中国工业体系、加速实现国家的工业化中承担了重要角色。虽是历经战火，百废待兴，但上海仍旧有着较好的经济基础和发展优势。当时，上海的工业产值虽有下降，但仍占全国的四分之一，其中沿一江一河布局的轻工业尤其发达，上海得天独厚的区位水路条件使其成为国内外贸易的重要枢纽。改革开放后，两岸的港口和老工业基地在经历了结构性调整后，迎来了鼎盛发展期。

20世纪80年代，上海市委、市政府逐步形成了以发展"第三产业"为核心的新思路，提出了"三二一"产业发展方针，而一江一河沿线的工业用地也随着发展浪潮迎来了更新和升级。20世纪90年代初浦东开发开放，陆家嘴金融贸易区迅速崛起，外滩对岸矗立起座座高楼，上海启动了跨黄浦江两岸的联动开发，并逐步走向国际大都市。黄浦江作为海派文化的原点，聚集了源源不断的资金、技术、人才和文化艺术，聚拢起上海的活力和潜力，构成海派文化开放性、国际性的精神传统。

苏州河沿线工业企业经历改造新生，随着工业的高速发展和人民日益增长的生活和物质需求，逐步显现出了居住卫生、水体污染等一系列环境和社会问题。1980年到1990年间，苏州河治理工作逐步开展。1996年，上海全面启动苏州河环境综合整治工程。持续开展的大规模旧区改造也进一步推动了苏州河沿岸景观风貌的提升。经过多年努力，至2000年，苏州河的干流水水质指标年平均值基本达到国家景观水标准，沿线逐渐出现绿地和广场，市民开始可以接近滨水地区。总体环境的改善与功能的置换和提升相结合，以新的城市公共功能替代衰退的产业，用环境、文化、景观形象见证上海新的经济增长模式。

进入21世纪，中央明确了上海未来发展新的战略目标，同时也对一江一河两岸地区的发展提出了新的课题。

A window on socialist construction

After 1949, socialist construction began in earnest. With the adjustment of the central government's industrial strategy, Shanghai assumed an important role in building the new Chinese industrial system and accelerating the country's industrialization. Although decades of warfare took a severe toll on Shanghai, the city still had a good economic foundation and development advantages. At that time, Shanghai's industrial output has declined, but still accounted for a quarter of the total of China. The light industry whose facilities were located along the River and the Creek is particularly sophisticated. Shanghai's unique river traffic conditions make the city an important hub for domestic and foreign trade. Since China embraced reforms and opening up, the wharves and old industrial bases on both sides of the river ushered in a new phase of development after their structural adjustment.

In the 1980s, the Shanghai Municipal Government developed a new growth strategy, giving impetus to the tertiary industry. This heralded the upgrading of the industrial land along the River and the Creek. The development of Pudong in the early 1990s, the rapid rise of the Lujiazui Financial and Trade Zone, the Bund and the mushrooming high-rise buildings proved to be a breath-taking scene. Shanghai launched a synergetic development program on both sides of the Huangpu River. That helped the city muscle its way towards an international metropolis. The Huangpu River, as the origin of Shanghainese culture, witnessed a continuous influx of capital, technology, talents, culture and art; it highlights the vitality and potential of Shanghai, constituting the open and international spirit typical of Shanghainese culture.

The enterprises along the Suzhou Creek underwent a heated wave of industrial development. The public's material needs also ballooned. However, these triggered a series of social and environment problems, escalating the issues of housing shortage and water pollution. The 1980s-90s saw the execution of a water treatment campaign. In 1996, Shanghai launched a holistic environmental improvement program for the Suzhou Creek. The sweeping campaign of renovation of old neighborhoods improved the landscape along the Suzhou Creek.

Years of efforts eventually paid off. By 2000, the average annual water quality index of the main stream of the Suzhou Creek had basically reached the national standard, green spaces and squares had appeared on its bank, and citizens had gained access to the waterfront. The improvement of the environment coincided with the upgrading of functions. New industries have replaced declining industries. The new environment, cultural events and landscape have proved the effectiveness of Shanghai's new model of economic growth.

At the beginning of the 21st century, the central government of China announced the new strategic goals for Shanghai's future development. It also raised new requirements on the development of Shanghai's waterfront.

1973年上海市土地使用综合现状图
Comprehensive map of land use in Shanghai in 1973

从1973年上海市土地使用综合现状图中可以看到：作为全国工业中心的上海，聚集着大量的工业用地，并且很大一部分是沿着一江一河两岸布局的，尤其是黄浦江的南北两端（今杨浦、虹口、徐汇和浦东新区），以及苏州河西端（今普陀和新静安区段）。

From the comprehensive land use map of Shanghai in 1973, we can see that Shanghai, as the industrial center of the country, has a large swath of land for industrial use, and a large part of it was along both sides of the River and the Creek, especially at the northern and southern stretches of the Huangpu River (in today's Yangpu, Hongkou, Xuhui districts and Pudong New Area), and the west stretch of the Suzhou Creek (in today's Putuo and New Jing'an District sections).

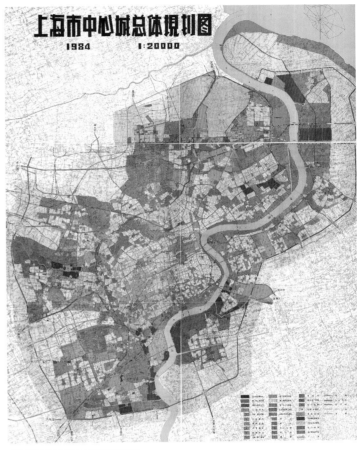

1984年上海市中心城总体规划图
Shanghai Central City Area Master Plan in 1984

进入80年代后，上海社会经济发展以"调整经济结构和振兴上海经济"为主，工业出现由中心城区向外扩散的趋势。1984年上海市中心城总体规划也体现了一江一河城市中心区段的土地更新策略，为后来的浦东开发开放、陆家嘴金融贸易区的崛起，以及上海跨黄浦江两岸的联动开发奠定基础。

After 1980, Shanghai's socio-economic development focused on "adjusting the economic structure and revitalizing Shanghai's economy". It became a tendency for industrial facilities to spread out from the central city to the outskirts. The Central City Area Master Plan 1984 also embodied the land renewal strategy of the central section of the city with one river. This laid the foundation for the later development and opening up of Pudong, the rise of Lujiazui Financial and Trade Zone, and the joint development of Shanghai across the Huangpu River.

人民城市的客厅

2002年，上海市政府启动黄浦江两岸综合开发，编制《黄浦江两岸地区规划优化方案》，并提出"人民之江"的发展目标。综合开发的核心思想就是还水于民，还岸于民，使其成为城市文化、社会和市民生活的中心，成为全体市民的资产。黄浦江的综合开发始于21世纪，从2002年到2007年，主要工作为释放沿江的土地和岸线空间，为今后新功能的引入做前期准备，动迁沿江大量工业厂房和仓储运输单位；从2008年到2010年，主要通过加速改造滨江环境和推进基础配套设施建设，为黄浦江畔世博会的举办以及滨江日后的综合开发提供环境设施配套；2011年到2015年，以开展功能项目建设为主，在沿江地区引入新功能。

同年，针对苏州河沿岸高楼林立、公共空间不足的问题，基于上海"双增双减"的发展要求，《苏州河滨河景观规划》提出要充分发挥苏州河在社会发展和环境建设中的作用，进一步挖掘苏州河两岸自然与人文历史景观，建成水质清洁、环境优美、气氛和谐的生活休闲区域。经过市政府的不懈努力，苏州河的运输功能被休闲、游览功能所取代，滨河景观逐步成为市民、游客驻足和流连忘返的城市公共空间。

2016年市委市政府提出，实现杨浦大桥至徐浦大桥之间的黄浦江核心段两岸45公里岸线贯通，打造健身休闲、观光旅游滨水空间。这项工程在2017年基本建设完成，引起热烈的社会反响。与此同时，2020年底，苏州河沿岸42公里滨水空间也基本贯通开放，沿河各区段特色鲜明、景观优美，各类观景平台、滨水公园、体育设施等活动节点，实现人民客厅的功能。

2019年1月批复的《黄浦江、苏州河沿岸地区建设规划（2018-2035年）》标志着新时期下一江一河正以建设全球影响力的世界级滨水区、体现上海卓越全球城市水平的标杆区域为目标。黄浦江沿岸定位为国际大都市发展能级的集中展示区，苏州河沿岸定位为特大城市宜居生活的典型示范区，以更高站位谋求规划升级，以更高标准推动建设实施，满足人民群众对美好生活的期盼。

A "living room" of the People's City

In 2002, the Shanghai Municipal Government launched the Comprehensive Development Plan on Both Sides of the Huangpu River, proposing to make the waterway a "People's River". The core idea of development is to return the river and its waterfront to the people, making it the center of the city's cultural, social and civic life and an asset for all citizens. The integrated development program began in the first decade of the 21st century. From 2002 to 2007, the main task was to repurpose the land along the river, to make room for its new functions, and to move a large number of industrial facilities along the river elsewhere. From 2008 to 2010, the main task was to accelerate the transformation of the riverfront environment and to promote the construction of infrastructure facilities on the waterfront for the upcoming World Expo, paving the way for future waterfront redevelopment. From 2011 to 2015, the focus was the construction of functional projects and the introduction of new functions along the river.

In the same year, the Suzhou Creek Waterfront Landscape Plan was promulgated to address the problems of crowded tall buildings and insufficient public space along the Suzhou Creek. The document seeks to leverage the advantage of the Suzhou Creek in social development and environmental construction. It proposes to maximize the benefit of the natural and cultural landscape of the Suzhou Creek, and build a sound environment and harmonious living and recreation space. Through the unremitting efforts of the municipal government, the transport function of the Suzhou Creek has been replaced by leisure and sightseeing, and the waterfront has gradually become an urban public space catering to both citizens and tourists.

In 2016, the municipal government proposed to open the 45 km waterfront on both sides of the Huangpu River - between the Yangpu Bridge and Xupu Bridges. The waterfront revitalization campaign aims to accommodate fitness and leisure activities, sightseeing and tourism. This project was basically completed in 2017, causing a warm social response. At the end of 2020, most sections of the 42 km-Suzhou Creek waterfront was also opened to the public. Each section of the waterfront has its distinctive feature and beautiful sights; the sightseeing platforms, waterfront parks, sports facilities have made the waterfront truly live up to the reputation as "the living room of the city".

"The Waterfront Development Plan for the Huangpu River and the Suzhou Creek (2018-2035)" approved in January 2019 heralds a new era. The River and the Creek are slated to be made world-class waterfront with global influence and a landmark to reflect Shanghai's excellence as a global city. The Huangpu River waterfront is expected to be a showcase of the development capacity of this international metropolis, while the Suzhou Creek waterfront will be a demonstration zone of the life of a mega-city. We pursue a high standard for construction, in a bid to carry forward high-quality construction up to people's expectations.

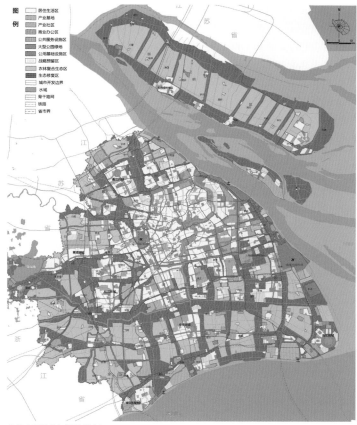
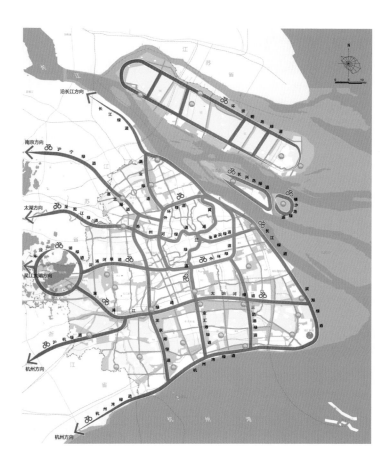

上海2035城市总体规划
Shanghai Master Plan 2035

上海2035总体规划更是充分落实"人民城市"的建设理念，通过更新改造，在一江一河沿线新增各类公园绿地、服务设施、商业办公，进一步提升一江一河的公共性和开放性。还水于民，还岸于民，使其成为全体市民的资产和活力客厅。

The Shanghai Master Plan 2035 seeks to fully implement "People's City" concept. Through the renewal and transformation, parks, green spaces, service facilities, commercial offices will emerge to further enhance the public nature and accessibility of the River and the Creek. Eventually, people can have access to the waterfront, so that it can become an asset of all the citizens, a dynamic "living room" that they demand and deserve.

上海2035总体规划以水为脉构建城市慢行休闲系统，在一江一河沿岸优化驳岸设计、增加公共空间，形成连续畅通的公共岸线和功能复合的滨水活动空间。

According to Shanghai Master Plan 2035, an urban non-motor trail system along Huangpu River and Suzhou Creek has been forming, composed of upgraded waterfront, shared space, pedestrian-friendly continuous pubic space and multi-functional zones.

理念与实践
Philosophy And Practice

1. 功能转型升级，谱写多彩乐章

一江一河是上海城市发展的"主动脉"和标志性区域，承载着城市最核心的功能，也勾勒出城市最经典的形象。自20世纪90年代浦东开发开放和2002年黄浦江两岸综合开发以来，一江一河的建设始终秉承"还水于民"的理念，通过持续的滨水区城市更新，实现了沿岸功能与形象的重塑。

2002年完成的《黄浦江两岸地区规划优化方案》中提出，"由生产型岸线转变为生活型岸线"的目标，以及"重塑功能、公众江岸、生态优先、再现风貌、彰显景观"五项策略，在后续的结构规划和详细规划中得以深化，并在多轮建设实施中有效贯彻落实。2019年完成的《黄浦江、苏州河沿岸地区建设规划（2018-2035年）》在继承"人民江河"的理念基础上，提出"发展为要、人民为本、生态为基、文化为魂"的指导思想，进一步发挥一江一河对提振城市经济能级、强化创新驱动、增强文化辐射力、修复生态环境的重要意义，对未来上海的城市建设发展起到引领和示范作用。

发展至今，黄浦江沿岸基本实现由工业、仓储为主的生产性功能向金融商贸、文化游憩、创新创意、居住生活等综合性功能的转变，沿岸景观形象、环境品质与设施配套不断提升，成为城市核心功能承载与标志性形象展现最重要的区域。沿岸形成一系列引领城市发展的重点片区。其中，陆家嘴、外滩、南外滩地区发展为金融集聚区；徐汇滨江以文化品牌打造为先导，培育科技创新、文化创意、创新金融等特色功能；世博会地区后续发展聚焦总部经济，在浦东世博A、B片区的建设中，探索公共空间与地下空间的一体化开发模式；前滩地区基于自然生态与城市组团的融合，打造总部商务、文化休闲、居住生活等多元功能复合的新型滨水城区。经过多年发展，苏州河沿岸也从早期水质治理迈向功能业态、景观形象、空间环境、基础设施的综合提升，逐步培育具有"生活化"特色的多元功能。沿线形成苏河湾金融文化等特色产业区、临空科创研发商务园区、M50文化创意区等重点区域。一江一河已成为城市功能的集聚带、城市文化的主阵地、公共活动的大舞台、生态文明的示范带、城市形象的展示区。

Writing a new chapter of function transformation and upgrading

The Huangpu River and Suzhou Creek are the main arteries that powers Shanghai's urban development. Their riverfronts are landmark areas of the city, carrying the city's core functions and serve as calling cards of the city. Since the development and opening up of Pudong in the 1990s and the comprehensive development of both sides of the Huangpu River in 2002, the construction of the River and the Creek has always adhered to the concept of "returning the waterways to the people". Through the continuous urban renewal of the waterfront, its functions and images have undergone tremendous changes.

"Optimization Planning of the Huangpu River Banks", a document formulated in 2002, proposes the objective of "Transforming the production-oriented waterfront into a living-oriented one", and puts forward five strategies, i.e., "changing the functions, opening waterfront to the public, prioritizing ecology, restoring past glory and presenting lovely sights". All these have been observed in subsequent structural and detailed plans and effectively implemented in several rounds of construction. "The Waterfront Development Plan for the Huangpu River and the Suzhou Creek (2018-2035)" completed in 2019 develops the concept of "People's River", stressed the importance of development, people's interest, ecology, and cultural flourishing. It requires practitioners should leverage the importance of the River and the Creek to boost the city's economic capacity, incentivize innovation, expand cultural influence and restore the ecological environment. All such efforts will play a leading and exemplary role in the future development of Shanghai's urban construction.

So far, the Huangpu River has basically changed its function from a working river to a waterway that accommodates finance, commerce, culture, leisure, innovation and creativity. The landscape, environmental quality and facilities along the river have been continuously improved, making the waterfront a showcase of the core functions and image of the city. A series of key sections have emerged to lead the city's development. Among them, Lujiazui, the Bund and the South Bund have been developed into financial hubs; the Xuhui section of the Huangpu River waterfront has established its branding as a cultural hub that also accommodate such functions as technological innovation, culture, creation and finance. The redevelopment of the sections A & B of the former Expo site focus on the headquarters economy, exploring the integrated development mode that involves both public spaces and underground spaces. Based on the integration of natural ecology and urban clusters, the New Bund area has been made a brand-new waterfront that takes on multiple functions, ranging from headquarters economy, business, culture, leisure, to residence. Over the years, the Suzhou Creek is not only a subject of water treatment but a testing ground of new functions. Along with the improvement of its aesthetic value, spatial environment and infrastructure, it has gradually cultivated multiple everyday life-related functions. Key features include the Suhe Bay Financial and Cultural Industrial Cluster, the M50 Creative Park. The waterfront has become a concentration space of urban functions, a cultural hub, an ecology demonstration zone and a showcase for the city's image.

陆家嘴
Lujiazui

南外滩
South Bund

上海市世博会地区会展及其商务区A片区规划
Bird's-eye view of Section A of Shanghai World Expo Site

以"绿谷"为概念，创造人与自然和谐的、突显世博记忆的绿色标志性空间；面向国际知名企业总部集聚区和具有国际影响力的世界级工作社区发展目标，提供商务、商业、文化娱乐、公寓式酒店等综合功能，打造24小时的活力城区；创造宜人尺度的建筑和城市开放空间，打造人车分流、安全舒适、活动连续的步行环境；探索整体开发和带方案出让的管理模式，绿谷街坊地下空间整体开发，上部建筑整体设计、建造。
Embracing the concept of a "Green Valley", Section A will be a green, landmark space that is harmonious between man and nature, which also evokes the memory of the World Expo. Aiming at the goal of gathering headquarters of internationally famous enterprises and world-class work community with international influence, Section A accommodates the facilities of commerce, commerce, cultural entertainment, serviced apartments, thus being a 24-hour dynamic urban area. It will also accommodate buildings and urban open space of reasonable volumes to create a safe and comfortable pedestrian-friendly environment. We will also explore the management mode of overall development, while the underground space of the Green Valley will be developed in an all-in-one design, and the same applies for the buildings above ground.

上海市世博会地区会展及其商务区B片区规划
Bird's-eye view of Section B of Shanghai World Expo Site

规划以环境宜人、交通便捷、低碳环保、具有活力的知名企业总部聚集区和国际一流的商务街区为目标，注重园内商务办公和配套商业、居住、餐饮、休闲、健身等功能的适度混合，满足各类人群的生活需求；注重延续世博空间意向，延续已形成的轴线、广场、绿廊，突显中国馆的标志性地位，形成整体融协的空间体系；强调人性化空间的整体营造，采取高密度、紧凑型街坊空间设计，形成连续的街道界面，鼓励建筑公共开放，构建立体公共活动体系。
The plan aims at building a gathering area enterprise headquarters and a world-class business community with pleasant environment, convenient transportation, low-carbon emission and vitality. It brings together office buildings and supporting business, residence, dining, leisure, fitness facilities to meet the needs of everyone. The existing Expo axis, square and green corridor will be preserved to stress the grandiosity of the China Pavilion and form an integrated space system. It gives emphasis to the people-oriented-ness of the space with the design of high-density and compact neighborhood, thus forming a continuous street interface, a three-dimensional public activity system which encourages buildings to be accessible.

2. 水岸贯通开放，共享活力空间

一江一河是城市最具代表性和吸引力的公共开放空间。在多轮规划与建设中，沿岸地区始终以打造开放、共享、活力、宜人的公共岸线为目标，从早期的滨江、滨河绿地建设，到近年来的滨水公共空间贯通开放建设，不断推进滨水公共空间体系的拓展和提升。目前，一江一河核心区段岸线已实现贯通开放，成为广受市民、游客欢迎的休闲游憩场所与展现城市魅力的水岸共享客厅。

通过滨水公共空间建设，打开封闭的水岸，构建贯通开放的滨水开放空间序列。自2002年黄浦江两岸综合开发启动以后，浦江沿岸陆续建成北外滩绿地、南园公园、后滩公园、前滩绿地等公共休憩空间，苏州河畔也建成了梦清园、九子公园、蝴蝶湾公园、长寿湾绿地、临空公园等。为了进一步加强滨水公共空间的连续性和开放度，黄浦江沿岸于2016年启动核心段"45公里"岸线贯通工程，并专门编制了《黄浦江两岸公共空间贯通开放规划》。

规划以"更开放、更人文、更美丽、更绿色、更活力、更舒适"为理念，采取开放封闭单位、架设二层连廊、建设水上栈道等方式打通沿江空间断点，将原先碎片化的滨江绿地开放空间连为整体，沿江设置总长度约150公里的漫步道、跑步道、骑行道三类贯穿全线的慢行活动线路，至2017年底实现杨浦大桥至徐浦大桥的滨江岸线空间贯通开放。苏州河沿岸于2018年启动岸线贯通工作，并于2020年底基本实现中心城段两岸42公里的公共空间全线贯通，应对空间相对狭小的现状特点，采用因地制宜、灵活多样的方式建成串联各个滨水公园绿地的游憩步道体系。加强滨水岸线与城市腹地的慢行联系，提升滨水公共空间的可达性。对沿江沿河道路进行改造，减少机动车道，从而削弱城市道路对于滨水空间的割裂，如2010年完成的外滩综合改造工程，将中山东路一部分地面机动车交通置于地下隧道，压缩地面车道，提升过街便利性并扩大滨水开放空间。依托垂直于水岸的街道、支流河道等建设慢行廊道，串联腹地轨交站点、重要公共服务设施与重要公共空间等，形成滨江至腹地的活力动线，如2020年完成的南京东路步行街东拓工程，将原本终止于河南中路的步行街向东延伸至中山东一路，形成连接外滩与人民广场的经典步行游览空间。

注重滨水公共空间的品质、功能塑造，以及多元化游憩服务设施配套。进行场地定制化设计，有效植入休闲、游憩、体育等功能，如徐汇滨江滑板公园、北外滩"魔都矩阵"极限运动场、前滩鳗鲡嘴儿

Creating a vibrant waterfront for everyone

The River and the Creek waterfront is the most representative and attractive public open space in Shanghai. Multiple rounds of planning and construction aimed to create an open, shared, vibrant and pleasant waterfront. It is a consistent goal, now as then, to expand and improve the waterfront. At present, the core sections of the River and the Creek waterfront have been opened to the public, becoming popular leisure and recreation spaces for the public and tourists. They have been civic spaces that showcase the city's charm.

Through construction, the previously separated inaccessible sections of the waterfront are linked up with the already open ones. Since the Huangpu River waterfront revitalization campaign started in 2002, the North Bund green space, Nanyuan Park, Houtan Park and New Bund Green Space have been built along the river, while the Mengqing Garden, Jiuzi Park, Hudiewan Park, Changshou Bend Green Space and Lingkong Park have also been built along the Suzhou Creek. In order to further enhance the continuity and openness of the waterfront, a special project was launched in 2016, to open up all sections of the 45 km main course of the Huangpu River and The Plan for the Opening of the Huangpu River Waterfront" was also prepared.

The plan seeks to make the waterfront "more open, more humanistic, more beautiful, greener, more energetic and more comfortable". It adopts the approach of making the waterfront sections in formerly not accessible institutions open to the public, erecting elevated corridors and constructing waterfront walkways to span the break-up points. The previously fragmented riverfront green spaces are linked up into a whole. A total length of about 150 kilometers of walkways, running and cycling trails are built along the river. By the end of 2017, the waterfront from the Yangpu Bridge to the Xupu Bridge was opened. The waterfront along the Suzhou Creek was opened up in 2018, and by the end of 2020, waterfront of 42 kilometers had been opened up along both sides in the city center. A jogging / cycling trail system has been built to link up the waterfront parks and green spaces in a flexible manner.

Our aim is to build slow traffic connections between the waterfront and the rest part of the city and enhance the accessibility of the waterfront public space. To this end, we remodeled waterfront roads to reduce motorized traffic, thus reducing the number of urban roads that make waterfront inaccessible. A prime example is the comprehensive reconstruction project of the Bund completed in 2010: It channels part of the surface motorized traffic on East Zhongshan Road to an underground tunnel, reduce the number of motor lanes on the ground. It is now more convenient to cross the street and access the waterfront. Slow traffic corridors are built along streets perpendicular to the waterfront and tributary waterways, linking rail stations, important public service facilities and public spaces in non-waterfront areas to form a dynamic route from the riverfront to the rest part of the city. For example, the East Nanjing Road Pedestrian Street expansion project completed in 2020 extends the pedestrian street from Middle Henan Road eastwards to East Zhongshan Road No. 1, forming a classic pedestrian space connecting the Bund to People's Square.

童乐园等。配套人性化的服务设施，建立包括便民服务、应急医疗、文体游憩、标识导引、智慧信息、安全预警、照明、无障碍在内的多类型设施体系。注重设施设计与建造的公共艺术性，如杨浦滨江人民城市建设规划展示馆、浦东滨江望江驿、徐汇滨江"水岸汇"驿站等，提供多元功能与集合式服务配套的同时，也成为景观特色亮点。

Our practice has focused on the quality and function of waterfront, and diversified recreational services and facilities. Customized space design, effective introduction of leisure, recreation, sports and other functions have become a standard practice. The Xuhui Riverfront Skateboard Park, the North Bund "Magic City Matrix" extreme sports field, the New Bund Eel Mouth Children's Playground all fall into this category. Visitor-friendly service facilities, covering convenient services, emergency medical care, cultural, sports and recreation, signage, intelligent information, security warning, lighting and barrier-free, are set up. We also emphasize on the aesthetic of design and construction of public art facilities, with such specimens as the People's City Construction Planning Exhibition Hall in the Yangpu section of the Huangpu River Waterfront, Riverview Service Stations in both Pudong and Xuhui sections of the Huangpu River waterfront. They perform a variety of functions and provide service support, and are also lovely features of the waterfront in their own right.

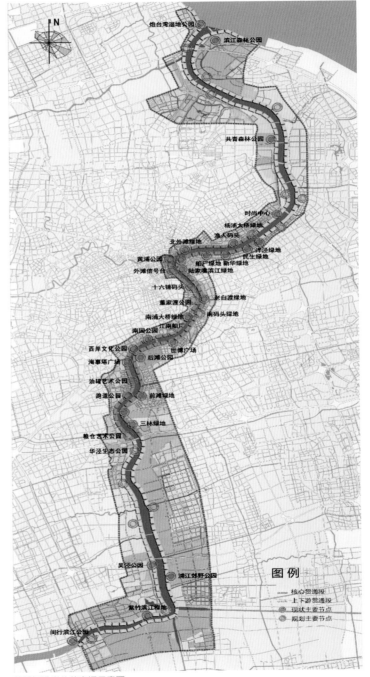

黄浦江沿岸公共空间示意图
Schematic diagram of the Huangpu River Waterfront

黄浦江两岸公共空间贯通开放规划
Plan for the Opening of the Huangpu River Waterfront

规划以"世界级滨水开放空间"为发展愿景，通过贯通黄浦江最核心的45公里岸线，建设开放共享的市民休闲空间。规划提出"开放、人文、美丽、绿色、活力、舒适"六项理念，在浦江两岸构建空间贯通、文化风貌、景点观赏、绿化生态、公共活动、服务设施六个系统。此外，在浦江两岸规划十个主题区段，形成沿江统一而各具特色的滨江空间序列，打造包括工业文明、海派经典、创意博览、文化体验、生态休闲、艺术生活等不同主题特色，可漫步、可阅读、有温度的精彩水岸空间。

With the development vision of "world-class waterfront open space", the plan will build an open and shared public leisure space by connecting the 45 km coastline at the core of the Huangpu River. The plan puts forward six concepts of "openness, humanity, beauty, green, vitality and comfort", and constructs six systems on both sides of the Pujiang River: space connection, cultural style, scenic spot viewing, greening ecology, public activities and service facilities. In addition, ten theme sections are planned on both sides of the Huangpu River to form a unified and distinctive riverside space sequence along the river, and create a wonderful waterfront space with different theme characteristics, including industrial civilization, Shanghai style classics, creative Expo, cultural experience, ecological leisure and artistic life, which can walk, read and have temperature.

3. 修复水绿生态，培育和谐生境

一江一河作为战略性的生态廊道，是城市生态安全得以保障的绿色主脉。多年来，以"人与自然和谐共生"为理念，探索在高密度超大城市中建立韧性平衡、自然亲和的水岸生态体系。通过滨水绿地建设、自然岸线营造、水体土壤污染治理等方式，不断拓展生态空间规模、提升生态品质，加强多维度生态系统修复和培育。沿岸生态环境显著提升，生态辐射效应日益增强。

以"增量、联网、提质"为导向，多方式增加沿岸生态空间规模，并建立"点、线、面"结合的生态网络结构。依水复绿，沿岸布局不同尺度的块状生态绿地和滨水公园，并通过连续的带状绿地，将这些生态斑块串联，构成完整的水绿生态廊道。中心城区段建成陆家嘴滨江绿地、南园公园、后滩公园、前滩绿地、虹桥河滨公园、梦清园、蝴蝶湾公园等多处集中绿地，世博文化公园等大型绿地空间正在加快建设；上下游郊区段建成炮台湾湿地公园、浦江郊野公园、三岔港滨江森林公园等大型生态空间。此外，横向设置楔形绿地与滨江绿带形成"梳"状绿化结构，并连接腹地绿网系统，开展张家浜楔形绿地等垂直于水岸的支流生态廊道建设，使滨水生态网络向城市腹地拓展和辐射。

以水体治理为重点，积极推进沿岸环境污染整治和生态修复。20世纪90年代苏州河综合治理工程启动至今已进入第四期工程，经过多年持续的水质治理，苏州河水质由早期的"黑臭水"改善为清澈干净的Ⅳ类水，该项目曾在2005年日本爱知世博会上获奥地利"能源全球奖（水资源组）"，并得到联合国人居署高度评价。黄浦江的水质也进行了多年治理，目前已达到Ⅱ～Ⅲ类水。此外，推进一江一河上游的生态复育，结合工业用地转型、村庄撤并等，大力开展上游生态涵养林建设和环境污染治理，构建稳固、安全的生态本底。

Restoring waterfront ecology and cultivating ideal habitats

As strategic ecological corridors, the River and Creek waterfronts are the green arteries to guarantee the city's ecological security. Over the years, following the concept of "harmony between man and nature", we have established a resilient and balanced waterfront ecosystem in a high-density mega-city. Through the construction of green, eco-friendly waterfront and the treatment of soil and water pollution, we have been expanding the scale of ecological space, improving ecological quality and strengthening the restoration and cultivation of multi-dimensional ecosystems. The waterfront ecological environment has been significantly improved and the ecological radiation effect is increasingly enhanced.

Guided by the principle of "increasing quantity, building network and improving quality", the waterfront ecological spaces has been expanding in various ways, and an ecological network has been established in an effort that can be described "linking up points into lines and lines into patches". The ecological green areas and waterfront parks of different scales are laid out along the waterways, and these ecological patches will be linked together to be connected, forming uninterrupted waterfront ecological corridors. At the city center, a number of centralized green spaces have emerged, such as the Lujiazui Green Space, Nanyuan Park, Houtan Park, New Bund Green Land, Hongqiao Riverside Park, Mengqing Park and Hudiewan Park. The construction of large green spaces such as the Expo Culture Park is now in full swing. The suburban sections of the River have already been built with large ecological spaces such as the Fortress Wetland Park, the Pujiang Country Park and the Sanchagang Riverside Forest Park. In addition, wedge-shaped green areas are created perpendicular to the waterfront green space, forming a network that reaches out to connect the green land at a larger area. Such tributary ecological corridors are set up perpendicular to the waterfront, such as the Zhangjiabang wedge-shaped green space.

With the emphasis on water pollution control, the project has promoted the remediation of environmental pollution and ecological restoration on the river banks. After years of continuous treatment effort, the water of the Suzhou Creek is no longer black and smelly, but has turned clear and clean, reaching the standard of Class IV. The water treatment program was awarded the Austrian "Energy Global Award (Water Resources Category)" at the 2005 World Expo in Aichi, Japan, and received high recognition from the UN-Habitat. The water quality of the Huangpu River has also reached Class II-III standard after years' treatment. In addition, we are promoting the ecological rehabilitation of the upper reaches of the River and the Creek. Taking advantage of the transformation of industrial land and the merger of villages, we carried out the construction of ecological forests and the treatment of environmental pollution in the upper reaches to build a solid and safe ecological base.

炮台湾湿地公园
Fortress Terrace Wetland Park

徐汇滨江绿地
Xuhui Section of Huangpu River Waterfront

长风滨河绿地
Changfeng Riverside green space

浦东杨浦大桥西侧绿地
Pudong Yangpu Bridge west green space

世博文化公园规划
General plan of the Expo Culture Park

规划以"生态自然永续、文化融合创新、市民欢聚共享"的新时代伟大公园为目标，在上海城市的中央活动区建设一座2平方公里的"都市绿肺"，构建黄浦江生态廊道上的重要核心。配置大量乔木形成高密度中心城的绿色森林，并打造湖区、湿地、疏林草地、密林，营造多样生境。此外，在公园中设置大型文体设施，突显文化特质，并延续"城市，让生活更美好"的世博理念。

Aiming at the Great Park of the new era of "ecological and natural sustainability, cultural integration and innovation, and citizens gathering and sharing", the plan will build a 2 square kilometer "urban green lung" in the central activity area of Shanghai, and build an important core on the ecological corridor of Huangpu River. A large number of trees are configured to form a green forest in the high-density central city, and create lake areas, wetlands, sparse forests, grasslands and dense forests to create diverse habitats. In addition, large-scale cultural and sports facilities will be set up in the park to highlight cultural characteristics and continue the Expo concept of "Better City, Better Life".

4. 传承历史记忆，新旧融合辉映

一江一河作为上海的母亲河，积淀了深厚的历史底蕴，是城市文化和精神的象征。沿岸丰富的历史遗存，承载了上海城市的金融贸易、港口运输、工业生产、社会生活的发展和变迁，编织着上海城市文化。在沿岸的建设过程中，充分保护和利用沿岸历史建筑、工业遗产等各类风貌资源，留驻历史、唤醒历史、续写历史，展现其蕴含的人文价值。一江一河已成为人们重拾记忆、感受和展望城市情怀与魅力的最佳场所。

以"应保尽保""整体保护"为原则，对一江一河沿岸的历史资源进行深入的梳理挖掘和严格保护。一江一河沿岸地区共有外滩、老城厢、提篮桥、龙华、愚园路5处历史文化风貌区，200余处优秀历史建筑，百余处文物保护点，以及多片工业遗产集聚区域。这些珍贵的历史风貌资源凝聚着上海城市发展的历史：老城厢残破的城垣、鳞次栉比的街市，依稀可辨明清邑城的景象；外滩建筑厚重的石柱、石墙，以及充满异域特色的精美雕饰，展露出近代远东都市的繁华；遍布江畔的工厂、港口，其岁月斑驳的痕迹折射出上海自近代以来作为港口工业城市的辉煌。在建设开发过程中，对各类历史建筑进行了严格的保护和妥善的修复，同时加强历史环境和空间肌理的协同保护，并将塔吊、船茇等历史构筑物保留下来且运用于历史性场景的景观塑造，在空间要素上最大限度地延续城市文脉、传承历史记忆。

重视历史文化积淀与当代新锐文化特质的兼收并蓄，运用多元方式对沿岸历史遗产进行合理的活化利用。强调历史空间的整体性塑造，形成多个文化风貌展示体验区：对经典的风貌区进行品质与功能提升，如外滩的历史建筑群进行功能置换，恢复金融、商贸等历史功能，并植入文化、休闲、酒店等新业态；不断探索工业遗产的保护性更新利用，把"工业锈带"转换为"生活秀带"，打造杨浦滨江工业文明展示带、徐汇滨江西岸文化长廊、南外滩老码头创意区、世博会地区城市最佳实践区等展现工业遗产特色的时尚文化风貌区域。基于各类历史建筑的空间适应性分析，植入商业、办公、创意、展示、演艺等功能，打造文化地标。例如浦东老白渡煤仓改造为艺仓美术馆，杨浦国棉十七厂改造为时尚体验中心，徐汇滨江北票码头建筑改造为

Keeping history alive through a fusion of the old and the new

As Shanghai's mother rivers, the Huangpu River and the Suzhou Creek are glittering cachets of the city's culture and spirit. The myriads of heritage sites along them have witnessed the sweeping changes that have occurred to Shanghai's finance, trade, port transportation, industrial production and social life. They are incarnations of the city's culture. In the process of redeveloping the waterfront, various landscape resources, such as historical buildings and industrial heritage, are well protected and utilized to honor the history of the waterfront and to showcase its cultural value. The River and the Creek areas have become the best places for people to call up their deep memories and savor the city's charm.

Based on the principle of "preservation as far as possible" and "holistic protection", the historical resources along the River and the Creek have been thoroughly researched and strictly protected. There are five historical and cultural districts along the River and the Creek: the Bund, the Old Town, Tilanqiao, Longhua and Yuyuan Road. They involve a total of more than 200 heritage buildings and more than 100 heritage sites and a number of industrial heritage clusters. These precious historical and landscape resources highlight Shanghai's urban evolution. The peeling walls of the old Shanghai town, closely packed neighborhoods bear traces left from the Ming and Qing dynasties; the towering stone pillars and walls of the buildings on the Bund and the fine carvings reveal the prosperity Shanghai in the early 20th century; the factories and ports that line the waterfront reflect the glory of Shanghai as a riverborne industrial city since the mid-19th century. During the construction and development process, various historical buildings have been strictly protected and properly restored, while the historical environment and the spatial fabrics have been enhanced and the historical structures such as tower cranes and ship barges have been preserved and used to be living memorials. They celebrate the city's cultural identity and bring history alive with spatial elements.

We need to pay attention to the integration of time-honored cultural resources with contemporary cultural traits, and use multiple approaches to carry out rational adaptive use of the heritage on the waterfront. We should emphasize the holistic approaches to redevelop historical spaces and produce multiple cultural landscape experience areas. We should upgrade these areas' quality and functions, such as repurposing the historical buildings on the Bund to restore their historical functions of finance and commerce, and introduce new functions, such as culture, leisure and hotels. We should explore the modes of conservation and renewal of industrial heritage, transforming the "industrial rust belt" into a "waterfront lifestyle showcase", creating trendy cultural showcase areas with industrial heritage elements. The Yangpu Riverfront Industrial Heritage Demonstration Belt, the Xuhui Waterfront West Bank Cultural Corridor, the South Bund Old Wharf Creative Zone, and the Expo Area Urban Best Practice Zone are all fine examples of this kind.

Based on the analysis of the spatial adaptability of various historical buildings, we can convert former industrial facilities into commercial spaces, offices, creative parks, exhibition halls and performing arts venues to create new cultural landmarks. For example, the Laobaidu coal warehouse in Pudong has been converted into the Shanghai Modern Art Museum, the 17th Cotton Mill in Yangpu District has been transformed into a fashion experience center, the Beipiao Wharf building in the Xuhui section of the Huangpu River waterfront

龙美术馆西岸馆，静安四行仓改造为抗战纪念馆。在此基础上加强多层次文化设施的建设，构建文化功能集群，深厚历史底蕴与时尚文化功能相结合，年代久远的历史遗存焕发出新的生命力。

has emerged as the Long Museum West Bund, and the Joint Trust Warehouse in Jing'an District has been made a war memorial. We should promote the construction of multi-level cultural facilities on this basis, build cultural clusters, and introduce new cultural functions to heritage sites and buildings, thus lending new vitality to the age-old waterfront.

上海市世博会地区城市最佳实践区规划
Planning of urban best practice area in Shanghai World Expo area

规划较为完整的保留了世博会城市最佳实践区的展览场馆，继承上海世博会的无形遗产，继续成为交流、分享、推广城市最佳实践的全球平台，不断演绎"城市，让生活更美好"的世博主题，依托存量建筑和设施，进行功能转换，打造成为集创意设计、交流展示、产品体验等为一体，具有世博特征和工业遗产特色的文化创意产业集聚区。

The exhibition venues in the urban best practice area of the World Expo are fully planned to inherit the intangible heritage of the Shanghai World Expo, continue to become a global platform for the exchange, sharing and promotion of urban best practices, constantly interpret the theme of the World Expo "Better City, Better Life", transform functions according to the existing buildings and facilities, and build into a collection of creative design, exchange, exhibition It is a cultural and creative industry cluster with the characteristics of World Expo and industrial heritage.

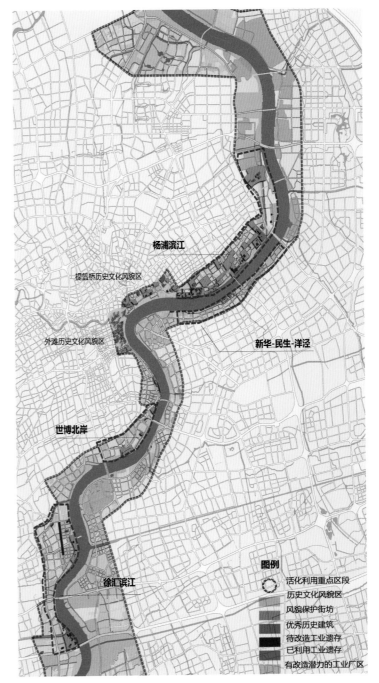

黄浦江沿岸历史遗产示意图
Schematic diagram of historical heritage along the Huangpu River

5. 完善基础设施，提升人本环境

一江一河是探索和实践城市基础设施建设先进理念的引领区和示范地。多年建设过程中，以安全、便捷、韧性、低碳为基本导向，引入人性化、景观化、智慧化的要求，沿岸形成较为完备的综合交通体系和市政设施体系。公共交通持续完善，跨江跨河通道不断加密，滨水慢行系统初步成形；防汛设施建设完备，安全应急、智慧信息等各类市政配套不断提升。

以公交优先、两岸联动、水陆衔接、注重体验为原则，建立多层次综合交通体系。针对一江一河地区开展关于越江隧道、轨道交通、水上码头的各类交通专项规划研究，相关建设在规划的指导下有序推进：黄浦江沿岸桥梁、隧道、地铁等跨江跨河交通成网成片，近江交通和越江交通状况不断改善，滨江空间的可达性日益增强，规划的38条越江桥隧中已建成20余条，规划的27条轨道交通中已建成线路达10余条，其中杨浦大桥、南浦大桥等桥梁还成为浦江之上的标志性景观。以兼顾游憩和通勤的复合功能为导向，构建包括越江轮渡、水上游船、公务船只等在内的多样化水上交通体系，近年来重点建设了以十六铺旅游码头为代表的多处水上游船码头，拓展水上游憩功能，并成为重要的水岸休闲服务节点。基于密路网、小街坊的建设，塑造人性化的空间格局与尺度，并结合一江一河滨水公共空间的贯通建设，构建沿岸慢行交通网络。苏州河沿岸因地制宜，开展滨河路慢行化改造，并建设普济路桥、西康路桥、宝成桥等多座慢行跨河桥梁，形成两岸一体的慢行空间体系。

市政设施建设突显人性化、景观化、集约化、智能化等要求，采取一江一河定制化的处理方式。防汛墙的设计与建设在满足防汛安全的前提下，综合考虑塑造景观美感和承载游憩活动，在滨水建设空间较为充裕的区段，将防汛墙与自然坡地、踏步台地等景观要素结合设置，形成宜人的亲水空间感受。对沿岸变电站、雨污水泵站、水闸等设施采取地下建设、隐蔽建设等景观化处理方式。例如，苏州河河口水闸、延安东路雨水泵站、世博通信基站等均采取隐蔽式处理，与周边环境融为一体。积极推进运用智慧设施与建设智慧场景，采用街道基站、小微基站、智能街具等设置方式，如北外滩滨江公共WiFi信息亭、苏州河长宁段智慧座椅等，探索智慧城市建设与智能化决策管理的新模式。

Improving infrastructure and creating people-friendly environment

Waterfront of the River and the Creek is a leading and demonstration area for exploring and practicing advanced concepts in urban infrastructure construction. In the construction process over many years, we have observed the principles of safety, convenience, resilience and low-carbon, and have introduced the requirements for people-friendliness, aesthetics and intelligence. As a result, we have established a sound, comprehensive transportation system and municipal facilities system on the waterfront. Public transportation has been steadily improved, the number of cross-river channels has increased, and the waterfront non-motorized traffic system has taken shape; flood control facilities are well built, and various municipal facilities such as safety, emergency and intelligent information are continuously upgraded.

A multi-level integrated transport system has been established based on the principles giving priority to public transport, linkage between the two banks of the River and the Creek, connection between land and water transportation, and emphasis on experience. Special planning studies on river-crossing tunnels, rail transport and river traffic terminals have been carried out for the areas, and the relevant construction has been implemented in an orderly manner under the guidance of the planning: bridges, tunnels and subways along the Huangpu River have formed a network; near-river and cross-river transport have been continuously improved, and the accessibility of the riverfront space has been increasingly enhanced. More than 20 of the 38 planned river-crossing bridges and tunnels have been completed, and more than 10 of the 27 planned metro lines have been completed, while such bridges as the Yangpu and Nanpu Bridges have become iconic sights.

With the composite functions of recreation and commuting in mind, a diversified water transportation system - including ferries, cruise ships and executive boats - has been built. In recent years, a number of cruise terminals, represented by the Shiliupu Terminal, have been built to adopt recreation function. They have become important waterfront leisure facilities. Based on the construction of a dense road network and small street blocks, a humane spatial pattern has been created, and a non-motorized traffic network is built along the waterfront in conjunction with the construction of waterfront public spaces. A number of non-motorized bridges have been built over the Suzhou Creek, including the Puji Road Bridge, Xikang Road Bridge and Baocheng Bridge. They have formed an integrated system of slow traffic spaces on both sides of the Suzhou Creek.

The construction of municipal facilities lives up to the requirements for people-friendly, aesthetics, intensification and intelligence. Many measures are tailored for the conditions of the River and the Creek waterfront. In addition to having the design and construction of flood control walls meet safety requirements, we have taken into account the aesthetics of the landscape and the conditions for recreational activities. Whenever there is abundant room in the waterfront, we plan the flood control walls, sloping plantation, stepping terraces and other landscape elements in an overall manner, in a bid to evoke an enjoyable feeing.

Facilities such as substations, rainwater and sewage pumping stations and sluices along the waterways are treated in a visually interesting ways by means of undergroundor or concealed construction. For example, the Suzhou Creek estuary sluice gates, the East Yan'an Road rainwater pumping station and the Expo telecommunication base station are all treated in a concealed manner, making them integrated into the surrounding environment. The use of smart facilities and the construction of smart scenes are adopted, e.g., street base stations, small and micro base stations, and smart street furniture are set up, such as the public WiFi kiosks along the North Bund waterfront and the smart seats in the Changning section of the Suzhou Creek waterfront. These are concrete efforts to explore new modes of smart city construction and intelligent decision-making.

杨浦大桥
Yangpu Bridge

南浦大桥
Nanpu Bridge

白莲泾步行桥
Bailianjing Creek Footbridge

苏州河口河闸及外白渡桥
Sluice and Garden Bridge (Waibaidu Bridge) at the estuary of the Suzhou Creek

6. 打造文化品牌，共建卓越水岸

一江一河是市民与游客感受城市魅力、共享欢乐幸福的"客厅"与"舞台"。依托贯通开放的水岸公共空间、优美迷人的景观环境、丰富多元的人文资源，在沿岸组织各类精彩纷呈的艺术、体育、节庆、游览活动，一江一河已打造成为上海最经典的文化和旅游品牌。

以"全域旅游"为理念，依托重点文化旅游资源策划一系列经典旅游线路和游览项目。例如以外滩、陆家嘴为核心，组织浦江水上游、外滩观光巴士游等，成为体验海派文化、观赏都市美景最具代表性的旅游项目。

定期策划举办具影响力的赛事与文化活动，以及滨江漫步、市民沙龙等亲民休闲活动，例如滨江马拉松赛、上海杯帆船赛、滨江音乐节、新年灯光秀、上海城市空间艺术季、苏州河文化艺术节、苏州河城市龙舟国际邀请赛等。其中，上海国际马拉松赛自1996年以来每年举办，多数在黄浦江畔进行。上海城市空间艺术季分别于2015年、2017年、2019年连续三次在徐汇滨江、浦东滨江、杨浦滨江举办，以开放、互动的艺术化形式将上海的城市建设成果展示给广大市民，已成为市民参与城市建设、感知城市艺术的品牌性、长效性共享共治平台。徐汇滨江集聚20余座文化场馆打造"美术馆大道"，每年举办超过100场文化活动，成为全球首发首秀首展新地标。

Burnishing cultural brands and building excellent waterfront

The waterfront of the River and Creek allows citizens and visitors to sample the charm and happiness the city has to offer. By tapping the open waterfront, the enjoyable environment, and the rich humanity resources, we can launch an exciting range of arts, sports, festivals and excursions events along the River and the Creek. The River and the Creek areas have become a sparkling cultural and tourism brand in Shanghai.

Based on the concept of "all-for-one tourism", a series of classic tourism routes and tourist attractions are planned based on key cultural tourism resources. For example, with the Bund and Lujiazui as the core, we have organized cruise tour on the Huangpu River and bus tours that pass the Bund, which have become representative tourism programs to dip into the culture of Shanghai and experience the glamour of the city.

We regularly plan and organize influential cultural events and leisure activities such as waterfront marathon, Shanghai Cup Regatta, riverfront music festival, New Year's light show, Shanghai Urban Space Art Season, Suzhou Creek Culture and Art Festival, International Suzhou Creek Dragon Boat Invitational Race, etc. Among them, the Shanghai International Marathon has been held annually since 1996, mostly on the waterfront of the Huangpu River. The Shanghai Urban Space Art Season was held in Xuhui, Pudong and Yangpu sections of the Yangpu River waterfront in 2015, 2017 and 2019 respectively, showcasing Shanghai's urban renewal achievements to the general public in an open, interactive and artistic form. The event has established its branding as a long-term joint governance platform for citizens to participate in urban construction and appreciate urban art. Xuhui section of the waterfront is home to more than 20 cultural facilities, creating an "Art Museum Avenue" that holds more than 100 cultural events each year. It has become a new destination for world's premiere shows and exhibitions.

西岸美术馆
West Bund

徐汇滨江美术馆大道
Xuhui River Waterfront Art Museum Avenue

发挥文化先导的作用，建设西岸美术馆、龙美术馆（西岸馆）、余德耀美术馆、油罐艺术公园等20余座文化场馆打造"美术馆大道"，每年集聚超过100场文化活动，成为全球首发首秀首展新地标。2019年中法最高级别文化交流项目"蓬皮杜五年展陈合作"开幕；2020年第七届西岸艺术与设计博览会规模位列全球第四；上海国际艺术品交易月以西岸为核心联动全城，上海国际艺术品交易中心、上海国际文物交易中心落户西岸，助力上海国际文化大都市建设。

To celebrate visual art, we have built more than 20 cultural venues such as the Westbund Museum, Long Museum West Bund, Yuz Museum and TANK Shanghai to create an art museum avenue. The area hosts more than 100 cultural events each year, becoming a new landmark for the world's foremost destination of premiere shows and exhibitions. The seventh West Bund Art and Design Fair was the fourth largest event of this kind worldwide in 2020; the Shanghai International Art Trade Month takes the West Bund as the core destination and have an impact on whole city, and the Shanghai International Art Trade Center and Shanghai International Cultural Relics Trade Center have also located their operation here, in a bid to make Shanghai an international cultural center.

2019上海城市空间艺术季
2019 Shanghai Urban Space Art Season

2019上海城市空间艺术季在杨浦滨江举办，主题为"相遇"，意喻艺术季所呈现的人与人的相遇、水与岸的相遇、艺术与城市空间的相遇、历史与未来的相遇，将激发更多美好生活、美好情感的相遇，成为人们不可磨灭的城市记忆。主展分为"城市空间艺术"和"规划建筑"两个版块，演绎滨水空间的话题。规划建筑版块主题为"再生"，立足于一江一河，聚焦"滨水空间为人类带来美好生活"，以"再生——水之魔力"为策展理念，将学术性、历史感、前瞻性以及公众参与融为一体。

Held in the Yangpu section of the Huangpu River waterfront, the 2019 Shanghai Urban Space Art Season interprets the theme of "Encounter". On this occasion, people can encounter each other, water can encounter the banks, art can encount urban space, and the past can encounter the future. All these can create more encounters with good life and blissfulness, which will become indelible urban memories. The main exhibition is divided into two parts: "Urban Space Art" and "Planning and Architecture", which focus on the waterfront topic. The theme of the latter part is "Regeneration", showing "how the waterfront of the River and the Creek brings a better life for humans." The curatorial concept of "Rebirth (Mutation) - Water, an Enigma" has both academic and historical depth, and embraces a forward-looking vision.

卓越未来
Future Of Excellence

上海的一江一河有着蜿蜒绵长的岸线形态、古今辉映的建筑界面、独具标识性的城市天际线，同时拥有世界级的商务办公、文化展览、休闲旅游等功能和精彩纷呈的艺术、体育、游览等水岸活动，充分体现了上海在历史文化的传承、当代建筑的创新、滨水空间的建设以及市政设施的配套四个方面的建设成就。

一江一河地区的综合开发中不但注重历史建筑的保护，也关注工业厂区、水系格局等历史肌理和环境的协同保护。滨水地区更是通过更新改造，建成一批具有引领作用和标志性的当代建筑，不但体现了绿色低碳的设计理念，更引领集约高效的城市发展模式。

保障滨水公共空间的全线贯通、安全亲水、活动丰富、舒适宜人是规划建设工作的重要原则。通过桥梁、隧道、轮渡码头、地铁等增强跨河联系和两岸联动。这些桥梁和码头本身也成为一江一河空间中重要的景观要素，与沿线各区域特色景观相辉映。随着上海《黄浦江、苏州河沿岸地区建设规划（2018-2035年）》的发布，一江一河未来将打造与卓越全球城市相匹配的世界级滨水区，也将成为上海开放度最高、为大众公平共享的城市客厅，成为城市活力的集聚地。它不但可以展示城市历史和人文风貌，更可以通过滨水贯通并构建向腹地延伸的公共游憩网络，大力增加公共服务配套设施，成为宜居宜业的共享空间。未来，一江一河地区必将更开放、更人文、更美丽、更绿色、更活力、更舒适。把最好的岸线资源留给市民，下大力气开放滨水公共空间，让市民出门就能享受到和谐优美的环境，这些都体现了一江一河建设中人民城市的重要思想，以及协同治理、创新实践的发展方向。

在新时期，上海将加快完成发展转型，"四个中心"的功能迈上新台阶，人民城市的建设迈出新步伐，为新时代"城市，让生活更美好"的乐曲谱写新篇章。一江一河两岸地区在上海发展战略中的重要地位将更加突显，同时也面临更高的要求和更激烈的竞争。只有坚持更高站位谋求规划升级、更高标准推动建设实施、更高水平落实协同管理，坚持国际视野，坚持持续创新，才能建设国际著名的世界级滨水区，在未来的发展中体现上海的全球竞争力，让人民群众拥有更多的获得感、幸福感、安全感。

Shanghai's Huangpu River and Suzhou Creek each has a long and winding shoreline, a fascinating architectural profile and a distinctive skyline. The banks are lined with world-class waterfront office buildings, cultural, leisure and tourism facilities, hosting a wonderful range of artistic, sporting and touristic activities. All these reflect Shanghai's achievements in its celebration of history and culture, innovation of contemporary architecture and construction of waterfront and supporting municipal facilities.

The integrated development of the River and the Creek areas not only focuses on the conservation of historical buildings, but also on the conservation of the historical fabric and environment, such as plants, roads and waterway system. The waterfront areas have also been transformed through the land regeneration to create a number of leading and iconic contemporary buildings. These structures will not only embrace green and low-carbon architectural designs, but also lead the way to an intensive and efficient urban development mode. In recent years, the redevelopment of waterfront along the River and the Creek has been highly appreciated by the public.

It is an important principle of the planning and construction work to ensure the connection, safety, abundance of activities and comfort of the waterfront. Cross-river connections and linkages between the two sides of the river can be enhanced through bridges, tunnels, ferry terminals and subways. The bridges and docks themselves have become important landscape elements of the riverfront, reflecting the distinctive landscape of these areas. With the release of Shanghai's "The Waterfront Development Plan for the Huangpu River and the Suzhou Creek (2018-2035)", world-class waterfront will be built along the Huangpu River and the Suzhou Creek to match the status of Shanghai as an excellent global city, and the two waterways will become the fully open and equitably shared urban civic areas, serving as gathering venues to showcase urban vitality.

They will not only showcase the city's history and culture, but will also create shared spaces for living and working through building continuous waterfronts and extending them to the rest parts of the city and introducing supporting facilities. In the future, the River and the Creek areas are slated to be more open, more beautiful, more humanistic, greener, more energetic and more comfortable. The effort to leave the best shoreline resources to the public and to open them to the public will enable residents to enjoy a harmonious and beautiful environment when they go out. All these reflect a commitment to a "People's City" and the strategy of collaborative governance and innovative practices.

In the new era, Shanghai will accelerate its process of seeking new development paths to push its construction of "four centers" to a new level. The campaign of constructing a People's City will take a further step, thus realizing the goal of "Better City, Better Life" in the new era. The position of the area on both sides of the River and the Creek will become more important in Shanghai's development strategy, and at the same time the waterfront will have to be up to higher requirements and meet fiercer competition. We need to be forward-looking in our planning, pursue a high standard for development, improve collaborative management, have an international vision and continue our innovation. Only by doing so, can we build a world-class waterfront; it can add to Shanghai's global competitiveness in the future and give its people a greater sense of benefit, happiness and security.

综述・卓越未来　OVERVIEW · FUTURE OF EXCELLENCE

黄浦江篇
HUANGPU RIVER WATERFRONT

闵行区 Minhang District

徐汇区 Xuhui District

浦东新区 Pudong New Area

宝山区 Baoshan District		
01	宝山滨江公共空间 Baoshan Yangtze Waterfront	92
02	上海长滩 Shanghai Long Beach	250

杨浦区 Yangpu District		
03	上海国际时尚中心 Shanghai International Fashion Center	106
04	绿之丘 Green Hill	114
05	杨浦滨江公共空间 Yangpu Section of Huangpu Waterfront	42
06	世界技能博物馆 World Skills Museum	118
07	明华糖厂 Ming Hua Sugar Refinery	120
08	灰仓美术馆 Ash Gallery	110

虹口区 Hongkou District		
09	虹口滨江公共空间 Hongkou Section of Huangpu Waterfront	60
10	上海白玉兰广场 Sinar Mas Plaza	172
11	世界会客厅 The Grand Halls	124

黄浦区 Huangpu District		
12	外滩公共服务中心 Bund Public Service Center	128
13	外滩 SOHO Bund SOHO	176
14	外滩金融中心 The Bund Finance Center	180
15	董家渡金融商业中心 Dongjiadu Financial Center	184
16	世博会博物馆 World Expo Museum	188
17	黄浦滨江公共空间 Huangpu Section of Huangpu Waterfront	62

徐汇区 Xuhui District		
18	上海世博会城市最佳实践区 Urban Best Practices Area of Shanghai World Expo	130
19	上海当代艺术博物馆 Power Station of Art	134
20	日晖港步行桥 Footbridge on Rihui Creek	262
21	龙美术馆西岸馆 Long Museum West Bund	144
22	徐汇滨江公共空间 Xuhui Section of Huangpu Waterfront	68
23	星美术馆 Start Museum	142
24	龙华港桥 Longhua Creek Bridge	268
25	余德耀美术馆 Yuz Museum	138
26	西岸美术馆 West Bund Museum	192
27	油罐艺术公园 Tank Shanghai	148
28	西岸智慧谷 West Bund AI Valley	196
29	西岸传媒港 West Bund Media Center	202
30	游客集散中心（三港线"港口渡口"） Tourist Distribution Center (Sangang Line Ferry Terminal)	264

浦东新区 Pudong New Area		
31	洋泾港步行桥 Yangjing Creek Pedestrian bridge	272
32	民生码头八万吨筒仓 80,000-ton Silos at Minsheng Wharf	152
33	民生轮渡站 Minsheng Ferry Station	276
34	上海浦东船厂 1862 Shanghai Pudong Shipyard 1862	156
35	浦东美术馆 Museum of Art Pudong	208
36	上海中心大厦 Shanghai Tower	212
37	东昌栈桥 Dongchang Elevated Passage	280
38	上海 JW 万豪侯爵酒店 Shanghai JW Marriott Marquis Hotel	216
39	上海艺仓美术馆及长廊 Shanghai Modern Art Museum and its Walkways	160
40	白莲泾 M2 游船码头 Bailianjing M2 Tourist Terminal	282
41	新开发银行总部大楼 New Development Bank Headquarters	220
42	浦东滨江公共空间 Pudong Section of Huangpu Waterfront	74
43	倪家浜桥 Nijiabang Bridge	286
44	上海宋城·世博大舞台 Shanghai Songcheng · World Expo Grand Stage	164
45	上海大歌剧院 Shanghai Grand Opera House	224
46	上海久事国际马术中心 Shanghai Juss International Equestrian Center	228
47	前滩国际商务区 New Bund International Business District	232

闵行区 Minhang District		
48	闵行滨江公共空间 Minhang Section of Huangpu Waterfront	98

黄浦江篇　HUANGPU RIVER WATERFRONT

HUANGPU RIVER SUZHOU CREEK
THE WATERFRONT SPACE & ARCHITECTURE IN SHANGHAI

滨水空间重塑
REVITALIZATION OF WATERFRONT

杨浦滨江公共空间 Yangpu Section of Huangpu Waterfront
杨树浦电厂遗址公园　Art Waterfront of Yangshupu Thermal Power Plant
上海第十二棉纺厂　Shanghai No.12 Cotton Mill
杨树浦六厂滨江公共空间　Six-Plant Joint Site in the Yangpu Section of Huangpu Waterfront
上海船厂段　Shanghai Shipyard Section of Huangpu Waterfront
杨浦滨江公共空间二期　Southern Section of Yangpu Riverfront, Phase II
杨浦滨江公共空间示范段　Demonstration Stretch of Yangpu Section of Huangpu Waterfront

虹口滨江公共空间 Hongkou Section of Huangpu Waterfront
上海国际航运服务中心景观设计　Landscape Design of Shanghai International Shipping Center

黄浦滨江公共空间 Huangpu Section of Huangpu Waterfront
南外滩滨水岸线公共空间（复兴东路轮渡站至南浦大桥）
South Bund Waterfront (From East Fuxing Road Ferry Station to Nanpu Bridge)
黄浦滨江公共岸线贯通工程（南浦大桥至日晖港段）
Huangpu section of Huangpu waterfront (From the Nanpu Bridge to the Rihui Creek)

徐汇滨江公共空间 Xuhui Section of Huangpu Waterfront
徐汇滨江党群服务中心　Xuhui Waterfront Community Service Center

浦东滨江公共空间 Pudong Section of Huangpu Waterfront
上海民生码头水岸景观　Waterfront at the Minsheng Road Wharf
黄浦江东岸滨江公共空间　Waterfront on the East Bank of the Huangpu River
东岸望江驿　Riverview Service Stations, East Bund
上海世博文化公园　Shanghai Expo Culture Park

宝山滨江公共空间 Yangtze River Waterfront in Baoshan
宝山滨江贯通工程（一期）　Yangtze River Waterfront in Baoshan, Phase I
上海吴淞炮台湾国家湿地公园　Shanghai Wusong Paotai Bay Wetland National Park

闵行滨江公共空间 Minhang Section of Huangpu Waterfront
兰香湖　Orchid Lake

设计·师说
MASTERS' WORDS

涤岸之兴，向水而生
章 明

自古以来"择水而居"作为一种栖居的理想，如同基因般根植于绵长不绝的社会进化链条之中。工业时代曾使得人们舍弃逐水而居的空间，为工业发展做出让步：水岸因其航运的便利被货仓码头所占据，成为重要的生产岸线，居住之地则被推移至远离水岸的城市腹地。城市生活与水岸渐行渐远。

随着城市转型和产业结构调整，城市区域的工业区纷纷面临关停和转型，全球范围内出现荒废闲置的滨水工业区。因此，自1960年代末开始，城市滨水区的再生逐渐受到关注并历经几十年的探索，其间催生出一大批从生产岸线到生活岸线迭代的成功案例，如伦敦金丝雀码头、西班牙毕尔巴鄂、美国巴尔的摩内港等。

时至今日，对于依水而建的城市而言，滨水空间依然是城市建成环境中最重要的部分之一，它不仅在整个公共空间体系中有着举足轻重的作用，关乎市民日常生活体验，也体现着城市治理与精细化管理的现代化水平。这也是上海在实现全球卓越城市的历程中强调"一江一河"滨水空间复兴的原因之一。作为杨浦滨江南段和苏州河黄浦区段的总设计师，率领团队参与其中，见证了这两条母亲河沿岸的蜕变。

"一江"之杨浦滨江南段：叠合生长，还江于民

自2002年始，黄浦江两岸综合开发成为上海市的重大战略。2010年世博会的举办，加速了相关工业企业的搬迁。2015年，黄浦江两岸45公里公共空间三年行动计划启动。

杨浦滨江段位于黄浦江东端，拥有长15.5公里——上海浦西中心城区最长岸线，曾是黄浦江历史上工厂开设最早、最密集的区域，是上海乃至中国工业的发祥地，曾被联合国教科文组织专家称为"世界仅存的最大滨江工业带"。曾经高速粗放的城市化进程，使得大拆大建成为一种理所当然的思维模式，它几乎完全抹去了原有场地上的历史痕迹，而以主观的逻辑来统一场所的景观特征。

这是一条扭转观念的改造之路。一个具有丰富历史积淀的场所蜕变为城市滨水公共空间，就需要在场所的存留痕迹中挖掘价值与寻求线索。要发掘场所的潜在价值与精神，就无法脱离场地上的各种物质存留，它们是场地中历史记忆最真实的映射。城市公共空间应当呈现出城市文化的时间厚度和延续性。

防汛墙、渔市货运通道和防汛闸门、趸船的浮动限位桩、老码头的地面肌理以及钢质栓船桩和混凝土系缆墩等都是克服众多困难保留下的成果。同时在改造实践中引入新元素，既保持着对既有环境的尊重，又以一种清晰可辨的方式避免与既有环境的粘连，与留存的部分形成比对性的并置关系。譬如场地中新建的两组交通复合体——集合了坡道、座椅、展示、种植等功能的纺车廊架，以及通过单一元素"水管"的组合变化形成的栏杆与灯柱系列。

最终杨浦滨江的总体设计在此思想基础上，将"还江于民"的宏大主旨落位到"以工业传承为线索，营造一个生态性、生活化、智慧型的杨浦滨江公共空间"的理念中。通过实施有限介入，低冲击开发的设计策略，实现工业遗存的"再利用"、原生景观的"再修复"、城市生活的"新整合"。通过水上栈桥、架空通廊、码头建筑顶部穿越、景观连桥等不同方式贯通道路上的六个断点，形成5.5公里连续不间断的工业遗存博览带，漫步道、慢跑道和骑行道并行的健康活力带以及原生景观体验带。全线包括上海船厂、上海杨树浦自来水厂、上海杨树浦发电厂等在内的"八厂一桥"作为工业遗存博览带中九段特色公共空间，各自结合其空间与景观条件实现叠合再生。让工业文明的记忆以具"有时间厚度和空间深度"的城市景观的方式丰富城市文化和融入城市生活。正如邦尼·费舍在《滨水景观的设计》一文中阐述的："没有一个滨水区完全相似于另一个，也不该是，设计应该承认每一座场景的内在特性"。

从最初公共空间示范段的艰难尝试，到5.5公里总体概念方案的一气呵成，再到2.8公里公共空间的全新亮相，直至5.5公里全线公共空间全面开放；从对工业遗存全面的甄别、保留与改造，到现代技术与材料的探索，再到水岸生态系统的修复、基础设施的复合化利用与景观化提升，杨浦滨江正是通过公共空间的复兴，改变了曾经"临江不见江"的城市空间结构。在改善城市公共服务、修复城市生态、促进周边区域发展与产业升级的基础上，激发了新一轮的城市更新，并将工业区原有的特色空间和场所特质重新融入城市日常生活之中，使之从人们记

忆中的"大杨浦"印象中蜕变而出，迎来新的身份认同。

这些尝试与突破，使得杨浦滨江公共空间再生受到业界和社会的高度关注与评价，对全中国范围内的大量产业转型期的滨水工业遗地建设提供了积极的示范和借鉴价值。

"一河"之苏州河沿岸：风情长卷，向史而新

继2018年"黄浦江两岸公共空间贯通开放工程"之后，上海启动"苏州河沿岸公共空间提升工程"，希望以此提升城市生活品质、加强城市竞争力。

有别于黄浦江两岸的开阔性和城市关联度，苏州河的滨水空间较为逼仄，但其尺度更为亲人。面对这条具有百年历史与丰富内涵的滨水岸线，设计希望既能够"眷顾历史"又可以映射未来，在"向史而新"的基础上提出"上海辰光，风情长卷"的总体定位，综合考虑历史、城市功能、城市肌理、特征性要素，让苏州河黄浦段形成具有差异性的水岸空间，展现"典雅精致的，有内容的，有记忆的，有活力的'海派风情博览带'"。

虽然苏州河黄浦段整体的腹地进深有限，但设计依然致力于公共空间步行体系、绿化体系、亲水体系等系统的全线贯通或提升，因此充分挖掘贯通性的要素，提出五体系协调共荣的策略，包括两岸联动的漫步道、多元的绿化带、因地制宜的驿站建构筑物、连续的"马赛克"铺装艺术带、高水平的多艺术展呈带。这五个系统成为苏州河滨河空间整体性的基础。

三公里的滨水岸线以河南路桥和乌镇路桥为界，划分为东段、中段、西段三个段落。每个段落中都依据空间特征保活化、利用、改造历史遗存的节点：如以历史回溯和景观一体化的两种策略设计一系列通透轻盈的绿化棚架构筑物来活化原划船俱乐部；利用原吴淞路闸桥桥墩建构一座可以眺望东方明珠和上海大厦的钢质长亭——介亭；通过趣味化的建筑改造以及开放边界等方式，营造共享的城市公园——九子公园；利用原半地下倒班房结构体打造多层次立体观景活动空间——"樱花谷"；将加油站与咖啡厅进行叠合再生的中石化一号加油站等等。

在苏州河全段贯通中，为了释放更多高品质的滨水公共空间，设计在满足安全性的前提下根据空间的需要对防汛墙做了多种亲水化改造：例如利用二级挡防汛墙形成亲水平台，通过打开防汛闸门的方式增加联通性，抑或利用插板式防汛墙让狭窄空间段拥有亲水的可能等。防汛墙的处理体现的是滨水空间设计从"挡水"到"亲水"的思路转变，让城市空间呈现出更多的人性关怀与城市内涵。

城市滨水空间复兴历经几十年的探索，如今已不再是单一功能的简单置换，更倾向于一种城市的多维复合模式，通过嵌入公共交通、重要节点、生态体系等，力求将滨水空间重新纳入城市的生长与更新体系中。相应的，在操作路径上更强调跨领域、跨学科的协同合作：跨越传统意义上城乡规划学、建筑学、风景园林学，甚至市政设计、艺术设计、工业设计、智能设计和活动策划等相关领域边界和知识壁垒，以适应功能复合的实际需要，营造活力共享的高品质公共空间。

LIVE AND THRIVE BY THE WATER
Zhang Ming

The ideal of "living by water" has long been rooted in public mind in the social evolution process since ancient times. In the industrial age, people had to live away from waterfronts to make room for industrial development: the riverfront was occupied by warehouses and wharves for the convenience of shipping; waterways were reserved for the use of production. Residential quarters were set back from the waterfront, leaving civic life alienated from the waterfront.

With urban transformation and industrial restructuring, industrial facilities in urban areas were closed down and took on new functions, resulting in an abundance of derelict and disused industrial waterfront around the world. Since the late 1960s, the regeneration of urban waterfronts has received attention and has been explored for decades. A number of successful cases of transformation have appeared, such as the Canary Wharf in London, Bilbao in Spain and the Inner Harbour of Baltimore in the USA.

Today, for river-crossed cities, their waterfronts remain one of the most important parts of their built environments. These waterfronts not only play a pivotal role in the overall public space system and closely related to living experiences of their citizens, but also reflect the level of urban governance and refined management. This is one of the reasons why Shanghai emphasizes the revitalization of its waterfront in the course of building a global city of excellence. As the chief architect of the south Yangpu section of the Huangpu River waterfront and the Huangpu District section of the Suzhou Creek waterfront, I led a team to participate in the projects and witnessed the transformation of the waterfronts of these two waterways.

The South Yangpu Section of Huangpu River Waterfront: Symbiotic Growth, People's River

The waterfront revitalization of the Huangpu River has been made a major strategy for Shanghai since 2002, and the relocation of related industrial enterprises accelerated before the 2010 World Expo. 2015 saw the launch of a three-year action plan for 45 km of public space on both sides of the River.

The Yangpu section of the waterfront is located at the eastern end of the Huangpu River and stretches a length of 15.5 km - the longest sections among all the districts in Shanghai's Puxi downtown area. It is the birthplace of industrial sector in Shanghai and even nationwide. The rapid, often haphazard, urbanization process has made wholesale demolition the default modus operandi, completely erasing the historical traces of the sites in question and dictating the landscape in a process largely divorced from considered planning.

It is a transformation that requires a new mindset. To transform a historically rich place into a waterfront open to the public, we need to conduct a search for values and clues in the surviving traces of the venues. To discover the potential value and underlying spirit of a place, it is impossible to ignore the tangible remnants left on the site. They are authentic source of the site's historical memory. Urban public space should preserve the profundity and continuity of urban culture.

The flood control wall, freight lanes and flood gates in the fishing market, the floating limit piles of the barges, the pavement of the old quay, the steel pegs and concrete bollard piers are the preserved despite myriads of difficulties. At the same time, new elements are introduced into the renovation practice, showing respect for the existing environment while avoiding immediate relevance. This brings a contrasting juxtaposition with the remnants. For example, the two new transport complexes on the site -- including ramps, seats, displays, planting and other functions in a spinning gallery, and a series of railings and lampposts formed by the combination of a single element, the "water pipe".

The overall design of the Yangpu Riverfront is based on this idea, with the grand theme of "returning the river to the people" being placed in the concept of "creating an ecological, living and intelligent Yangpu Riverfront public space with industrial heritage as a thread". By implementing a design strategy of limited intervention and low-impact development, it achieves the "reuse" of industrial heritage, the "restoration" of the native landscape and the "new integration" of urban life. The six breakpoints on the road are connected by means of water bridges, elevated corridors, crossing on top of quay buildings and landscaped bridges, creating a 5.5 km continuous industrial heritage exposition zone, a health and vitality zone with walking, jogging and cycling paths, and a native landscape experience zone.

The eight factories and one bridge, including the Shanghai Shipyard, the Shanghai Yangshupu Water Treatment Plant and the Shanghai Yangshupu Power Plant, have been regenerated as nine sections of distinctive public space in the Industrial Heritage Expo Zone, each taking into account their spatial and landscape conditions. The memory of industrial civilization is enriched and integrated into the city's culture and life in the form of an urban landscape with "temporal and spatial depth". As Bonnie Fisher explains in Designing the Waterfront Landscape: "No waterfront is exactly like another, nor should it be; design should recognize the intrinsic character of each scene".

From the initial difficult attempts of the demonstration section of public space, to the completion of the 5.5 km overall concept plan, the 2.8 km of public space, the full opening of the entire 5.5 km public space; from the comprehensive screening, preservation and transformation of industrial heritage, to the exploration of modern technology and materials, to the restoration of the waterfront ecosystem, the composite use of infrastructure and landscape enhancement. The Yangpu section of the Huangpu River waterfront has changed the spatial structure of the city, which was once "no river in sight", through the revitalization of public space. The revitalization of public space has transformed the spatial structure of the city, which was once a "river without a view", by improving public services, restoring the ecology of the city and promoting the development and industrial upgrading of the surrounding areas, stimulating a new round of urban regeneration and reintegrating the characteristic spaces and places of the industrial area into the daily life of the city, transforming it from the memory of "Yangpu" to a new identity. The new identity of the industrial area has been reintegrated into the daily life of the city.

These attempts and breakthroughs have led to the regeneration of Yangpu section of the Huangpu River waterfront being highly regarded and evaluated by the professionals and the general public alike, providing a positive demonstration and reference for the construction of a number of waterfront industrial heritage sites in China during the industrial transformation period.

The Suzhou Creek Waterfront: Splendid Scenes of Shanghai

Following the "Huangpu River Crossing Project" in 2018, Shanghai launched the "Suzhou Creek Public Space Enhancement Project" in the hope of improving the quality of urban life and strengthening the city's competitiveness.

Unlike the Huangpu River waterfront that is both wide and closely related with the rest of the urban areas, the Suzhou Creek waterfront has limited space, but there is an intimate element in it nonetheless. The design seeks to straddle the past with the future. To interpret the grand idea of "Suzhou Creek: Splendid Scenes of Shanghai", the design is to take into account history, urban functions, urban fabric and special features, so that the Huangpu section of the Suzhou Creek waterfront is still distinctive, presenting a living museum of the social history of Shanghai, with abundant splendid scenes.

Although the overall depth of the non-waterfront area of the Huangpu section of the Suzhou Creek is limited, the design is still committed to building a continuous pedestrian system in the public space, greening system and waterfront platform system, exploring the elements of the five systems and improving their coordination. Concrete measures include linking up walkways on both sides of the Creek, diversified green belts, locally adapted convenience stations, continuous "mosaic" paving art zones, and a high-level multi-art exhibition zone. These five systems contribute to the integrity of the Suzhou Creek waterfront.

The three-kilometer waterfront is divided into three sections: the eastern, middle and western sections, separated by the Henan Road Bridge and the Wuzhen Road Bridge respectively. In each section, facilities of historical heritage are revitalized, utilized and transformed according to their spatial conditions: For example, a series of light and airy green scaffolding structures are designed to suggest the architectural traces of the former rowing club by integrating historical retrospection and landscape; Pavilion In-Between, a steel pavilion with a view of the Oriental Pearl and the Shanghai Mansions is built using the piers of the now defunct Wusong Road Sluice Bridge.

Through interesting architectural modifications and open boundaries, a shared urban park, Jiuzi Park, has been created; a multi-level three-dimensional viewing space, Cherry Blossom Valley, has been created using the former semi-underground shift house structure; and a petrol station and café have been overlaid and regenerated in Sinopec No.1 Petrol Station.

In order to give public a larger area of high quality waterfront along the Suzhou Creek, the design of the flood control wall has been adapted to meet the needs of the space, such as using two flood control walls -- one high and one low -- to create a waterfront platform, opening the flood gates to increase connectivity, or using the insert flood control wall to create access to water in limited spaces. The flood control walls are a reflection of the shift in the design of waterfronts from "keeping off water" to "embracing water", giving the urban space a human touch and generosity that characterizes the city.

After decades of exploration, the revitalization of urban waterfront is no longer simply a matter of function change, but a multi-dimensional model that seeks to re-integrate waterfront into the growth and regeneration of the city by introducing public transport, important facilities and ecological systems and so on. Accordingly, there is a greater emphasis on cross-disciplinary and interdisciplinary collaboration: crossing the traditional boundaries and knowledge barriers of urban and rural planning, architecture, landscape architecture, and even municipal design, art design, industrial design, intelligent design and event planning. The ultimate purpose is to adapt to the practical needs of functional complexity and create a vibrant and shared high-quality public space.

杨浦滨江公共空间
YANGPU SECTION OF HUANGPU WATERFRONT

兼具自然野趣和工业特色的景观环境
Unique landscape environment with natural and industrial characteristics

图片说明：遗址广场 / Square

杨树浦电厂遗址公园
Art Waterfront of Yangshupu Thermal Power Plant

项目地点：杨浦区杨树浦路2800号
设计单位：同济大学建筑设计研究院（集团）有限公司原作设计工作室
建设单位：上海杨浦滨江投资开发有限公司
设计时间：2015.5-2019.2
建成时间：2020.3
用地面积：36000m²

Location: No. 2800, Yangshupu Road, Yangpu District
Designer: Original Design Studio (A subsidiary of Tongji Architectural Design (Group) Co., Ltd.)
Constructor: Shanghai Yangpu Riverside Investment and Development Co., Ltd.
Design Period: May 2015 - February 2019
Date of Completion: March 2020
Site Area: 36,000m²

从污染严重的火力发电厂到生态共享的滨水岸线

杨树浦电厂遗址公园前身为建于1913年的杨树浦发电厂。这座曾经的远东第一火力发电厂在上海的城市发展中扮演了重要的角色，但大量燃煤也造成了空气污染，影响生态环境。随着黄浦江两岸公共空间贯通开放工程的启动，电厂关停，开始实施生态和艺术改造，从封闭的"闲人免入"的生产岸线，向文化、生态、共享的生活性滨水开放空间转型，修复环境污染，塑造场所精神，嵌入城市律动，实现"还江于民"。

"向史而新"、原生态的景观设计理念

电厂遗址公园的工业遗构改造设计始于两个锚点。一个是充分利用场地遗存特有的空间体量和形式：高105米的烟囱、江岸上的鹤嘴吊、输煤栈桥、传送带、清水池、储灰罐等作业设施，展开公共空间营造，这是塑造场所精神的出发点。另一个是采用有限介入、低冲击开发的策略，在尊重原有厂区空间和原生形态的基础上进行生态修复和改造。保留原有的地貌状态，形成可以汇集雨水的低洼湿地。植物配植以原生草本植物和耐水乔木池杉为主，同时配以轻介入的钢结构景观构筑物，形成别具自然野趣和工业特色的景观环境。

遗迹花园叠合的原真

介入场地时，输煤栈桥中靠近江岸的两座转运建筑连同一座办公楼已拆除。就着拆除的基坑挖掘出三方池塘，暴露出基础与钢筋，平整场地构成遗址广场，以展现原先建构筑物存在的痕迹。在场地最西侧，利用挖掘的土方堆出一座小山坡，与池塘构成负向的景观。

从无边水池看向江边塔吊 / Crane viewed from the wall-less pool

From thermal power plant to green space

The Yangshupu Power Plant Park is converted from the now defunct Yangshupu Power Plant, a British-funded facility dating from 1913. Although the plant—the largest in the Far East—once played a critical role in Shanghai's urban development, it also caused severe air pollution by burning coals, and engendered the surrounding environment. In 2015, the power plant was shut down for ecological repairing and artistic renovation. The project aims to transform the isolated and closed industrial waterfront to shared ecological and cultural open space, "returning the riverside to all the citizens". It has three targets: repairing environmental damages caused by coal-burning, forging the genius loci of the site and implanting the public space into the whole city.

Design idea: "Looking back for transforming anew" and returning to nature

The reuse of industrial structures serves as the starting point of forging genius loci. The site is left over with 105-meter-high chimney, huge crane-mouth hanging along the riverside, trestle works that used to carry the coal, clean water pond, damp and dry ash tanks and etc., which boasts spectacular spaces and impressive forms. The rearranging of public space began with comprehending the huge structures as well as the original industrial technological process. On the other hand, applying a strategy of limited intervention and low impact design, the team designed a restoration system of the ecological environment based on the spatial condition of the factories and the local eco-system. The design kept the topology and inserted an ecological wetland which collects rainwater. The plants for vegetation are mainly reeds and water-resistant arbor, which are commonly seen in local eco-system. The tectonics of passageways features a light steel structure that minimizes impact to the environment during construction and in visual aspects places an interesting juxtaposition of industrial and natural landscape.

Layered authenticity of the ruin garden

When the team first came to the site, two transportation structures and one office building were already torn down. We decide to seize the momentum and dig up three ponds upon the destructed foundation ditches of the three buildings, exposing the basement and steel bars. In the meantime, the site is leveled to become a ruin piazza. To the west, the earth dug up is used to formulate a low slope, composing a negative landscape to the ponds. Service rooms such as facility and public toilets are hidden under the slope made of cast-in-place concrete, whose framework are corrosion rust steel plate, the same as the pond fence. Three coal scuttles that were dismantled somewhere else are placed on one of the ponds, forming a visual focus like sculptures. The transportation belts along the riverside are also reserved. Its half-circular rubber bands are replaced with steel ones, which are meant to be covered with earth and with plants in it.

建造过程 / Building process

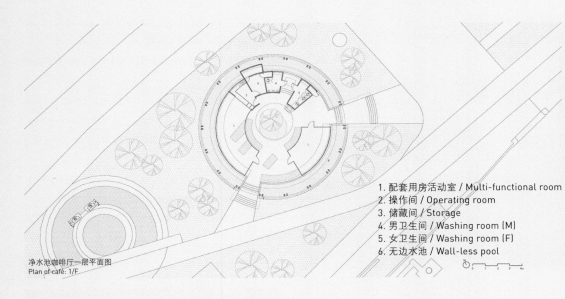

1. 配套用房活动室 / Multi-functional room
2. 操作间 / Operating room
3. 储藏间 / Storage
4. 男卫生间 / Washing room (M)
5. 女卫生间 / Washing room (F)
6. 无边水池 / Wall-less pool

净水池咖啡厅一层平面图
Plan of café: 1/F

咖啡厅内庭院
Courtyard of café

生态与休闲的双生花：净水池与咖啡厅

电厂作业设施中有一组储水、净水装置，拆除两个圆形储水池等上方结构后留下两个基坑。设计保留其中一处基坑作为雨水花园，另一处改建为咖啡厅。雨水花园种植芒草等净水植物，池中铺有鹅卵石，大雨时调蓄降水、滞缓雨水排入市政管网。

咖啡厅则在基坑上盖劈锥拱，点式细柱落在以同心圆方式形成的外圈基础上，上部的穹顶在顶部打开，引入自然光，同时也暴露出后方标志性的烟囱。人们坐在下凹的净水池咖啡厅中，既能透过柱间的开口瞥见一旁净水池做成的雨水花园，又能抬头望见高耸入云的烟囱，想见这片场地曾经的忙碌。

Clear Water Pool and the Cafe

There is a set of installations for water storage and cleaning, of which only two round water ponds are preserved. They have been demolished but the structural foundations remain. The design intends to turn one into a rainwater garden, and to transform another into a café. The foundation that is supposed to be a rainwater garden is planted with water-cleaning plants such as miscanthus sinensis, and has cobblestones covering the bed of the foundation. It can adjust the falling water and release pressure on the municipal drainage. The foundation that is meant to be a café works as a formal type. It is covered with a round conoid arch, standing on the round basement that is enlarged as a concentric circle to the foundation. The center of the arch is open to the sky, introducing natural light into the café and inviting the huge chimney into view. One can see the clear water reservoir while drinking coffee through the openings of the arch, recalling their fading past.

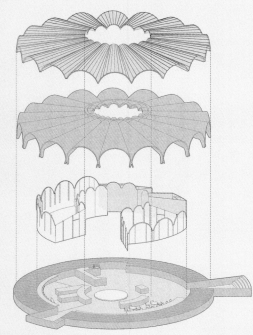

直立锁边铝合金屋面
Aluminum alloy roof with vertical locking edges

混凝土壳体
Concrete shell

超白玻璃幕墙
Ultra-clear glass curtain wall

混凝土基座
Concrete pedestal

净水池咖啡厅爆炸轴测图
Axonometric view of the café

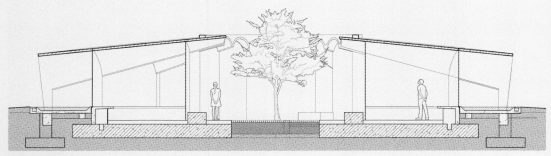

净水池咖啡厅剖面图
Sectional drawing of the café

净水池咖啡厅剖透视
Perspective drawing of the café

全景鸟瞰
Panoramic view

慢行步道
Footway

上海第十二棉纺厂
Shanghai No.12 Cotton Mill

项目地点：杨浦区北至安浦路、南至黄浦江、西至隆昌南路、东至腾越路
设计单位：上海大观景观设计有限公司
合作设计单位：上海聚隆绿化发展有限公司
建设单位：上海合生滨江房地产开发有限公司
设计时间：2018.1
建成时间：2019.5
用地面积：26700m²

Location: The block bordering on Anpu Road (north), Huangpu River (south), South Longchang Road (west), Tengyue Road (east)
Designer: DA landscape
Partner: Shanghai Julong Greenland Development Co., Ltd.
Constructor: Shanghai Hopson Riverfront Real Estate Development Co., Ltd.
Design Period: January 2018
Date of Completion: May 2019
Site Area: 26,700m²

基地原为日商创办的大康纱厂（1912年），后为国营上海第十二棉纺厂（以下简称十二棉）。过往，棉纺厂将棉花加工成人们可使用的纺织物品，如今滨江公共空间的开放将工厂逆转为人们可以使用的绿色植物空间，延续场地与人们生活的关联性。设计方案以"编织"为概念，融合场地中最显著的两种肌理，即代表原有工厂的斜线与代表沿江慢行动线的水平线条。方案通过新的防汛体系置换出更多的绿色空间，形成连续的滨江林荫步行网络，并嵌入活动场地与休憩设施，创造多样化的滨江体验。
在景观配置方面，作为杨浦滨江南段唯一没有大面积高桩码头的岸线，良好的覆土种植条件带来充足的绿量和生态自然的风貌。场地选用特殊配方的卡其色透水性地面材料，颜色和质感与真实的土壤非常接近，回应十二棉因生产行业领先的优质卡其布而被称为"卡其大王"的辉煌历史。

The site was originally occupied by the Japanese-run Dah Kong Cotton Spinning Milldating to 1912 and its successor, the Shanghai 12th Cotton Mill. Previously, the mill processed cotton into fabircs, but now the reuse campaign has turned it into a home of green plants, perpetuating the bond between this site and humans. The designer tries to "weave" two textures with the diagonal lines that characterize the original cotton mill and the horizontal line that suggest the prommonade along the river. By means of stepped flood control walls, the new design has produced an uninterrupted riverside network of walk paths.
This is only waterfront section in Yangpu without high-piled wharves, where ideal planting soil has brought a big swath of greenery. Khaki-colored water-permeable paving materials are used, as their color and texture are akin to the soil. It is a fitting reminder of the cotton mill that was once famous for the quality Khaki fabrics it produced.

鸟瞰图 / Bird's eye view

绿林迷径 / Green maze

卡其乐园 / Khaki Park

结合慢行系统设置无动力游乐设备空间，延续编织的主题，适合不同年龄段儿童的使用，与防汛体系地形的高低组合构建具有参与度和趣味性的活动空间。中部的纱泉广场利用高差形成层层跌落的阶梯水池，塑造联系江边与后方地块的水景轴线。水池边互动型模拟摇纱机、可控制高度的喷泉，水池内以纺织机零件为原型设计水车游乐装置，为市民提供丰富的主题互动式游览体验。十二棉的历史遗存在滨江空间改造前已消失，因此公共空间设计的重心放在人的体验和活动上，嵌入历史线索，在离黄浦江最近的地方，营造一片以绿色为主，趣味、可游玩的空间，为工业锈带加入轻松、柔和的生活质感。

The space with unpowered amusement equipment also addresses the theme of "weaving", and it is reserved for children of various ages. The flood control wall, with steps at different heights, offers a user-friendly experience. The Shaquan (meaning: fountain of yarn) Plaza in the middle also forms scattered heights with stone steps. The "yarn spinning wheels" by the pool can control the height of the water gushed from the fountain. The spinning wheel-shaped amusement devices beckon to the public in an interesting way. Before the renovation campaign, the historical relics of the cotton mill on the waterfront had already disappeared without a trace, so the public space design has focused on user experience and interactive activities. A green, fun and enjoyable space has emerged in the former cotton mill, making the industrial rust belt a place of relaxation.

纱泉广场
Yarn fountain square

阶梯水池
Stepped pool

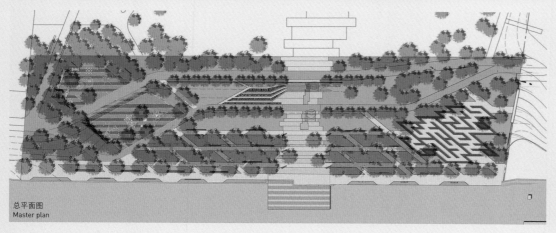
总平面图
Master plan

江边林荫道
Riverside footway

上海制皂厂-皂梦空间
Shanghai Soap Factory: Dream Space

上海制皂厂外景
Exterior view of Shanghai Soap Factory

电机辅机厂（西厂）-共生构建
Auxiliary Equipment Works (west): flora and landscape

杨树浦六厂滨江公共空间
Six-Plant Joint Site in the Yangpu Section of Huangpu waterfront

项目地点：杨浦区杨浦大桥宁国南路至隆昌南路范围内滨江公共空间
设计单位：致正建筑工作室、刘宇扬建筑事务所、大舍建筑设计事务所
合作设计单位：悉地（苏州）勘察顾问有限公司（设计审核）、上海浦海求实电力新技术股份有限公司（电力规划及供电设计）、上海邮电设计院有限公司（室外弱电设计）
建设单位：上海杨浦滨江投资开发有限公司
设计时间：2017.9-2019.8
建成时间：2019.10
用地面积：142588m²

Location: Waterfront from South Ningguo Road to South Longchang Road, Yangpu District
Designers: Atelier Z+, Atelier Liu Yuyang Architects, Atelier Deshaus
Partners: CCDI (Suzhou) Exploration & Design Consultant Co., Ltd. (Design review), Shanghai Puhai Electric Power High Technology Co., Ltd. (Power planning and power supply design), Shanghai Post and Telecommunications Design Institute Co., Ltd. (Outdoor weak electricity)
Constructor: Shanghai Yangpu Riverside Investment and Development Co., Ltd.
Design Period: September 2017 - August 2019
Date of Completion: October 2019
Site Area: 142,588m²

电站辅机厂西厂鸟瞰
Auxiliary Equipment Works (west): Aerial view

堆煤场景观带
Landscape area of coal yard

项目为国棉九厂、电站辅机厂东厂与西厂、上海制皂厂、杨树浦堆煤场及杨树浦煤气厂六个厂区的滨江公共空间更新和改造。设计始终坚持"有限介入，低冲击改造；保野性、保生鲜、保自然、保慢活；大主题线索串联，强调每个厂区场地特性"的策略。通过片段式地保留厂区界墙，以景观手法构建空间节点，传承杨浦百年工业的肌理；修复历史建筑风貌，保留厂房结构框架，置入新的空间，结合生态景观，提升体验互动，使其焕发新活力；保留场地的植被风貌，定制景观，培育野趣盎然的生态群落；连通高桩码头，结合防汛墙后移，形成多层次、连续、亲水的杨浦滨江空间。希望空间的再创造与它的历史相关联，并将其转化为一种空间特质。

This Six-Plant Joint Site revitalization project was done on the premises of six former industrial facilities, namely, the Shanghai No.9 Cotton Mill, the East and West Sections of the Shanghai Power Station Auxiliary Equipment Works, the Shanghai Soap Factory, the Yangshupu Coal Field and the Yangshupu Gas Works. The redevelopment observed the principle of "limited intervention, preservation of special features, slow process but uncompromised quality." The project, although subject to a single theme, preserves the special features of each facility. Some parititition walls between different facilities remain, and special approaches of landscape design are adopted to pay homage to the century-long industrial glory. To restore heritage buildings, the plan has preserved the structures in the original plants while new landscapes are created to appeal to the public. The vegetational features of the joint site are preserved and the landscape is innovatively designed, with ecological community having rich wild interest cultivated. The high-pile wharves are connected to each other, while the flood control walls stepping back from the waterfront combine to form a continuous, multi-layered Yangpu section of Huangpu Waterfront.

生态草阶
Grassed steps

煤气厂-边园
Riverside Park in gas works

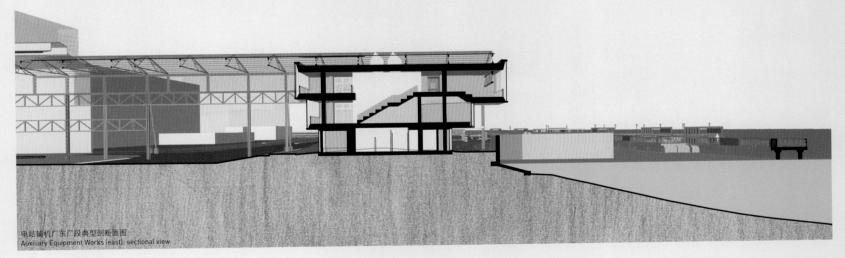

电站辅机厂东厂段典型剖断面图
Auxiliary Equipment Works (east): sectional view

上海船厂段
Shanghai Shipyard Section of Huangpu Waterfront

项目地点：杨浦区西起秦皇岛路游船码头站，东至杨树浦水厂
设计单位：上海优实建筑设计事务所
合作设计单位：上海优德达城市设计咨询有限公司（方案）、上海市政工程设计研究总院（集团）有限公司（码头结构）
建设单位：中国铁建港航局集团有限公司
设计时间：2016.11-2017.4
建成时间：2017.9
用地面积：31756m²

Location: From Qinghuangdao Road Cruise terminal to Yangshupu Waterworks
Designer: URBANDATA., Inc.
Partners: URBANDATA (Scheme design), Shanghai Municipal Engineering Design Institute (Group) Co., Ltd. (Wharf structures)
Constructor: China Railway Construction Port & Navigation Administration Group Co., Ltd.
Design Period: November 2016 - April 2017
Date of Completion: September, 2017
Site Area: 31,756m²

上海船厂浦西区记录并见证了上海乃至中国造船业近代以来百余年的发展史，具有重要的历史意义。

提炼历史及工业文化特征，在创新的设计里，传承历史、文化的独特性，同时展示出现代感、科技感，打造市民喜闻乐见的城市水岸空间。在公共空间重塑的过程中，设计师以结构性保护厂地原有肌理为重点，造访历史见证者，考察船舶造修工艺及生产过程；通过系统性梳理船厂历史文脉，挖掘其工业遗存特征并予以保留，通过空间设计方法，建立历史作业场景和当代滨水都市空间活动之间的内在关联，在传承和保护的基础上，打造出富有特色的船厂滨江公共空间，把项目落实到城市开放空间体系中。

从工业水岸到人文水岸的转化，突出"以人为本"的理念。研究人性化尺度，将旧有巨大的工业布局转换为舒适宜人的生活布局，让人与建筑、景观、环境对话，打造兼具艺术性、娱乐性、教育性和展示性的公共空间，构建集休闲、健身、观光、交流等功能于一体的城市水岸，构建船厂工业历史与现代生活的对话，浦西水岸与浦东水岸的对话。项目竣工后，在"三道"贯通的基础上，运动驿站、船排广场咖啡厅、坞门栈道、装焊广场、百草花园均成为市民的打卡地和时尚演艺的集聚场所。

With a history of more than 100 years, the Puxi plant of the Shanghai Shipyard has witnessed the evolution of Shanghai, especially that of Chinese shipbuilding. History is an essential element pertaining to the property.
A challenge to the designer is how to respect the history while cultivating a sense of style. During the revamping process, the designer, aims to preserve original features so the staff made extensive investigation related to shipbuilding and reached out to witnesses, studying, repair and manufacture process. Besides, advanced methods are applied to the space design.
The revitalization campaign has converted the former industrial facility to a space that combines art, entertainment with education, featuring the functions of recreation, fitness, sightseeing, and networking. The property can offer trails for walking, jogging and cycling; its gym, shipyard square cafe, wharf gate walkway, assembly and welding square and herb garden are destinations for leisure seekers.

船排空间
Shipyard square

百草花园
Peaceful park

船厂段鸟瞰
Bird's eye view of shipyard section

杨浦滨江公共空间二期
Southern Section of Yangpu Riverfront, Phase II

项目地点：杨浦区安浦路（丹东路、宁国路两个轮渡站之间）
设计单位：上海大观景观设计有限公司
合作设计单位：上海聚隆绿化发展有限公司、同济大学建筑设计研究院（集团）有限公司原作设计工作室
建设单位：杨浦区浦江办、上海杨浦滨江投资开发有限公司
设计时间：2016.8
建成时间：2017.10
用地面积：67000m²

Location: Anpu Road (between Dandong Road and Ningguo Road Ferry Station), Yangpu District
Designer: DA landscape
Partners: Shanghai Julong Greenland Development Co., Ltd., Original Design Studio (A subsidiary of Tongji Architectural Design (Group) Co., Ltd.)
Constructors: Huangpu River Administration Office of Yangpu District, Shanghai Yangpu Riverside Investment and Development Co., Ltd.
Design Period: August 2016
Date of Completion: October 2017
Site Area: 67,000m²

阶梯型长凳和露天剧场
Benches and amphitheater

基地历史上曾集聚祥泰木行（1902年）、怡和冷库（1920年）、明华糖厂（1924年）以及旧上海鱼市场（1946年），在后工业时代留存大量空置厂房与码头，成为市民与江岸的阻隔，也切断了沿江的步行动线。基地所面临的另一个挑战是防汛体系。作为人口超过2400万人的高密度城市，上海在黄浦江两岸执行严苛的千年一遇的防汛标准，这使得大部分的防汛墙顶部要比周边地坪高出2~3米，导致视线无法看到江面。

祥泰木行旧址
Site of Xiangtai Wood Firm

The 6.7-hectare plot is a 30-to 100-meter-wide stretch of hinterland that once housed the Xiangtai Wood Firm (dating from 1902), Jardine Cold Storage (dating from 1920), Minghua Sugar Factory (1924) and the Old Shanghai Fish Market (1946). In the post-industrial era, vacated buildings and abandoned wharves have made the waterfront inaccessible and also interrupted the otherwise continuous riverfront prommonade. Another challenge was posed by the flood control wall. As Shanghai has implemented a high flood control standard along the Huangpu River, most of the flood control walls are two meters above the ground, making it impossible for the public to see the river.

杨浦滨江公共空间二期鸟瞰
Bird's eye view of Yangpu Riverfront Phase II

保留码头起重机
Dock crane remained

因此，设计提出"看台与舞台"的构想，将原来单一的防洪墙改造成两级系统。第一级防汛墙顶部降低到与现状高桩码头地面高度相同，形成连续的场地空间，成为融合旧有遗存、承载城市生活的舞台，以留白的方式鼓励多元化的使用，同时亦满足百年一遇的防汛要求；第二级防汛墙采用千年一遇的标准，位置后退20～30米，形成6%坡度的面江休憩空间，形成融入景观地形和种植的看台，服务游客和市民。以黄浦江沿岸美丽的城市天际线为背景，杨浦滨江南段二期公共空间从人的视角出发，最终落实到人的活动，过去以生产为主的工业岸线在更新后实现了日常休憩与城市活动的弹性使用，重振活力的水岸真正回归于公众，也给未来留下了更多可能。

The solution, "Stand and Stage", turns the original single-wall flood control system into a two-wall one. The top of the first wall is as lower as the ground of the existing high-piled wharf to form a continuous space. It is the STAGE that integrates built heritage into urban life, giving incentive to diversified use while leaving undeveloped space. This flood control wall can also defend once-in-a-century floods nonetheless. The second wall is set 20-30 meters back from the first to brace for once-in-a-millenium floods. The waterfront space, with a 6% slope, makes for a STAND integrated into the green space.

With Shanghai's breathtaking skyline in the background, the project has taken into account the viewing effect and activities of the general public. The waterfront, once for the service of the industrial sector, has now become a place for an enjoyable assortment of events.

江边夜景
Charming night view

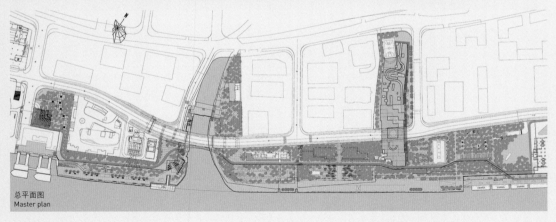

总平面图
Master plan

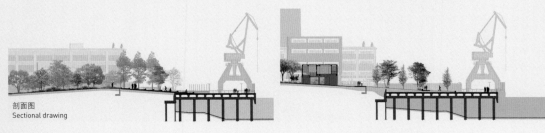

剖面图
Sectional drawing

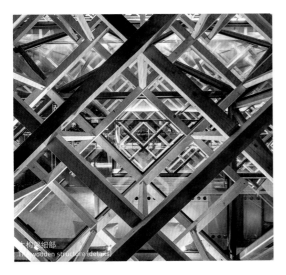
木构架细部
The wooden structure (details)

雨后小屋
The hut after raining

人人屋
House for Everyone (Ren Ren Wu)

项目地点：杨浦滨江杨树浦港东侧
设计单位：同济大学建筑设计研究院（集团）有限公司原作设计工作室
建设单位：上海杨浦滨江投资开发有限公司
设计时间：2018.3-2018.5
建成时间：2018.7
建筑面积：72m²

Location: The east flank of Yangshupu Creck
Designer: Original Design Studio (A subsidiary of Tongji Architectural Design (Group) Co., Ltd.)
Constructor: Shanghai Yangpu Riverside Investment and Development Co., Ltd.
Design Period: March - May 2018
Date of Completion: July 2018
Floor Area: 72m²

人人屋是杨浦滨江南段公共空间的一处滨水驿站。它是向每一位市民敞开的，提供休憩驻留、日常服务、医疗救助的温暖小屋，故取名"人人屋"。

源自场地历史的材质特征

杨浦滨江百年工业的历史背景，使每一段江岸都能追溯到一处历史悠久的工业厂址，人人屋所在区域是祥泰木行的旧址。据老产业工人回忆，直到20世纪90年代还有大量直径1米以上的木材在此区域运输、加工、分解。基于这一历史背景，设计之初便确定了采用钢木结构的策略，希望温润的木材能够唤起人们对这段滨江历史的回顾与感知。

激活场地空间的功能介入

人人屋所在的滨水区段为塔吊演艺区。以塔吊为背景、以码头为舞台、以绿地条椅为看台，夜晚可以放映露天电影和观看各种民间文艺演出。人人屋作为滨水驿站，提供直饮水、医疗急救、全息沙盘、微型图书室等服务，完善空间的日常服务功能。当该区域举行文艺表演时，驿站也可作为舞台的辅助用房，完善整个区段的功能设定。

快速建造的预制拼装体系

木结构构架的设计以人字形落地杆件为基本单元，不断重复并相互支撑形成整体空间结构，这也是取名人人屋的另一层含义。木杆件为云杉胶合木，空间结构模数为800mm，截面尺寸为60mm×60mm。构件标准化、工厂预制、现场拼装，在极其紧张的工期下依然保证了人人屋的较高完成度。

作为表征的外置结构系统

以相互连接的细密杆件代替尺度较大的梁柱体系，以共同作用的系统代替独立作用的系统，这一结构策略形成一个有厚度的朦胧界面，从而缩减了建筑体量对滨水空间的侵占，并使得内部与外部保持一种连续的、独特的空间体验。同时，外置的木构架成为滨水景观之中的重要元素，折射出木构建筑独有的温润之光。

西侧夜景
People enjoying night peace in the hut

The House for Everyone is a lovely jewel in the southern section of the Yangpu section of Huangpu waterfront. It is a community service center where every citizen can take a rest or receive medical assistance.

History-inspired materials

As Yangpu was a century-old industrial base of Shanghai, each section of the waterfront was formerly occupied a century-old industrial facility--the site of the House for Everyone included. It was formerly the site of the China Import & Export Lumber Company, a firm whose history goes back to 1902. According to the memories of old workers, blocks of wood with a diameter of more than one meter were shipped and processed here well into the 1990s. With this in mind, the designer is determined to use steel and wood as the materials to recall its past function.

New function to activate the space

The venue is reserved for performing art, an ideal place for open-air movies and performances in the evening, with cranes in the background, the pier as the stage and the benches in the green space as the stand. As a waterfront service station, the House also provides drinking water, first-aid kits, a holographic projection of riverfront urban model and a micro-library. When a performance is in progress, the House can also serve as an auxiliary room.

The prefabricated and fast assembly system

The design of the wooden frame uses herringbone rod as the basic unit which is shaped from the Chinese character for 'person' (ren). This makes another reason why people call it Ren Ren Wu. Each unit is repeated and mutually supported to form the overall structure. The wooden rods are made of spruce plywood with the modulus of the spatial structure as 800mm and the cross-sectional dimension as 60 x 60mm. The standardized components, prefabrication and fast on-site assembly enable the property to be completed in a very tight schedule.

The exterior structural system

Widely applying interconnected thin rods instead of beams and columns, the property uses a joint system in favor of traditional separate structure. It reduces the protion of the building that occupies the waterfront space, and maintains a unique, consistent spatial experience of the exterior and the interior. At the same time, the external timber frame becomes an important element in the waterfront landscape. The building lends a warm touch to the surroundings with wood as the material of choice.

水厂栈桥和保留的系缆墩
Elevated passage and the mooring dolphin remained

杨浦滨江公共空间示范段
Demonstration Stretch of Yangpu Section of Huangpu Waterfront

项目地点：杨浦滨江通北路至丹东路段
设计单位：同济大学建筑设计研究院（集团）有限公司原作设计工作室
合作设计单位：上海市政工程设计研究总院（集团）有限公司、上海市城市建设设计研究总院、上海东方罗曼城市景观设计有限公司、中交上海港湾工程设计研究院有限公司、上海宏辉港务工程有限公司、上海宏辉港务工程有限公司
建设单位：上海杨浦滨江投资开发有限公司
设计时间：2015.7-2016.7
建成时间：2016.7
用地面积：38000m²

Location: The stretch of Huangpu River Waterfront from Tongbei Road to Dandong Road
Designer: Original Design Studio (A subsidiary of Tongji Architectural Design (Group) Co., Ltd.)
Partners: Shanghai Municipal Engineering Design Institute (Group) Co., Ltd., Shanghai Urban Construction Design & Research Institute, Shanghai Dongfangluoman Urban Landscape Design Ltd., CCCC Shanghai Harbour Engineering Design & Research Institute Co., Ltd., Shanghai Honghui Harbour Engineering Co., Ltd.
Constructor: Shanghai Yangpu Riverside Investment and Development Co., Ltd
Design Period: July 2015 - July 2016
Date of Completion: July 2016
Site Area: 38,000m²

既存工业环境的挖掘和延续

杨浦滨江示范段也是杨浦滨江公共空间的启动段，设计注重既存工业环境的挖掘和延续，将老码头上遗留的工业构筑物、刮痕、肌理作为最真实、最生动、最敏感的记忆进行保留，使人们感受到老码头与黄浦江对岸陆家嘴CBD之间的差异，展现时间厚度，延续城市文化。

Honoring industrial heritage

As the demonstration stretch is a pilot project of the waterfront revitalization campaign in Yangpu, the designer has preserved the structures, scratches and fabrics as the physical reminders of the past. A stark difference can be felt between the old wharf and the Lujiazui CBD: such is the hybrid culture of Shanghai.

同水厂建筑相结合的坡道
Accessible slopes

夏日傍晚纳凉的周边居民
People living nearby

雨水湿地、钢结构栈桥和历史建筑
Greenland, water, steel elevated passage and historical buildings

钢索花园内的廊桥和凉亭
Steel elevated passage and pavilion in water

1、2号码头间搭建的钢栈桥
Steel elevated passage connecting wharfs 1 and 2

钢索坡道廊架和奔跑的小孩
Long steel corridor and lovely girl

原生态的景观设计理念

在尊重原有厂区空间基础和原生形态的基础上，采用有限介入、低冲击开发的策略，进行生态修复改造。防汛墙后腹地较大区域原本是一片低洼积水区，水生植物丛生，设计保留原本的地貌状态，形成可以汇集雨水的低洼湿地。大雨时能起到调蓄降水、滞缓雨水排入市政管网的作用。通过设置水泵和灌溉系统，湿地中汇集的水还可用于整个景观场地的浇灌之用。

基础设施的景观化利用

贴近黄浦江的杨树浦水厂对于公共空间贯通来说，既提出了挑战也提供了机会。利用水中基础设施的结构作为栈桥的结构基础，实现断点的贯通。同时将该基础设施纳入景观设计的范畴，人们可以欣赏到难得一见的角度和景象，也为景观设计和公共活动增添新的内涵。

The ecology-minded landscape design

Applying a strategy of limited intervention and low impact design, the designer designed a restoration system of the ecological environment based on the spatial condition of the factories and the local eco system. The hinterland behind the original floodwall is a puddle of water with many aquatic plants. The designer has kept the original topology and made it a wetland that could collect rainwater. The solution not only alleviates the pressure on the drainage system, but makes it serve an irrigation purpose with the use of a pump-powered water circulation system.

Reuse of industrial facilities

The Yangshupu Waterworks poses a challenge but also provides an opportunity to connect different sections of the Huangpu River waterfront. When the old facilities serve for the purpose of landscape design, they offer a new and inspiring perspective for visitors.

总平面图
Master plan

广场上保留的栓船桩
Mooring dolphins remained

虹口滨江公共空间
HONGKOU SECTION OF HUANGPU WATERFRONT

全景鸟瞰
Panoramic view

东侧鸟瞰
Bird's eye view from the east

上海国际航运服务中心景观设计
Landscape Design of Shanghai International Shipping Center

项目地点：虹口区公平路36号
设计单位：华建集团建筑装饰环境设计研究院有限公司
建设单位：上海银汇房地产发展有限公司
设计时间：2012.5
建成时间：2017.6
用地面积：64000m²

Location: No. 36 Gongping Road, Hongkou District
Designer: Shanghai Xian Dai Architectural Decoration & Landscape Design Research Institute Co., Ltd., Arcplus Group PLC
Constructor: Shanghai Yinhui Real Estate Development Co., Ltd.
Design Period: May 2012
Date of Completion: June 2017
Site Area: 64,000m²

西块港池实景
River bank in the west

上海国际航运中心与外滩金融街区、浦东小陆家嘴金融贸易区呼应，形成"黄金三角"格局。滨江景观与西侧的上海港国际客运中心相连，拥有长达780米的黄浦江岸线。中心围绕航运交易与商务需求，强化服务配套功能，形成以游艇港池为特色的商办滨水公共空间。

滨水景观设计以游艇港为主题，充分考虑实用性，坚持以人为本、可持续发展的规划建设理念。中心融合江景、码头、水岸、绿地及周边环境，既提供了世界级商业展示场地，又为附近市民、都市白领及国内外游客带来一处集观赏、休憩、运动、游玩等多功能于一体的绿色滨水天地，展示黄浦江两岸"世界会客厅"的风采。

The Shanghai International Shipping Center faces the Lujiazui Financial Hub on the other side of the Huangpu River and the Bund a few kilometers downstream. Connected with the Shanghai Port International Cruise Terminal on the west, the 780-meter-long waterfront focuses on the needs of shipping and trade, especially those of yachts.

The landscape design is people-oriented to accommodate the primary function of the property as a yacht wharf. It brings the river view, the wharf, the river bank, the green space and the environment into a whole. In addition to being a world-class complex, it also boasts a green space for the recreation of residents, office workers and tourists.

甲板空间
Waterfront space skin to timber deck

黄浦滨江公共空间
HUANGPU SECTION OF HUANGPU WATERFRONT

南侧鸟瞰
Bird's eye view from the south

开阔的滨江步道
Riverside footway

南外滩滨水岸线公共空间
（复兴东路轮渡站至南浦大桥）
South Bund Waterfront (From East Fuxing Road Ferry Station to Nanpu Bridge)

项目地点：黄浦区南外滩段滨江区域（复兴东路轮渡站至南浦大桥段）
设计单位：德国gmp建筑师事务所
合作设计单位：德国魏斯景观建筑（景观设计）、施莱希工程设计咨询有限公司（玻璃雨棚结构设计）、上海市政工程设计研究总院（集团）有限公司（施工图）
建设单位：上海外滩滨江综合开发有限公司
设计时间：2011.5-2013.12；2016.12-2018.6
建成时间：2018.12
用地面积：36300m²

Location: South Bund, Huangpu District (from East Fuxing Road Ferry Station to Nanpu bridge)
Designer: gmp Architects
Partner: WES (Landscape), Schlaich Bergermann Partners (Structure of glass canopy), Shanghai Municipal Engineering Design Institute (Group) Co., Ltd. (Constrution)
Constructor: Shanghai Bund Investment Group
Design Period: May 2011 - December 2013; December 2016 - June 2018
Date of Completion: December 2018
Site Area: 36,300m²

曾经的码头——南外滩区域是上海最早形成的城区之一，以生产、轮渡功能为主，建筑多为当地商人行会的仓库和会馆，缺少连接滨江的公共空间。在滨水空间贯通前，这里鲜有居民活动。老城区与江岸之间只有少量历史痕迹，如何与历史建立连接，如何把漫步道的设计与宽广的江景结合起来，如何营造相较外滩更私密的空间氛围，是设计的难点。

Formerly the seat of Shanghai's foremost passenger wharf, this area was historically one of the oldest urban districts in Shanghai, dominated by warehouses and guildhalls of local merchants. Before the waterfront revitalization, very few public activities took place here, and the neighborhood bore few traces of history. It was quite a challenge for them to bring history alive, integrate the promenade into the river view to evoke an intimate atmosphere.
The design extends the linear urban fabric of the old Shanghai town to the waterfront, creating a signature "wavy-shape" glass canopy that dominate the promenade.

北侧鸟瞰
Bird's eye view from the north

滨江步道夜景
Night scene

设计将老城厢的线性城市肌理延续到水岸，创造具有标志性的波浪形岸线，蜿蜒在城市与江景之间，提供步移景异的体验。区域内东西向街道的尽端就是滨江空间的"入口"，拾级而上即可欣赏宽阔江景。各入口处的步行道局部拓宽，形成面向城市和江面的"阳台"，入口之间的步行道缩窄，入口上方设扇形玻璃雨棚，增强辨识度，在夜间兼具照明功能。

设置低于步行道高度的亲水绿岛空间。波浪形岸线沿路一侧种植树木，为长椅提供阴凉，靠江的绿岛种植芦苇和草丛，增加岸线的动感。沿着栏杆、长椅和绿岛弧形轮廓的线性照明设计遵循并突出了波浪形。南外滩段滨水空间提供给广大市民和游客使用，成为城市的"大客厅"。

At the end of each perpendicular street is an entrance to the waterfront. People can come to enjoy the river view. The entrance is equipped with a fan-shaped glass canopy, which is illuminated at night to make it recognizable.
In the area between the entrances, the promenade swings back, creating "green islands" by the river. These "islands" are lower than the promenade. The green islands along the riverbank are planted with reed and grass to bring a dynamic touch. The lighting design also follows and emphasizes the wave-shape. Since the completion of the redevelopment, the waterfront space of the South Bund has been a favorite destination of tourists and locals alike.

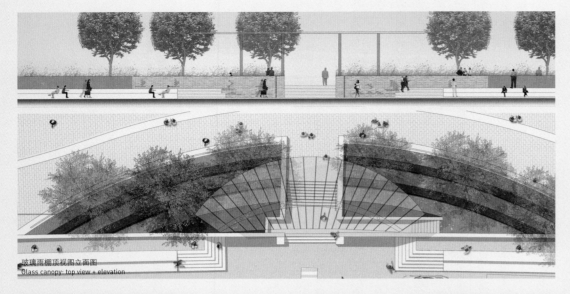

玻璃雨棚顶视图立面图
Glass canopy: top view + elevation

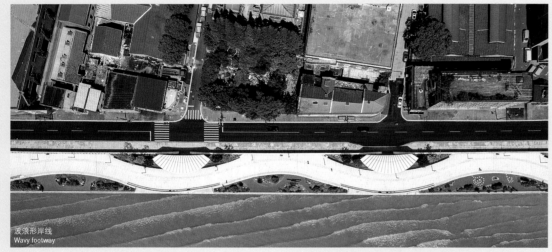

波浪形岸线
Wavy footway

总平面图
Master plan

入口处扇形玻璃雨棚
Fan-shaped glass canopy at entrance

岩石园步道
Footway in rock park

黄浦滨江公共岸线贯通工程（南浦大桥至日晖港段）
Huangpu Section of Huangpu waterfront (From the Nanpu Bridge to the Rihui Creek)

项目地点：黄浦区东至南浦大桥，西至日晖港，北沿苗江路，南临黄浦江
设计单位：华建集团建筑装饰环境设计研究院有限公司
建设单位：上海市黄浦区市政工程管理所
设计时间：2017.1-2017.8
建成时间：2017.11
用地面积：25.8hm²

Location: Area enclosed by the Nanpu Bridge, Rihui Creek, Miaojiang Road and the Huangpu River, Huangpu District
Designers: Shanghai Xian Dai Architectural Decoration & Landscape Design Research Institute Co., Ltd., Arcplus Group PLC
Constructor: Shanghai Huangpu Municipal Engineering Administrative Office
Design Period: January - August 2017
Date of Completion: November 2017
Site Area: 25.8 hm²

滨江景观与南浦大桥
Visible Nanpu Bridge

月季花园
Chinese rose park

项目以原江南造船厂三座船坞为中心向两侧展开，与对岸的中华艺术宫和奔驰文化中心隔江相望。设计追溯已消逝的"里滩"时代，因地制宜构建"一带、三道、七园"的景观格局，即一条带状开放空间，三条慢行道路（漫步道、跑步道、骑行道）与七个主题公园（杜鹃园、月季园、岩石园、春园、秋园、药草园、草趣园）。场地原先基本看不到江面、岸线亦不可达。

设计以公共性、公益性和安全性为主要原则，在不改变防汛墙高度的前提下，拓宽场地作为市民慢生活的空间，真正做到"还江于民"。通过制订"334"的三道设计准则，即漫步道和跑步道至少宽3米，骑行道至少宽4米，保证三道始终可以双向通行。结合不同路段和地形设计慢行系统的路线、宽度与坡度，串联沿江重点区域和重要节点，穿行于绿树、花草及休憩空间之间。此外，设计还保留原世博灯具、世博场馆排队时间记录铭牌等细节，传续场地记忆。

The waterfront stretches for 3.5km, with the three dockyards of the former Jiangnan Shipyard in the middle, opposite to the China Art Museum and Mercedes-Benz Arena across the Huangpu River. Given the topography, the landscape design can be summarized into "one belt, three trails and seven gardens": A belt-shaped open space, three trails (walking path and running and cycling trails) and seven theme gardens, namely the Gardens of Rhododendron, Rose, Rock, Spring, Autumn, Herbs, and Grass). The formerly inaccessible waterfront is now open to the public.
Although the height of the flood control wall remains unchanged, a wide waterfront platform is formed for leisure activities. Connecting major features and facilities, the two-way slow-traffic system includes a three-meter-wide walking path, a three-meter-wide jogging trail and a four-meter-wide riding trail. The route, width and slope of the slow-traffic system are determined according to the topography, while green space and recreation space are arranged along the way. The lighting fixtures and the electronic sceens used in the Shanghai World Expo are preserved as memorabia.

苗江路琴键春园
Qinjian Park on Miaojiang Road

琴键春园鸟瞰
Bird's eye view of Qinjian Park

保留"远望1号"
"Yuanwang 1" remained

滨江草甸景观
Riverside green land

徐汇滨江公共空间
XUHUI SECTION OF HUANGPU WATERFRONT

北侧鸟瞰
Bird's eye view from the north

景观小品
Ornaments

项目地点：徐汇区龙腾大道沿线
设计单位：北京东方易地景观设计有限公司
合作设计单位：上海市园林设计研究总院有限公司、上海市政工程设计研究总院（集团）有限公司、中交第三航务工程勘察设计院有限公司、上海市城市建设设计研究院(集团)有限公司
建设单位：上海徐汇滨江开发投资建设有限公司
设计时间：2008.5-2017.5
建成时间：2017.12
用地面积：约800000m²

Location: Longteng Avenue, Xuhui District
Designer: Beijing East Design Company, Inc.
Partner: Shanghai Landscape Design & Research Institute Co., Ltd., Shanghai Municipal Engineering Design Institute (Group) Co., Ltd., Cccc Third Harbor Consultants Co., Ltd., Shanghai Urban Construction Design & Research Institute (Group) Co., Ltd.
Constructor: Shanghai Xuhui Waterfront Development Investment and Construction Co., Ltd.
Design Period: May 2008 - May 2017
Date of Completion: December 2017
Site Area: About 800,000 m²

龙腾大道鸟瞰
Bird's eye view of Longteng Avenue

这里曾是20世纪民族工业的摇篮，"铁、煤、油、砂、粮"集聚，工业岸线封闭，沿江分布着上海铁路南浦站、龙华机场以及大量工厂、仓库和码头，为城市留下许多包括大跨度厂房、办公建筑和特色构筑物在内的工业遗产。以2010上海世博会的举办为契机，滨江沿岸自2008年起启动公共环境和基础设施建设。遵循"规划引领、文化先导、生态优先、科创主导"的发展原则，坚持高起点规划、高水平建设，持续推进岸线改造和城市更新。

The Xuhui section of Huangpu waterfront used to be dominated by production and logistics facilities, notably the Nanpu Railway Station and the Longhua Airport. The factories, warehouses and wharves have proved to be industrial heritage for the city, including large-span plants, office buildings and characteristic structures.

南侧鸟瞰
Bird's eye view from the south

该项目是黄浦江水岸空间的重生设计，是生态恢复与资源再利用的重要实践。设计以"上海滨江"为理念，分级设置防汛墙，建设绿色堤防，由景观大道至沿江亲水平台形成阶梯式、多层次的活动空间，提供多样化的观江体验。骑行道、跑步道、漫步道三道全线贯通，规划有轨电车贯穿南北，串联节点广场和各类活动场地，促进水、绿、人、文、城融合发展。借鉴国际一流滨水区开发经验，汇聚全球卓越水岸开发智慧，徐汇滨江成为可以驱车看江景的景观大道、多层次沿江公共活动空间、充满人文活力的滨水岸线。

This design seeks to activate the Huangpu River waterfront by means of ecological restoration and resource reuse. Dubbed the "Shanghai Corniche", the project builds two flood control walls rising to different heights and green embankment, forming a multi-level stepped space from the landscape avenue to the lookout platform. The waterfront has provided trails for walking, jogging and cycling, and has reserved space for north-south running tram tracks that connect the square and various gathering spaces.

滨江雪景
Romantic snow

立面图
Elevation

保留塔吊
Crane remained

Entrance (S)

北侧人视
Entrance (N)

徐汇滨江党群服务中心
Xuhui Waterfront Community Service Center

项目地点：徐汇区瑞宁路99号
设计单位：上海波城建筑设计事务所有限公司
合作设计单位：华建集团建筑装饰环境设计研究院、上海美术设计有限公司（布展）、哈塞尔国际设计公司（景观）
建设单位：中共上海市徐汇区委组织部
设计时间：2020
建成时间：2021
建筑面积：1560m²

Location: No. 99, Ruining Road, Xuhui District
Designer: Shanghai Bocheng Architectural Design Firm Co.,
Partners: Shanghai Xian Dai Architectural Decoration & Landscape Design Research Institute Co., Ltd., Arcplus Group PLC, Shanghai Art-Designing Co., Ltd. (Exhibition space design), HASSELL (Landscape)
Constructor: Organization Department of Shanghai Xuhui District Committee
Design Period: 2020
Date of Completion: 2021
Gross Floor Area: 1,560m²

主入口大厅
Main entrance

二层展厅
Exhibition hall on 2/F

一层展厅
Exhibition hall on 1/F

一层多功能厅
Multifunctional hall on 1/F

徐汇滨江党群服务中心是上海市委组织部与徐汇区委、区政府共同打造，作为新时代上海滨江党建创新实践基地，是一座集党的教育、党群服务、党员实践、党建创新为一体的枢纽型载体，更是集成各类服务功能的综合性阵地。

建筑呈现迎风而立、勇立潮头的姿态，由江面看来，如同一叶漂浮的轻舟，轻盈活泼，给予江边运动休闲的市民亲近之感。屋顶采用银白色的波纹板，整体造型向上折起，从卢浦大桥望去，犹如一被风拂起的书页；外立面穿孔板设计为水滴形孔洞，并通过参数化设计成为水波纹造型，错折相接，犹如江风吹起的党旗形成的褶皱呼应水岸滨江，树立庄严的党群中心形象。秉承把最好的资源留给人民这一原则，设置水岸露台、滨水步梯等高品质的开放空间，由建筑向水岸逐渐延伸，自然地过渡到滨江空间，有效串联景观大道和沿江开放空间。

室内设计以"一江一河"为设计线索，加入现代科技手段，通过3D浮雕水网、滨江掠影、水岸掠影、L形立体多维空间讲述水岸故事；一层挑空空间形成多功能水岸讲堂，设置可升降巨幅党旗，可作为党课、沙龙、论坛的活动空间，也为市民提供开放式阅读空间；设置咖啡吧、"第三卫生间"、母婴室、跑者驿站、"爱心加油站"等爱心服务的功能，切实体现"人民城市为人民"的设计初衷。

The Xuhui Waterfront Community Service Center is jointly built by the Organization Department of Shanghai Municipal Party Committee, the Xuhui District Party Committee and the District Government. As an innovation riverside practice base of CPC affairs in the new era, it covers a wide range of functions: providing party affairs education, community services, configuration for party members practice, and testing ground of innovative programs.

The building showcases a gallantry of braving wind and tide. Seen from the Huangpu River, it looks like a lightly floating boat, evoking an intimate sense for leisure-seekers along the river. Its upward-tilting roof, made of silver-white corrugated panels, makes it look like a wind-flipped page corner of a book when viewed from the Lupu Bridge. The perforated panels on the facade have teardrop-shaped holes; with the parametric design, they assume the shape of rippling waves, resembling a waving party flag kissed by the breeze. Such a spatial form echoes the surrounding waterfront environment and establishes a solemn image of the CPC. In response to the call of "leaving the best resources to the people", the Center provides high-quality open spaces, such as waterfront terraces and steps. The architecture extends from the river to the waterfront, effectively bringing the landscaped avenue and the open space into a whole.

The interior design focuses on the theme of "River and Creek"; modern technologies present the splendid scenery with 3D means in the L-shaped space. The double-height space on the first floor is a multi-functional lecture hall where a giant party flag that can be lifted and lowered; it can be used for party-themed lectures, salons and forums, and is also an open reading space for the general public. Other configurations include a cafe, a "third bathroom", a family room, a service station for runners, and a medical service center, among others. All these embody the goal of "a people's city for the people".

浦东滨江公共空间
PUDONG SECTION OF HUANGPU WATERFRONT

从西段眺望陆家嘴
View of Lujiazui from the west

上海民生码头水岸景观
Waterfront at the Minsheng Road Wharf

项目地点：浦东新区洋泾港至民生轮渡站
设计单位：刘宇扬建筑设计顾问（上海）有限公司
合作设计单位：上海市政工程设计研究总院（集团）有限公司（水工及结构）
建设单位：上海东岸投资（集团）有限公司
设计时间：2016-2017
建成时间：2018
用地面积：27191.5m²

Location: Between Yangjing Creek and Minsheng Wharf, Pudong New Area
Designer: Atelier Liu Yuyang Architects
Partner: Shanghai Municipal Engineering Design Institute (Group) Co., Ltd. (Hydraulic Engineering & Structure)
Constructor: Shanghai East Bund Investment (Group) Co., Ltd.
Design Period: 2016-2017
Date of Completion: 2018
Site Area: 27,191.5m²

贯通桥细部
Details of bridge

民生艺术广场
Art square

项目着眼于城市空间的整体性和延续性，通过低线慢步道、中线跑步道、高线骑行道"三线贯通"的总体设计，创造多样的慢行空间及丰富的游赏体验。设计以"艺术+日常+事件"为目标，采用"旧骨新壳、旧基新架、新旧同生"策略，在保留原有产业遗存空间特质的前提下，自东向西依次置入林木绿坡区、滨水步廊区、中央广场区、架空贯通道等活动空间，彼此叠加。

整体流线的规划组织适应日常及庆典两类活动的需求。为化解防汛墙在街道与水岸之间形成的阻隔，设计从剖面入手，结合地形和基础设施，用绿坡、步道、台阶、广场等方式，重组组织场地标高关系，新旧糅合，实现场地空间、景观和动线的连续性，改善水岸空间的体验感。这是一个融合当代城市功能与后工业景观的转化过程，是一个连接水岸断点与强化游憩体验的景观基础设施。设计通过独特的空间营造策略，打造黄浦江东岸新旧景观共生的水岸空间。

The project focuses on the integrity and continuity of the urban space, and creates a non-motorized traffic infrastructure in which the walkway is at the low elevation, the jogging trail at the middle elevation and the cycling trail at a high elevation. The design aims at "Art + Daily + Event" and adopts the strategy of "combining old contents with new appearance". While preserving the spatial characteristics of the original industrial heritage, it has introduced green slopes, a waterfront prommonade, a central plaza and an overhead passage way.

The overall design seeks to meet the needs of both daily and festival activities. The spaces between the river bank, the flood control wall and the street are arranged with green slopes, walkways, steps and plazas according to the topography and facilities. With a changing elevation, the site blends the old with the new, integrating contemporary urban functions into post-industrial landscape. It links up interrupted waterfront and enhances recreational experiences. Through a unique spatial creation strategy, the design creates a waterfront space on the east bank of the Huangpu River where old and new landscapes coexist.

贯通桥螺旋坡道
Spiral road

西段鸟瞰
Bird's eye view from the west

贯通桥
Bridge

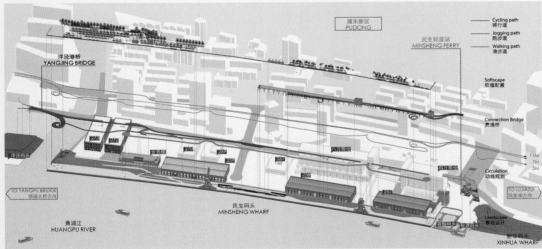

轴测爆炸图
Axonometric view

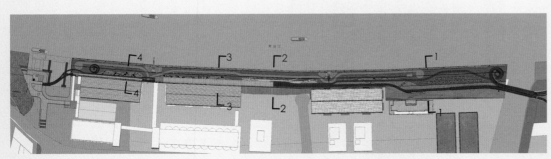

总平面图
Master plan

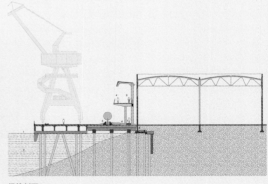

场地剖面1-1
Section 1-1

场地剖面4-4
Section 4-4

东段景观
Landscape in the east

鸟瞰图
Bird's eye view

黄浦江东岸滨江公共空间
Waterfront on the East Bank of the Huangpu River

项目地点：浦东新区黄浦江沿岸
设计单位：华建集团建筑装饰环境设计研究院有限公司
合作设计单位：法国岱禾景观事务所（TER）、同济大学建筑设计研究院（集团）有限公司原作设计工作室、上海城建市政工程（集团）有限公司等
建设单位：上海东岸投资（集团）有限公司、上海市申江两岸开发建设投资（集团）有限公司
设计时间：2016.12
建成时间：2019.3
用地面积：582000m²（其中后滩公园地块215000m²、世博公园地块139000m²、白莲泾公园地块116000m²、南码头地块44000m²、南栈地块68000m²）

Location: East Bank, Pudong New Area
Designer: Shanghai Xian Dai Architectural Decoration & Landscape Design Research Institute Co., Ltd., Arcplus Group PLC
Partners: Agenceter, Original Design Studio (A subsidiary of Tongji Architectural Design (Group) Co., Ltd.), Shanghai Urban Construction Municipal Engineering (Group) Co., Ltd., etc.
Constructors: Shanghai East Bund Investment (Group) Co., Ltd., Shanghai Shenjiang Liang'an Development Construction Investment (Group) Co., Ltd.
Design Period: December 2016
Date of Completion: March 2019
Site Area: 582,000m², including the Houtan Park (215,000m²), Expo Park (139,000m²), Bailianjing Creek Park (116,000m²), South Wharf (44,000m²), and South Stack (68,000m²).

别有洞天的下沉广场
Sunken plaza

南栈绿地
South Stack Green Space

南栈绿地的设计重点是漫步道与运动道的有机结合。为了与南码头的标高衔接，同时避免大量堆土对南浦大桥造成影响，设计采用空腔抬高地面，通过骑行道、跑步道与高桩码头相接。空腔形成三个鼓起的地形就像三只眼睛凝视着黄浦江，屋顶上各色草花争芳斗艳。通过建设连续而贯通的滨江走廊，极大地提升了世博区域的景观品质，加强了城市的连续性和可达性。江河与城市重新连接，新的绿道系统休闲、生态、充满活力。

The design of green space focuses on the organic combination of the walking path and the sports trails. The elevation of the viaduct connecting the north side of the South Stack Project to the South Wharf is 10.2 meters, 5.7 meters higher than the ground. The close distance from this green space to the Nanpu Bridge makes it not suitable for piling soil, so the ground is lifted in to form a cavity and the cycling trail and jogging trails are gradually connected with the high-pile wharf from the roof. The cavities form three bulging terrains, just like three eyes staring at Huangpu River. All kinds of grass flowers on the "eyes" are vying for attention against each other. By creating a continuous riverfront corridor, the landscape of the former Expo site has been greatly improved and the accessibility of this stretch of waterfront has been enhanced. The new greenway is a destination of leisure, both ecological and brimming with vitality.

黄浦江之畔
Beautiful eyes of Huangpu River

草坪上的覆土空腔
Lifted ground for cavity

白莲泾公园
Bailianjing Creek Park

世博云桥夜景
Expo Cloud Bridge

公园位于白莲泾汇入黄浦江的河口处，由白莲泾滨江绿地、世博村滨江绿地、白莲泾河道两侧绿地及沿江码头四部分组成。白莲泾公园段以"穿行于活力水岸，享受现代雕塑艺术之美"为主题，通过道路系统梳理、局部节点改造，强化与城市的联系，解决现状空间功能单一、可达性差的问题，提供体验滨水活力景观与感受雕塑艺术的穿行路径。设计利用现有码头，营造亲水休闲的低线滨水漫步空间；梳理林下空间，打通江景视线，营造林中穿行的中线慢跑道；利用现有地形，架设高架桥梁，营造视野开阔的高线骑行道。
新建M2码头，打造立体交通模式，解决游船和滨江休闲人流的不同交通需求；新建世博云桥连接亩中山水园和奔驰中心北侧滨江空间，打通滨江步道在此区域的断点，使白莲泾公园滨江绿地成为更具活力的市民休闲娱乐新空间。

The Bailianjing Park seeks to present an exciting collection of modern sculptures along the waterfront. The effort of re-planning its road system has strengthened its bond with the surrounding urbanscape and also addressed the problems of single function and inadequate accessibility of the space. The walking path affords visitors an opportunity to sample the dynamic waterfront landscape and inspired sculptures on their walk. The existing wharf will be transformed into a low-elevation prommonade. The grove will be remodeled to afford an unblocked view of the Huangpu River, and a midline walking path will be planned to cut through the grove. The existing terrain will be made use to build a viaduct, and create a high-line riding path with a broad vision.
A three-dimensional traffic hub at the M2 wharf will be built to meet the different traffic needs of cruise ships and pedestrians along the prommonade. The newly-built Expo Cloud Bridge connects the Muzhong Landscape Garden and the waterfront on the north side of Mercedes-Benz Arena, and spans the breakup point on the riverfront footpath, which makes the Bailianjing waterfront space gain more dynamicsas a destination for leisure and entertainment.

总 视图
Top view

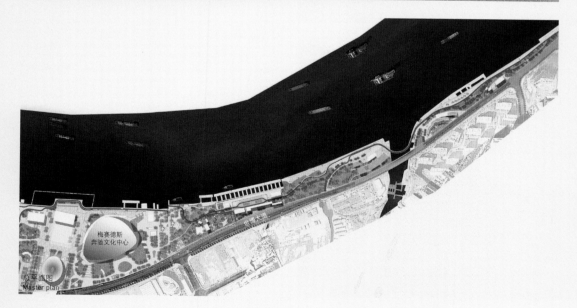
总平面图
Master plan

鸟瞰图
Bird's eye view

南码头
South Wharf

设计兼顾轮渡码头和消防支队两方面功能，采用竖向分层分流的交通流线组织，立体分隔公共空间人流与轮渡站过境人流。在滨江和腹地两条高线并行，凌空双桥形成立体交通，不仅满足了观景需求，也保证了使用安全。中心草坪、雨水花园、轮渡广场以及桥下篮球场共同构成舒适宜人的滨江环境。

The design takes into account the functions of both the ferry station and the fire brigade, and adopts a vertical hierarchical traffic flow organization to separate the people flow in the public space from the passengers at the ferry station. The elevated walkway and cycling trail not only meets the viewing needs, but also ensures safety. The central lawn, the rainwater garden, the ferry square, and the basketball court under the bridge jointly form a lovely riverside environment.

跑步及骑行桥
Bridge for running and cycling

山体东立面
East elevation of rockery

保留架空弧形桥
Retained overhead arc bridge

世博公园
Expo Park

世博公园北临黄浦江，南至世博大道，西起后滩公园，东至世博园区庆典广场。公园的原址是上海第三钢铁厂和江南造船厂，公园于2010年首次建成开放。此次景观提升，梳理了公园原有道路系统，提升了绿化空间品质，将其打造成集城市绿肺、科教文化、户外观演和庆典演艺等功能于一体的都市森林。

The Expo Park is framed by the Huangpu River in the north, the Expo Avenue in the south, the Houtan Park in the west, and the Expo Square in the east. The site had been formerly occupied by the Shanghai No. 3 Iron and Steel Works and the Jiangnan Shipyard before the Expo Park was opened in 2010. This landscape enhancement has rearranged the road network in the park and improved the quality of the green space, turning it into an urban forest that integrates the functions of greenification, education, open air performance and festivity events.

和兴仓库前广场
Square in front of Hexing warehouse

穿行在树林中的贯通道
A passageway through the grove

后滩公园
Houtan Park

公园北临黄浦江，南至世博大道，东接世博公园，西至倪家浜，场地原为浦东钢铁集团和后滩船舶修理厂所在地。公园于2010年首次建成开放，是一个具有湿地保护、科普教育功能的城市公园。此次景观提升，在满足贯通的前提下，保留公园的湿地生态、鸟类科普教育和承载城市活动的功能定位，通过局部节点改造，强化"最绿岸线"的公共空间定位。

The park is bordered by Huangpu River in the north, Expo Avenue in the south, Expo Park in the east, and Nijiabang in the west. The site was formerly the location of the Pudong Iron and Steel Group and Houtan Ship Repair Factory and was converted into a park in 2010. The latest landscape redevelopment retains many original functions of the Houtan Park, including wetland ecology, ornithology education and accommodation of festivity events.

鸟瞰图
Bird's eye view

野趣小径
Tranquil footway

后滩公园跑步道
Running track in Houtan Park

总体顶视图
Top view

19号望江驿鸟瞰
Aerial view of Service Station 19

6号望江驿
Aerial view of Service Station 6

东岸望江驿
Riverview Service Stations, East Bund

项目地点：浦东新区黄浦江贯通工程东岸滨江公共空间内
设计单位：致正建筑工作室
合作设计单位：上海思卡福建筑工程有限公司
建设单位：上海东岸投资（集团）有限公司（1-4#、6-17#、19-22#）、上海陆家嘴（集团）有限公司（5#、18#）
设计时间：2017.8.11 - 2018.8.7
建成时间：5#（北滨江）望江驿 2017.9.25；18#（前滩休闲公园）望江驿 2017.12.22；1-4#、6-17#、19-22#望江驿 2018.8.15
建筑面积：130m² (5#)；146m² (18#)；151m² (1-4#、6-17#、19-22#)

Location: Inside the East Bund Riverside Space, Pudong New Area
Designer: Atelier Z+
Partner: Shanghai SKF Construction Co., Ltd.
Constructor: Shanghai East Bund Investment (Group) Co., Ltd. (#1-4,6-17,19-22) / Shanghai Lujiazui (Group) Co., Ltd. (#5,18)
Design Period: August 11, 2017 - August 7, 2018
Dates of Completion: September 25, 2017 (#5); December 22, 2017 (#18); August 15, 2018, (#1-4, 6-17, 19-22)
Gross Floor Area: 130m² (#5), 146m² (#18), 151m² (#1-4,6-17,19-22)

东岸望江驿是位于上海黄浦江贯通工程东岸滨江公共空间内的一系列服务驿站，22个驿站分布在从杨浦大桥到徐浦大桥的东岸贯通带里，每公里一个沿江排布，为市民提供休憩停留空间和公共卫生间。为了平衡极短工期和我们对于完成品质、空间体验的最大诉求的矛盾，我们设计了一个标准统一的建筑形制，并根据不同的场地条件和地形特征，归纳了数种有差异的落地类型，在基础结构形式和场地标高关系上各不相同。建造上采用以胶合木结构为主的钢木混合体系，实现质量可控的超快速建造。利用当代施工企业在施工组织、结构优化及细节处理上具有的自我协调能力，通过设计与施工的高度整合和合理切分，达成极短工期内空间品质及综合效应的最大化。

线性的贯通"运动"空间增加了事件发生和公众互动的可能性，而22个点状的望江驿则为其补充了"停留"的休憩空间。望江驿作为贯通工程中必不可少的城市基础设施，摆脱了基础设施曾经隐匿、冷峻、严肃的形象，通过与景观、地形的多维整合，形成日常、自主、有活力的公共空间，增强了浦江贯通的场所体验。

这一系列位于上海市中心最具有公共性的滨江空间的微小驿站给了我们机会探讨建筑与风景的关系，以及微观场地与宏观公共空间及城市标志物的关系。希望驿站在以平易近人的氛围为市民提供支持和服务的同时，能够强化场地自身的特性，从而让建筑有机会成为风景的放大器。"望江驿"这一命名正突显了驿站的双重述求。

每一个望江驿都由两部分构成：一侧是相对封闭的公共卫生间，另一侧是开放通透的、布置有信息导览和发布、阅读书架等服务设施的公共休息室。两部分的左右分布原则首先是考虑驿站所处局部场地中保证休息室的开放性与面江视线的最大化，其次是在宏观尺度上考虑休息室的对外视线对于小陆家嘴中心区的关照。这两部分之间是一条穿越建筑的有顶通廊，连接背江一侧的骑行道和面江一侧的跑步道和漫步道，其中布置有自动售卖机、储物柜、冷热直饮水、共享雨伞机等便民设施。而整个沿江一侧包括休息室的侧面都是深广的檐下平台空间，靠墙设有通长的坐凳供市民小坐，无形中放大了市民的观江休憩空间。

驿站简洁方正的平面轮廓和相对复杂的半螺旋直纹曲面单向檩条木结构仿铜铝镁锰板屋面间形成鲜明的对比。从临江侧看，驿站像是一个微微架空在场地上的出檐深远、屋顶轻盈起翘的大凉亭。反坡屋檐下伞状放射形布置的木檩条成为整个休息室内外的视觉焦点；檩条汇聚处自然形成一个三角形的天窗，一半在休息室内，一半在通廊上，将幽暗的屋顶深处照亮，强化了空间的进深感。从背江面看，驿站的屋顶分为高低不同、但都向内倾斜的左右两半，特别在正中靠近通廊的休息室屋顶被压至接近视平线的最低点。这种特意压低的尺度加强了背江面的进入感，将人由居中的通廊引向江景。

The 22 East Bund Riverview Service Stations are located in the Huangpu waterfront on the east bank of the river between the Yangpu and Xupu Bridges, with a distance of one kilometer from one another. The stations are both venues for taking rest and public toilets. As it is almost impossible to build the stations in conventional design and with conventional construction procedures, the designer uses a consistent building shape and several stereotypes based on different site conditions and topographical features; they are differerent in basic structural forms and the site elevations. The construction adopts a steel-and-wood hybrid system mainly composed of laminated wood to meet the demand of short construction period and high quality. With the benefit of contemporary structural optimization and detail processing method, the constructor finished the task through high-level coordination with the design firm.

While the linear distribution of the stations catalycizes the interactions between them and the public, the 22 stations are like points to provide welcome respite to the public. As indispensable urban facilities in the waterfront revitalization project, the Riverview Service Stations stay clear of the stereotyped images typical of similar facilities in the past: secluded, cold and serious. Through the integration with the landscape and the terrain, the stations appear as user-friendly dynamic public space, optimizing the space experience of the waterfront.

Although moderate in size, these stations in the iconic waterfront at the heart of Shanghai explore the relationship between architecture and landscape, and the relationship between small facilities and the macro-scale public spaces and urban landmarks. The designer expects that the stations can be architecturally interesting sites compatible to the waterfront scenery while providing service to the citizens.

Each station consists of two parts: a relatively closed public toilet on one side and an open, transparent public lounge with service facilities such as bulletins and magazine racks. While working on the layout, we prioritize a panoramic river view from the lounge. Secondly we want to make Small Lujiazui appear at the center of the view. A corridor brings these two parts together, connecting the cycling lane at a distance from the Huangpu and the running and walking trails along it. Vending machines, lockers, hot and cold drinking water machines, rack of shared umbrellas and other convenience facilities are available along the corridor. Along the river, there is a wide, roof-covered platform that covers a part of the lounge. Benches are set up near the wall to give convenience to the visitors.

The square planar contour of the station is in sharp contrast to the relatively complicated half-spiral straight line curved surface single-direction purline roofing structure made of copper imitated Al-Mg-Mn Alloy sheeting. Viewed from the riverside, the station resembles a large pavilion with far-extending eaves and a lithe crimped roof, which is slightly elevated from the site. The umbrella-shaped radial purlines under the counter-slope eaves in the north have become a visual focus. The purlines converge to form a natural triangular skylight, one half in the lounge and the other half above the vestibule, illuminating the dark roofrecesses and strengthening the space depths. Viewed from the back of the river, the station roof is divided into two halves in the east and west, which have different heights but both tilt inward. Especially, the roof at the corner of western-section stair case in the center has been depressed almost to the lowest point of visual horizon. This specially depressed dimension has strengthened the entering sense of the side back to the river, guiding visitors from the central vestibule to the riverfront.

希望望江驿能够形成对公众的身体性引导和对空间自由使用的鼓励。人们可以随着三角天窗光亮的指引，穿过低矮亲切的通廊走向江岸，屋顶在配合身体的运动渐次升高，通廊尽端外的江景在视野中缓缓显现，冲破高敞的屋檐；来到望江檐廊上豁然开朗中视线向下，江面在粼粼波光中水平向展开，与江边或散步或奔跑的人影共同构成流动的风景。两侧身后廊下的长凳会吸引人们安坐下来，悠闲地观赏江景。隔着闪烁的江面，对岸的城市或隐在眼前的树丛之后，或在人们面前水平展开，有一种特别的宁静。当然，当人们走下檐廊，下到漫步道或亲水平台，在那里，浦江两岸壮丽的城市天际线一览无余。

The designer hopes that the station can be visitor-friendly and enable the public to make free use of the space. People can walk through the lovely corridor to the river bank with the guidance of the triangular-shaped skylight. The roof gradually rise up, the river view at the end of the vestibule will also raise in your field of vision, breaking through the high and spacious eaves in the north. Standing on the open platform and viewing through the trunks at the bottom, you can see the glistening river surface is stretching horizontally, constituting a flowing scene with figures walking or running along the riverside. The benches on both sides of the vestibule will attract people to sit idly and appreciate the river scene leisurely.

12号望江驿
Service Station 12

18号望江驿
Service Station 18

9号望江驿内景
Service Station 9

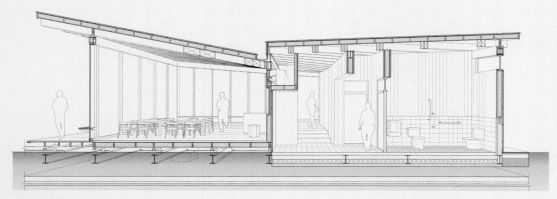

4号望江驿东西向剖透视
Section perspective of Service Station 4

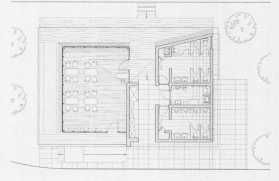

4号望江驿平面图
Plan of Service Station 4

86

2号望江驿鸟瞰
Bird's eye view of Service Station 2

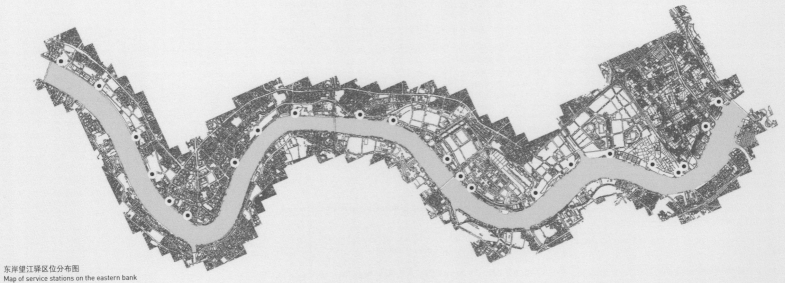

东岸望江驿区位分布图
Map of service stations on the eastern bank

上海世博文化公园
Shanghai Expo Culture Park

项目地点：浦东新区西北部毗邻黄浦江，东至卢浦大桥—长清北路，南至通耀路—龙滨路
设计单位：华东建筑集团股份有限公司、上海市园林设计研究总院有限公司、杭州园林设计院股份有限公司、上海市城市建设设计研究总院（集团）有限公司
建设单位：上海地产（集团）有限公司
设计时间：2017.5-2020.5
建成时间：2023.12
用地面积：1541950m²

Location: Enclosed by the Huangpu River in the northwest, the Lupu Bridge and Changqing North Road in the east, and Tongyao-Longbin Road in the south
Designers: Arcplus Group PLC, Shanghai Landscape Design & Research Institute Co., Ltd., Hangzhou Landscape Design Institute Co., Ltd., Shanghai Urban Construction Design & Research Institute (Group) Co., Ltd.
Constructor: Shanghai Land (Group) Co., Ltd.
Design Period: May 2017 - May 2020
Date of Completion: December 2023
Site Area: 1,541,950m²

申园鸟瞰
Bird's eye view of Shanghai Garden

以"生态自然永续、文化融合创新、市民欢聚共享"为设计理念，以"世界一流城市中心公园"为设计目标，世博文化公园内保留改造法国馆等四个世博会场馆，构建四大功能组团，同时新建世博花园、申园、双子山、上海温室、世界花艺园等设施。项目是上海完善生态系统，提升空间品质，延续世博精神，加快建设卓越全球城市的重大举措之一，打造上海中心城区最大的沿江公园绿地。

城中有景，景中有城的整体架构

"造山"——最高48米的人造山体，与连绵起伏的余脉地形，环绕整个公园，阻隔城市的喧嚣，形成面向浦江的山势。

"引水"——U形水体成为缝合各个功能区的核心，利用后滩已有水利设施，实现自然流动。

"成林"——特色鲜明的七彩森林与水体，覆盖整个公园，高达80%以上的绿地率，成为城市中心的新绿肺。

"聚人"——世博保留场馆、温室、江南园林、大歌剧院、马术谷等引人瞩目的现代建筑，将丰富多彩的城市生活融入自然。

特色鲜明，互融互通的功能布局

整体布局围绕世博环区、人文艺术区、自然生态区展开，与后滩公园区景观融合设计，形成四大功能组团。世博环区位于公园东北角，以世博会保留场馆为核心，包含世博花园、舞动广场、静谧林三个片区，传承世博文化记忆，打造文化高地，提供文化交流场所与创新平台。人文艺术区位于公园西侧，包含江南园林、音乐之林、大歌剧院、世界花艺园、马术谷五个片区，片区以人为本，共享开放，打造便于市民前往并乐于驻留的高雅艺术生活圈。自然生态区位于公园南侧，包含温室花园、双子山两个片区，以生态为先，蓝绿网络渗透，完善生态格局，重塑自然生境，促进生物多样性，打造城市生态修复典范。

The Expo Culture Park is located at the heart of the of Pudong section of Huangpu waterfront, framed by the Huangpu River in the northwest, the Lupu Bridge-Changqing North Road in the east, Tongyao-Longbin Roads in the south, covering a total site area of 1.54km². It has convenient traffic network, accessible by a several roads and rail transit lines. The design of the park seeks to build an ecosystem to enhance natural sustainability, cultural integration and innovation, and public enjoyment. The designers expect to create a world-class city center park. Four original Expo pavilions have been preserved to be parts of the park. At the same time, the Expo Garden, Shanghai Garden, Twin Rockeries, Shanghai Greenhouse and World Gardens are thrown up. The property aims to improve Shanghai's ecosystem and spatial quality, while perpetuate the Expo spirit and accelerate the process of making Shanghai shine on the world map. It is slated to be the largest riverfront park and green space in downtown Shanghai.

Overall structure: Lovely scene of the city, and making the city in the scenery

"Hill building": A rockery 48 meters high

The whole park is surrounded by undulating residual terrain, which blocks out the noise of the city and builds an artificial "hill" along the Huangpu River.

"Introducing water": The U-shaped creek connects each functional area. The existing water conservancy facilities at Houtan can enable the water to flow naturally.

"Afforestation": The forest with distinctive features will cover the entire park. With a green coverage rate of over 80%, the park will be a new vast green belt in the city center.

"Gathering place" - The remarkable modern buildings, such as the former Expo pavilions, warm house, Jiangnan garden, Grand Opera House, and the Equestrian Valley, will lend the vigor of urban life to the natural environment.

Functional zones with distinctive features, integration and mutual connectivity

The overall layout lays empasis on three mainzones, namely, the Expo Ring Zone, the Cultural and Art Zone and the Natural Ecozone. These, together with the landscape of the Houtan Park, form the four functional zones. The Expo Ring Zone is located in the northeast of the park, where the preserved Expo pavilions are the key features. It has three major functional areas: the Expo Garden, Dance Square and Quiet Forest. They will be the cultural spring heads as the Expo pavilions once were, where cultural exchange and creativity events will be held. The Cultural and Art Zone is situated in the west of the park, including a Jiangnan Garden, the Music Forest, the Grand Opera House, the World Floricultural Garden and the Equestrian Valley. The zone aims to allow visitors to savor its multiple cultural resources. The Natural Ecozone in the south of the park includes two major functional features, namely, the Glasshouse and the Twin Rockeries. Ecology is designated as its most prominent feature: With the improvement of the environment, biodiversity will be restored along with the reemergence of natural habitats.

世博文化公园整体鸟瞰
Bird's eye view of Expo Garden

申园
Shanghai Garden

申园地处世博环区与人文艺术区之间，临后滩，面湖对山，形成北山南湖、东园西苑的空间布局，是世博文化公园中具江南园林特色的园中园，意在打造"虽由人作，宛自天开"的江南山水园林景观。北山通过起伏的山脉形成厚实的山林背景；南湖以大水面为主，广种荷花，岸边设水榭，形成空灵雅致的环境；东园以假山跌水为特色，水岸点植松树、鸡爪槭，形成极富古典韵味的山水画卷；西苑通过堤岛的设计呼应传统园林中一池三山的山水模式，并最终形成放逸洒脱的自然风景。

The Shanghai Garden is a part of the Expo Cultural Park tucked between the Cultural and Art Zone and the Expo Ring Zone, facing a lake and a rockery. It has distinct Jiangnan (south part of the Yangtze Delta) features, best known for its prinstine beauty. In the planning of Shanghai Garden, the spatial layout is hill in the north and lake in the south, park in the east and garden in the west. The hills in the north form a thick hill forest background through rolling hills. The lake in the south is dominated by large water surface, with lotus flowers planted in the lake and waterside pavilions set on the shore, with an ethereal and elegant environment formed. The park in the east is characterized by falling water in rockeries, with pines and acer palmatum planted on the waterfront, forming a landscape painting full of classical charm. The garden in the west echoes the landscape pattern of one pool and three hills in traditional gardens through careful design of the dike island, thus a natural scenery that is free and elegant is formed finally.

申园外景
Shanghai Garden

总平面图
Master Plan

上海温室
Shanghai Glasshouse

地处人文艺术片区，以"祥云"为概念，以原上钢三厂厂房保留构架为基底，采用铝合金上弦与钢拉索形成新型张弦组合结构体系，将曲线流畅通透的新建温室建筑与工业历史沉淀的构架有机交织，形成新与旧、工业与自然、传统与未来、陆地与岛屿的对比统一。园内规划海市沙洲、云上森林、云雾峡谷三个主题场馆，打造奇花异卉争奇斗艳的万花盛景。

The property is located in the Humanities and Arts section of the former Expo site. Built on the ground plot of the former Shanghai No. 3 Steel Plant, it has preserved its old workshop buildings. Embracing the concept of "auspicious cloud", the building adopts aluminum alloy winding and steel cable to form a new tension-string combination structure system, which organically interweaves the new greenhouse with the industrial heritage building, forming a contrast and unity between new and old, industry and nature, the past and the future, land and island. The garden houses three theme pavillions, namely, Mirage Sandbank, Forest in the Cloud and Cloud Gorge, where thousands of exotic and strangely scented flowers present a magnificent feast for the eyes.

温室鸟瞰图
Bird's eye view of Shanghai Glasshouse

温室夜景
Tranquil Shanghai Glasshouse

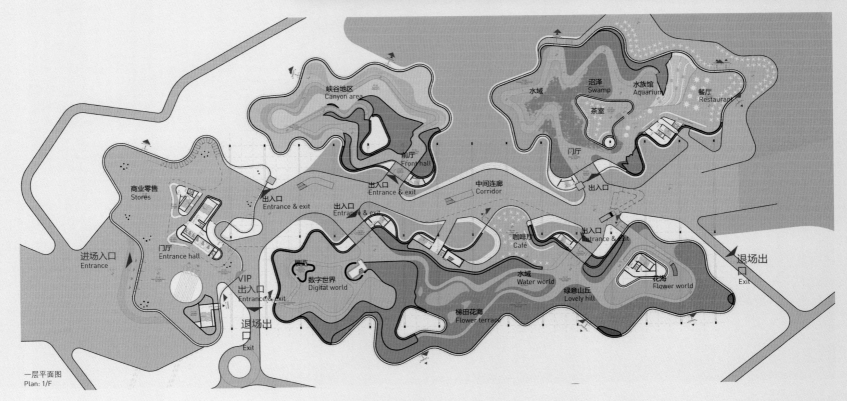

一层平面图
Plan: 1/F

宝山滨江公共空间
YANGTZE RIVER WATERFRONT IN BAOSHAN

鸟瞰图
Bird's eye view

宝山滨江贯通工程（一期）
Yangtze River Waterfront in Baoshan, Phase I

项目地点：宝山区吴淞口路滨江
设计单位：上海创霖建筑规划设计有限公司
合作设计单位：上海大观景观设计有限公司、上海泽柏景观设计有限公司
建设单位：上海宝建集团
设计时间：2020.3-8
建成时间：2020.10
用地面积：80000m²

Location: Waterfront at the estuary of Tangtze River, at Wusong Mouth, Baoshan
Designer: Shanghai Total Architectural Design & Urban Planning Co., Ltd.
Partners: DA landscape, Shanghai ZEB Landscape Design Co., Ltd.
Constructor: Shanghai Baojiian Group
Design Period: March – August 2020
Date of Completion: October 2020
Site Area: 80,000m²

改造前
Before modification

项目位于长江与黄浦江交汇处，改造前滨水岸线由混凝土防洪堤和防汛通道构成，滨江腹地以居住社区和公园绿地为主，岸线内最具标志性的建筑是吴淞口国际邮轮港。在上海打造全球卓越水岸，将黄浦江两岸公共空间贯通并向公众开放的背景下，宝山区着力将以防汛防洪为主的滨水区作为城市转型的重点。

Located at the estuary where the Huangpu River meets the Yangtze, this stretch of waterfront was formerly composed of concrete flood dike and channels, and behind them was a hinterland dominated by residential complexes and parks and green spaces. The iconic building in this stretch of waterfront is the Wusong Mouth International Cruise Terminal. To respond to the government's call to build an uninterrupted waterfront of the Huangpu River for the public, the Baoshan District focuses on converting flood control facilities into a waterfront that can entertain visitors and residents. Tapping into the resources of rivers, cruise ships and regional

充满自然野趣的漫步空间
Footway in nature

波浪花园
Wave-shaped park

设计结合江河、邮轮与宝山区域文化，以"宝山之舷"为主题，针对性提出自然韧性、趣味体验、公众融合三个策略。自然韧性用以应对滨水地区的生态敏感性以及全球气候变化带来的影响，建设自然型的水岸风貌；趣味体验旨在创造与邮轮文化相呼应的休闲互动体验；公众融合关注与周边社区的连接，为市民及游客提供便捷、优质、多元的滨水生活方式。

设计在不变动原有两级防汛体系的前提下，对现状提出相应提升方案。一级平台以波浪花园为主题，形成水陆过渡带，减少原有硬铺比重，增加自然野趣的绿植，形成花园式漫步空间；二级平台以城市吧台为主题，防浪墙作为工程遗存保留，结合宽厚的墙体特征，融入多种城市生活场景，实现远望江景近眺花园，形成类似船舱甲板的滨江空间体验。

---

culture at large, the waterfront revitalization campaign features the theme "Baoshan Gunwale". It proposes three principles: Fostering the resilience of nature, providing fun experience and engaging the public. By fostering the resilience of nature, the waterfront can regain an ecological sensitivity and cope with the impact of global climate change. The effort of providing fun experience aims to create an experience that corresponds to the cruise terminal. By engaging the public, the waterfront can have bond with the communities, providing citizens and tourists with a convenient, high-quality and diversified waterfront experience.

While following the mandatory two flood control wall system, the design has proposed a solution with platforms at two levels, one high and one low. The level-I platform, dubbed the "Wave Garden", forms a transitional zone between land and water; it reduced the proportion of hard pavement while increasing the coverage of green plants. The result is a garden-like promonnade. The level-II platform features nightlife as the theme, and makes use of the wave wall that has been preserved. Lifestyle facilities are introduced, allowing visitors to enjoy the view of the river and the garden together. The intended waterfront experience is akin to being on a ship deck.

平台上的休闲活动
Interesting activities

总平面图
Master plan

防汛体系
Flood prevention structure

总体鸟瞰
Bird's eye view

滨江湿地
Riverside wetland

上海吴淞炮台湾国家湿地公园
Shanghai Wusong Paotai Bay Wetland National Park

项目地点：宝山区塘后路206号
设计单位：上海市政工程设计研究总院（集团）有限公司
建设单位：上海市宝山区绿化和市容管理局
设计时间：2004.5-2011.6
建成时间：2007.5（一期）；2011.10（二期）
用地面积：106.6hm²

Location: No. 206 Tanghou Road, Baoshan District
Designer: Shanghai Municipal Engineering Design Institute (Group) Co., Ltd.
Constructor: Shanghai Municipal Baoshan District Landscaping and Appearance Bureau
Design Period: May 2004 - June 2011
Date of Completion: May 2007 (Phase I); October 2011 (Phase II)
Site Area: 106.6hm²

上海吴淞炮台湾国家湿地公园位于宝山区，长江与黄浦江的交汇口，总面积106.6公顷，其中湿地面积63.6公顷，湿地率60.3%，沿江岸线约2250米。公园以环境更新、生态恢复、文化重建为设计理念，以废弃地更新为契机，结合滩地保护与改造，将亲水空间融入湿地景观，将竖向设计融入防洪规划，将军事防御融入景观设计，实现人类、自然、军事、生态的和谐共生。

矿坑花园
Mine garden

古炮
Ancient fort

Shanghai Wusong Paotai Bay Wetland National Park Taking environmental renewal, ecological restoration and cultural reconstruction as the design concept, taking the renewal of abandoned land as the opportunity, combined with beach protection and transformation, the park integrates hydrophilic space into wetland landscape, vertical design into flood control planning, and military defense into landscape design, so as to realize the harmonious symbiosis of human, natural, military and ecology.

吴淞导堤
Wusong jetty

环境更新——更新废弃地，还岸于民

这里原来是长江的滩涂湿地，建国初期因战备需要，由废钢渣陆续回填形成43公顷陆域部分，钢渣平均深度8m，生态环境恶劣。为改善生态环境，弘扬炮台湾的历史与文化，发挥黄浦江与长江交汇处独一无二的地理优势和自然风貌，市、区两级政府共同规划建设公园。设计充分利用原有钢渣地的地形地貌，利用钢渣坑的深度设计矿坑花园，利用堆起的钢渣山设计假山等景观，利用陈钢渣作为道路、广场等硬化场地的路基，利用钢渣作为造型土，上覆种植土，满足种植条件，合理平衡土方，使钢渣不外运避免了二次污染，将岸线还给人民大众，实现人与长江零距离。

生态恢复——恢复吴淞口生态，改善环境

保留原生滩涂湿地，将防洪堤后退式设计，使滨水前沿的标高降低，塑造滨江景观步道、滨江湿地，扩大了湿地范围，并成为冬季候鸟迁徙停留的一个重要场所。现在公园内动植物种类丰富，有各类植物359种，鸟类超144种，常见鸟类有白鹭、斑嘴鸭、银鸥、家燕、乌鸫、白头鹎等，特色植物有薹草、芦苇、野茭白等。精准的设计定位，也为2016年5月完成国家湿地公园的验收，打下了坚实的基础。

文化重建——重拾场地精神，重塑文化

历史上吴淞口作为重要的军事要地，军事文化突出，两江交汇口的独特资源，造就了河口文化，钢渣堆场的工业遗迹地，体现了工业文明。充分挖掘传承场地文化，唤起人们对这片土地的追忆，并增加人们的国防意识，形成上海地区一个具有爱国主义教育特色基地，共同来维护宝山炮台湾这片具有深厚历史文化内涵的土地。

公园自建成开放以来，深受市民喜爱，被评为"国家AAAA级旅游景区""全国科普教育基地""上海市五星级公园"，并先后获得"改革开放30周年上海市建设成果银奖""中国人居环境范例奖""2010年IFLA亚太区第七届风景园林管理类杰出奖"等奖项。

Environmental renewal - Renew the abandoned land and return it to the people

It used to be a tidal flat wetland of the Yangtze River. In the early days of the founding of the people's Republic of China, due to the needs of war preparedness, waste steel slag was successively backfilled to form a land area of 43 hectares. The average depth of steel slag was 8m and the ecological environment was poor. In order to improve the ecological environment, carry forward Taiwan's history and culture, and give full play to the unique geographical advantages and natural features of the intersection of the Huangpu River and the Yangtze River, the municipal and district governments jointly plan to build the park. The design makes full use of the landform of the original steel slag land, the depth of the steel slag pit is used to design the mine garden, the piled steel slag mountain is used to design the rockery and other landscapes, the old steel slag is used as the subgrade of hard sites such as roads and squares, the steel slag is used as the modeling soil and covered with planting soil to meet the planting conditions, reasonably balance the earthwork, so that the steel slag is not transported outside and avoid secondary pollution, return the coastline to the people and realize zero distance between people and the Yangtze River.

Ecological Restoration - Restore the ecology of Wusongkou and improve the environment

Retain the original beach wetland, design the flood dike backward, reduce the elevation of the waterfront front, shape the riverside landscape trail and riverside wetland, expand the scope of the wetland, and become an important place for migratory birds to migrate and stay in winter. At present, the park is rich in animal and plant species, including 359 kinds of plants and more than 144 kinds of birds. The common birds are egrets, spotted billed ducks, silver gulls, domestic swallows, blackbirds and Bulbul. The characteristic plants are Scirpus, reed and wild water bamboo. The precise design positioning also laid a solid foundation for the completion of the acceptance of the National Wetland Park in May 2016.

Cultural reconstruction - Regain the spirit of the venue and reshape the culture

Historically, Wusongkou, as an important military place, has prominent military culture. The unique resources of the intersection of the two rivers have created the estuary culture and the Industrial Relics of the steel slag yard, reflecting the industrial civilization. Fully excavate and inherit the site culture, arouse people's memory of this land, increase people's national defense awareness, and form a base with patriotic education characteristics in Shanghai to jointly maintain Baoshan gun Taiwan, a land with profound historical and cultural connotation.

Since its completion and opening up, the park has been deeply loved by the public. It has been rated as "national AAAA tourist attraction", "national popular science education base" and "Shanghai five-star Park". It has successively won the "Silver Award for Shanghai construction achievements in the 30th anniversary of reform and opening up", "China living environment model award" and "the seventh outstanding award of landscape architecture management in IFLA Asia Pacific region in 2010".

闵行滨江公共空间
MINHANG SECTION OF HUANGPU WATERFRONT

总体鸟瞰
Bird's eye view

兰香湖
Orchid Lake

项目地点：闵行区紫竹高新技术产业开发区
设计单位：上海市水利工程设计研究院有限公司
合作设计单位：艾奕康环境规划设计（上海）有限公司（概念方案）
建设单位：上海紫竹半岛地产有限公司
设计时间：2016.9-2018.6
建成时间：2020.6
用地面积：370000m²

Location: Zizhu High-Tech Park, Minhang District
Designer: Shanghai Water Engineering Design & Research Institute Co., Ltd.
Partner: AECOM(Concept)
Constructor: Shanghai Zizhu Peninsula Properties Co., Ltd.
Design Period: September 2016–June 2018
Date of Completion: June 2020
Site Area: 370,000m²

湖中小岛
Island

总平面图
Master plan

兰香湖位于上海市闵行区吴泾镇境内，是由紫竹国家高新区建设的大型人工湖。

吴泾镇在历史上有一邢窦湖，湖以邢、窦二姓所居而得名，当时文人雅士汇聚于此，且有莺湖十景等绝胜风光。到了16世纪初，水面收窄成现在的樱桃河，绵延7公里，流经紫竹高新区内的大学校区和研发基地后，汇入浦江。鉴于湖已成河，为了延续历史，高新区在浦江第一湾选址了一处大型人工湖，原为水滴形状，后为契合高新区的科技人文特色，演变为"钥匙形"湖体，寓意着打开智慧大门的钥匙。如今碧湖波平、大学环绕、人才云集，恍如当年盛景再现。

湖名"兰香"，与"紫竹"呼应，以空谷幽兰和山间青竹比喻上海科创中心建设砥砺前行的那些高科技人员和大学师生们。是他们高度的创新意识和能力，以及孜孜不倦的攻坚克难精神，推动着紫竹高新区不断换挡提速，高质量践行科创中心重要承载区的历史使命。"深谷幽香、清雅淡泊"，正是他们心灵和精神的真实写照。

从水系功能上看，它是承担整个高新区调蓄功能的重要水体；从景观功能上看，它目标打造具有国际品质的开放空间；从地区功能上看，它是带动地区年轻活力的体育休闲旅游场所；从生态系统上看，它编织了一张拓展滨江绿带融入腹地的自然文化网络。多元化的功能满足从孩子到老人，居民到研发人员，创业者到大学生的各种休闲娱乐需求，宛如开启紫竹健康新生活的一把"钥匙"。

Located in Wujing Town, Minhang District, Orchid Lake is a large artificial lake built by the Zizhu High-tech Park.
There used to be a Xingdou Lake in Wujing town. Named after the Xing and Dou clans, the lake was a gathering place for literati and scholars, nearby there were "ten top sights." At the beginning of the 16th century, the lake shrank to be the Cherry River. Today the river stretches 7 kilometers, flowing through the university campus and R & D base in the Zizhu High-tech Park before it finally meets the Huangpu River. To remember the historical lake, the High-tech Park built one at the first bend of the Huangpu River since its source. The lake was originally in the shape of a water droplet, and later changed into the shape of a key, implying that the high-tech in the park is a key to wisdom. Now the blue lake is surrounded by science insitutions.

The lake is named, "Orchid", echoes with "purple bamboo", the literal meaning of the lake. It compares the intellectual pioneers here to the orchids in the valley and the bamboo on the mountains. Their innovativeness and commitment fuels the progress of the High-Tech Park.
The lake performs the water conservancy function of the entire high-tech park. As a piece of landscape design, it aims to create an open space with international quality. It is also venue of sports and recreation activities that instill vigor into the region. In terms of ecosystem, it extends the waterfront green space to the hinterland. It can meet the diversified leisure and recreation needs of different demographic groups. This key-shaped lake is slated to the new life in Zizhu.

HUANGPU RIVER SUZHOU CREEK

THE WATERFRONT SPACE & ARCHITECTURE IN SHANGHAI

历史建筑激活
RENOVATION OF HISTORICAL BUILDINGS

上海国际时尚中心　Shanghai International Fashion Center

灰仓美术馆　Ash Gallery

绿之丘　Green Hill

世界技能博物馆　World Skills Museum

明华糖厂　Ming Hua Sugar Refinery

世界会客厅　The Grand Halls

外滩公共服务中心　Bund Public Service Center

上海世博会城市最佳实践区　Urban Best Practices Area of Shanghai World Expo

上海当代艺术博物馆　Power Station of Art

余德耀美术馆　Yuz Museum

油罐艺术公园　Tank Shanghai

龙美术馆西岸馆　Long Museum West Bund

星美术馆　Start Museum

民生码头八万吨筒仓　80,000-ton Silos at Minsheng Wharf

上海浦东船厂1862　Shanghai Pudong Shipyard 1862

上海艺仓美术馆及长廊　Shanghai Modern Art Museum and its Walkways

上海宋城·世博大舞台　Shanghai Songcheng World Expo Grand Stage

黄浦江畔工业历史的当代风景

柳亦春

黄浦江是中国近代工业的摇篮。在工业时代，依托于黄浦江所展开的生产、运输等一系列工业活动造就了一个生产型城市的切面。码头、工厂、塔吊林立，这些工业活动及其建筑印迹是上海城市发展中现代性文化的重要组成部分。随着后工业时代城市功能的更新，诸多从城市生产活动中退场的工业建筑逐渐停用，继而废弃。它们中有的会保留改造，更多会被拆除。在黄浦江两岸，随着2010年上海世博会之后快速展开的城市更新，人们意识到保留更多工业建筑的空间价值与文化价值。曾经依水而建的工厂、仓储、码头，如今摇身变为美术馆、艺术中心或者文化型商业综合体，成为上海黄浦江岸线贯通中具有独特年代记忆与人文精神的公共空间，它们已经成为上海当代城市文化生活中亮丽的风景。

回顾近十年来黄浦江岸线的城市更新，这其中涌现的优秀建设案例，至少有四个方面的特点值得归纳与总结。

因地制宜地保护与更新

经历了20世纪90年代大拆大建的快速发展阶段，上海开始理性思考建筑遗产的保护，至今已经初步建立分级保护制度、保护机构与保护模式，在历史建筑的分级分类、修缮改建等方面均形成较为成熟的城市管理办法。对于历史建筑中的工业遗产，也给予特别的关注与相应的保护策略。事实上，黄浦江岸线上各类工业活动各有特点，对于遗产的保护并不局限于其中的建筑本体，构成工业活动的场址也是工业遗产的一部分。这其中既有特色鲜明的龙门吊、船坞、储油储气罐、储粮储煤的筒仓等，也包含看似普通的一些工业构筑物、运输构架等。尽管很多普通的工业建筑和构筑物并未列入历史建筑或遗产名录，很多建筑师在更新设计中仍能有意识地、最大限度地保留并因地制宜地加以利用。

比如，对龙美术馆西岸馆场地中的煤料斗、西岸艺术中心上海飞机制造厂组装车间的保留利用，都是在建筑完工之后，其中的工业建筑才得以列入历史保护建筑名录。而艺仓美术馆中的老白渡码头遗构、杨浦边园的煤气厂码头的卸煤挡墙，并未列入任何保护名录，仍能得到巧妙的保护和利用。现在名为杨浦滨江"绿之丘"的原烟草仓库，因为与规划道路的矛盾甚至是原本规划拆除的建筑。这些得到留存的一般遗迹，有时更能体现建筑在废弃的过程中，其中萦绕不变的文化意义。在杨浦边园的码头与岸线之间的"废墟"园林中，生物学家还发现了在黄浦江其他岸线近乎绝迹的生物物种。可见保护的内容是多方面的，但绝不是对更新使用的束缚，因地制宜地保护是一种价值的延续，也是全新的建筑所无法创造的。

适应性再利用

"适应性再利用"是由国际古迹遗址理事会（ICOMOS）澳大利亚国家委员会制定的《巴拉宪章》中特别为工业遗产提出的，指对某一场所进行调整使其容纳新的功能，而这个功能对于现有建筑与空间必须是恰当的。为既有的工业建筑"寻找恰当的用途"是一个非常重要的保护策略，它意味着要根据既有的建筑空间特点去安排新的功能与用途。无论是曾经的南市发电厂经由世博会未来馆再改造为上海当代历史博物馆，还是上海飞机制造厂组装车间改为西岸艺术中心，龙华机场飞机库改为余德耀美术馆，抑或民生码头八万吨筒仓临时作为上海城市艺术季主场馆，都是根据现有空间特点为其寻找合适的功能，以便较小程度地干预既有建筑，既减少建筑垃圾的产生，也更好地保护原有建筑特色。

新与旧的共生

新与旧的关系是所有遗产保护与更新的核心命题，这关乎历史记忆在当代的延续。在黄浦江岸线更新中涌现的众多案例，呈现了多种不同的新旧关系。龙美术馆西岸馆是其中的一个重要代表。场地中原北票码头运煤的煤料斗卸载桥，因功能需要不断重复呈阵列，受混凝土煤漏斗的启发，龙美术馆的设计通过当代清水混凝土技术与重复组合的"伞拱"墙体结构，既满足了展览功能的需求，又创造了美妙的光线与空间。新建筑的构成与煤漏斗的构成形成一组类比关系，寓旧于新，共同形成徐汇滨江北票码头旧址上崭新的城市空间。位于老白渡煤炭码头旧址的艺仓美术馆，通过悬吊结构拓展展览空间，同时将原有码头的粗砺煤仓藏在一个外表为阳极氧化铝板的新建筑之中，煤仓既成为新建筑内的展品，也作为展览空间使用。码头上的运煤构架通过新的悬吊钢结构，将原有的混凝土框架组合为新结构的一部分，形成高架的观景步道和步道下的便民服务空间，新与旧共生为一个新的整体。

公共性与景观性

黄浦江贯通工程中，除了历史与记忆，另一组最重要的关键词莫过于公共性与景观性。黄浦江由生产岸线转变为生活岸线，最重要的就是公共空间的塑造。在浦江两岸的更新案例中，对于公共性与景观性的塑造也各有特色。龙美术馆保留的煤漏斗成为城市连接江边的公共通道；艺仓美术馆通过层层外挑的平台，人们仍能穿过闭馆后的美术馆，甚至可以层层向上抵达它的屋顶平台，这也使得原本封闭的煤仓建筑具备了开放的景观性。

民生码头八万吨筒仓通过外挂的玻璃自动扶梯，解决了旧工业建筑封闭的外观与新文化建筑应有的开放性之间的矛盾，黄浦江的景色被得以带入筒仓建筑；"绿之丘"更是将原来的烟草仓库变为一个立体的绿化庭园；杨浦边园也将一处小小的运煤码头变为一处旱冰场与内外两重尺度的公共风景园林。它们都使曾经的工业印迹悄然进入上海的日常生活环境，这是上海城市更新中城市公共性的重要文化表现。

Huangpu Waterfront:
A living incarnation of Shanghai's industrial past
Liu Yichun

The Huangpu River is the cradle of the modern Chinese industrial sector. During the industrial era, production and transportation relied on the Huangpu River, creating a production-oriented city. The operation of the wharves, factories, cranes and their architectural contours are an important part of Shanghai's urban development. With the renewal of the city's functions in the post-industrial era, many of the industrial buildings lost their original functions and were gradually abandoned. Some of them will be preserved and renovated, while more will be demolished. On both sides of the Huangpu River, with the rapid urban regeneration that has taken place since the 2010 Shanghai World Expo, people have realized the spatial and cultural value of retaining industrial buildings. The factories and warehouses that were once built by the water are now converted into art galleries, art centers or cultural commercial complexes, and have become public spaces with a unique memory and spirit in Shanghai's Huangpu River waterfront revitalization campaign, glowing in the city's contemporary cultural life.

In retrospect of the urban regeneration of the Huangpu River waterfront in the last decade, there are at least four aspects that merit our attention.

Site-specific Conservation and Regeneration

After a period of rapid development in the 1990s when the city underwent a massive wave of demolition and construction, Shanghai began to think rationally about the conservation of its architectural heritage. The city has since established a graded conservation system, set up conservation institutions and measures, and has developed a mature approach to the grading, classification, repair and remodeling of historic buildings. Special attention has also been paid to the historical industrial facilities, and corresponding conservation strategies have been formulated. As a matter of fact, various types of industrial activities along the Huangpu River are different from one another, and the conservation of heritage is not limited to the buildings themselves, but also to the sites that accommodated the industrial operations. These include distinctive gantry cranes, shipyards, oil and gas storage tanks, silos for grain and coal storage, as well as seemingly ordinary industrial structures and transport structures. Although many of these common industrial buildings and structures are not listed as heritage buildings, many architects still seek to preserve the structures to the maximum and utilize them site-specifically in their regeneration designs.

For example, the coal hopper on the site of Long Museum West Bund, and the West Bund Art Centre's retained use of the assembly workshop of the Shanghai Aircraft Manufacturing Plant, were both industrial structures that were only listed for historic preservation after the projects were completed. The remaining structures on the Laobaidu Wharf in the Shanghai Modern Art Museum and the coal-unloading retaining wall of the coal factory wharf in the Riverside Passage are not on any conservation list but have been tactically preserved and utilized. The former tobacco warehouse, now known as the "Green Hill" on the Yangpu riverfront, would have been otherwise faced with demolition because of its conflict with a planned road. It is sometimes the ordinary historical sites that have been preserved that can better reflect the fundamental cultural significance during the fading process of architecture. In the garden of ruins between the docks along the Yangpu section of Huangpu waterfront, biologists have also found biological species that are nearly extinct along other parts of the Huangpu River waterfront. It is clear that conservation can be multidimensional, but it does not impose restriction on regeneration, and that site-specific conservation is a continuation of values that are not found in brand-new building.

Adaptive Reuse

Adaptive reuse is a concept specifically proposed for industrial heritage in the Barra Charter of the Australian National Committee of the International Council on Monuments and Sites (ICOMOS). It refers to the adaptation of a site to accommodate new functions that are appropriate to the existing building and space. "Finding appropriate uses" for existing industrial buildings is a very important conservation strategy, assigning new functions and uses according to existing spatial features of the space. The transformation of the former Nanshi Power Plant into the Expo Future Pavilion and then the Shanghai Museum of Contemporary History, the transformation of the assembly workshop of the Shanghai Aircraft Manufacturing Plant into the West Bund Art Centre, of the aircraft hangar at the Longhua Airport into the Yuz Art Museum, or the effort to convert the silo at the Minsheng Wharf into the main venue for the Shanghai Urban Spatial Arts Season, all of these are attempts to find appropriate functions for existing spaces with minimum interference. This approach can both reduce construction waste and better preserve the original architectural features.

The Symbiosis of the Old and the New

The relationship between the new and the old is central to all heritage conservation and regeneration projects. It is about perpetuating past memory in the present day. Many cases that have emerged in the regeneration of the Huangpu River waterfront present different relationships between the old and the new. The Long Museum West Bund is an important example of this. The coal-unloading bridge of identical hoppers of the former Beipiao Wharf formed an array for functional needs. Inspired by the concrete coal hoppers, the design of the Long Museum uses contemporary fair-faced concrete techniques and a cluster of "umbrella vault" wall structures to meet the needs of an exhibition space while creating a wonderful play of light and shade. The new building and the coal hoppers become an analogy, rejuvenating the legacy, jointly forming a new urban space at the old Beipiao Wharf of the Xuhui section of the Huangpu Waterfront. Located on the former site of the Laobaidu Coal Wharf, the Shanghai Modern Art Museum expands its exhibition space with a suspended structure. It hides a coal bunker from the original wharf in an anodized aluminium exterior. The coal bunker becomes both an exhibit and an exhibition space in the new building. The coal channels of the wharf is combined with the original concrete frame as part of the new structure by means of a new suspended steel structure, giving rise to both an elevated walkway and also a convenient service space underneath, where the new and the old coexist as a new whole.

The Public and Aesthetic Nature

In addition to preserving history and memory, the most important aspects of the Huangpu River Waterfront Revitalization project are its public and aesthetic nature. The public nature of the project is highlighted by the transformation from a production-oriented waterfront to a life-oriented one. In the case of the regeneration of both sides of the river, the expressions of the public and aesthetic nature are diversified. The Long Museum retains a coal hopper bridge that becomes the passage leading to the riverfront; the Shanghai Modern Art Museum and Walkways can still be entered via layers of cantilevered platform when the museum is closed, and visitors can even go to its roof terrace, which affords the originally closed coal bunkers a panoramic view.

A glass escalator at the former 80,000-ton silo of the Minsheng Wharf has resolved the contradiction between the closed nature of the old industrial building and the openness expected of a cultural facility. Visitors can command a vista of the Huangpu River from the former silo. The Green Hill has been transformed from a former tobacco warehouse into a three-dimensional green garden. The Riverside Passage in Yangpu has also transformed a small coal wharf into a roller-skating rink and a public landscape garden with multiple layers of scales. All of these are important cultural expressions of the public nature of Shanghai's urban regeneration, as the traces of the former industrial facilities have quietly been woven into the fabric of the city's daily life. This is a phenomenal cultural expression of the city's public nature in the urban regeneration of Shanghai.

上海国际时尚中心
SHANGHAI INTERNATIONAL FASHION CENTER

沿江侧鸟瞰
Bird's eye view along the bank

项目地点：杨浦区杨树浦路 2888号
设计单位：华建集团华东建筑设计研究院有限公司、法国夏邦杰建筑设计事务所（概念方案）
建设单位：上海纺织控股有限公司
设计时间：2009.8 - 2011.5
建成时间：2012.5
建筑面积：140000m²
用地面积：89333m²

Location: No.2888 Yangshupu Road, Yangpu District
Designer: East China Architectural Design & Research Institute Co., Ltd., Arcplus Group PLC, Arte-Charpentier-Architectes-Urbanistes (Conceptual Design)
Constructor: Shangtex Holding Co., Ltd.
Design Period: August 2009 - May 2011
Date of Completion: May 2012
Gross Floor Area: 140,000m²
Site Area: 89,333m²

立面细部
Details of evelation

入口广场
Entrance

前身为始建于1922年的裕丰纱厂，中华人民共和国成立后更名为国营第十七棉纺织厂。随着城市经济转型，纺织厂房空置衰败。方案梳理了建筑的工业遗产价值，以原有纺织功能为特色，赋予其新的现代功能：以时尚精品仓为主，辅以时尚秀场、休闲餐饮等配套设施。设计遵循最小干预原则，采用了保存、修复、重建及改造性再利用等技术措施，着重呈现建筑原有的质感和肌理。

The Shanghai International Fashion Center sits on the premises of the former Toyo Spinnning Co., Ld. dating from 1922, a facility renamed the Shanghai No.17 Cotton Textile Mill after 1949. When Shanghai redefined its economic functions in the 1990s, the textile industry lost its former luster, leaving many textile miles languishing on the verge of bankrupcy. The renovation plan confirmed the historical value of the premises and proposed to change its function to realize its potential inurban life. The architects have preserved the buildings on the basis of minimum intervention, and all their effort has emphasized the principle of PRRA: Preservation, Restoration, Reconstruction and Adaptive reuse.

立面的韵律
Rhythmed facade

建筑立面以保存为主、改建为辅。尊重建筑既有立面风格，以安全为底线，逐一判断现状墙体情况。对较好的墙体尽量予以保留，对破损严重的则采用修旧如旧的手法予以替换，使建筑从空间、体量、色彩上都保持了原有的重要元素，延续工业遗韵，突显独特气质。新建、改建和修缮部分的建筑采用与保留建筑类似的风格，尊重并保留原有空间肌理，完整留存下这片上海市区最完整、最具规模的锯齿形厂房建筑群。项目既保存了珍贵的历史记忆，又为城市注入新的空间元素，激发了城市活力。百年老厂旧貌焕发新颜，彰显城市文脉延续的魅力。

Most elevations of the original buildings are preserved. The new and renovated parts valued the enduring charm of the original buildings, preserving the largest and most complete cluster of plants with swatooth-shaped roofs. The design has injected new space and elements into the urban environment while preserving memories of its past.

街巷空间
Well-designed street

屋面肌理
Delicate roof pattern

总平面图
Master plan

时尚秀场剖面图
Section of fashion stores

西北侧鸟瞰
Bird's eye view from the northwest

灰仓美术馆
ASH GALLERY

西南侧远景
Bird's eye view

灰罐间的连廊
Ash Gallery

项目地点：杨浦区杨树浦路 2800 号
设计单位：同济大学建筑设计研究院（集团）有限公司原作设计工作室
建设单位：上海杨浦滨江投资开发有限公司
设计时间：2018.6
建成时间：2019.10
建筑面积：3840m²

Location: No.2800, Yangshupu Road, Yangpu District
Designer: Original Design Studio (A subsidiary of Tongji Architectural Design (Group) Co., Ltd.)
Constructor: Shanghai Yangpu Riverside Investment and Development Co., Ltd.
Design Period: June 2018
Date of Completion: October 2019
Gross Floor Area: 3,840m²

从塔吊下望向建筑
View from the crane

原为杨树浦发电厂最东端码头上的干灰储煤灰罐，现以艺术展览为主，采用灵活、开放的改造策略。方案用两个景观平台将原先独立的三个灰罐连成一体。采用朦胧界面的处理手法，改造原先15米通高的封闭灰仓。整个空间设计成一组完全公共的漫游路径，从底部的混凝土框架一直盘绕至灰仓顶部。方案以一种类似插入城市的模式，插入六个功能未确定的空间和一组折跑楼梯，在布置艺术品后形成艺术讨论和公共漫游紧密咬合的空间模式。

The Ash Gallery is on the site of the former the Yangshupu Power Plant, once the foremost thermal power plant in the Far East. It was converted from the dry ash tanks at the east of the wharf in a redevelopment project starting in 2015. The redevelopment has given the property a new flexibility and openness. Via inserting two platforms, the three isolated ash tanks are connected to each other. The original 15-meter-high steel surfaces of the tanks are tore down and replaced by thin horizontal bands, forming an ambiguous interface. The art space is designed as a public promenade, curling from the entrance carved by concrete framework to the top platform. Six boxes with a set of folding staircases are plugged in the tanks, whose functions are to be shaped by the behaviors of the visitors.

灰罐内部空间
Interior space of ash tower

建筑近景
Details

建筑近景
Details

建筑近景
Details

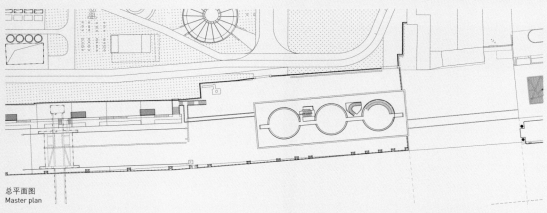
总平面图
Master plan

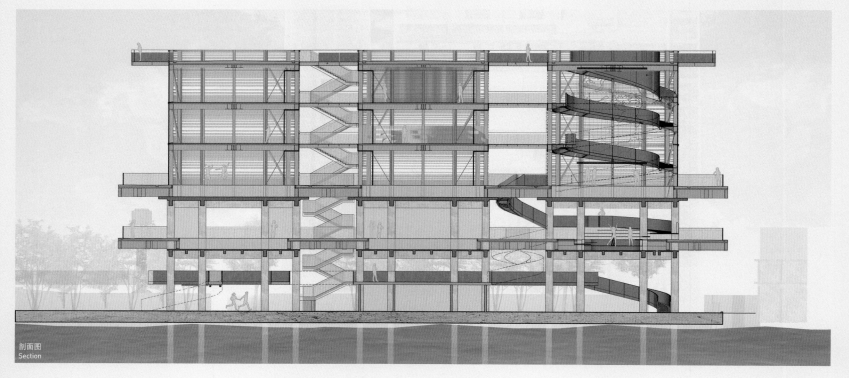
剖面图
Section

西南侧近景
Southwestern part

绿之丘
GREEN HILL

总体鸟瞰
Bird's eye view

观江眺望平台·连续的标高
Sight-Seeing Platform-a continuous gradation

项目地点：杨浦区杨树浦路 1500号
设计单位：同济大学建筑设计研究院(集团)有限公司原作设计工作室
建设单位：上海杨浦滨江投资开发有限公司
设计时间：2018.7
建成时间：2019.10
建筑面积：17500m²

Location: No.1500, YangshupuRoad, Yangpu District
Designer: Original Design Studio (A subsidiary of Tongji Architectural Design (Group) Co., Ltd.)
Constructor: Shanghai Yangpu Riverside Investment and Development Co., Ltd.
Design Period: July 2018
Date of Completion: October 2019
Gross Floor Area: 17,500m²

从 北侧草坡望向建筑
Northern part

绿之丘原是六层的烟草仓库，为钢筋混凝土框架板楼。设计利用滨江一线的区位优势，打造集市政基础设施、公共绿地和配套服务于一体的城市滨江综合体，"盘活"了这座工业建筑。巨大的、阻挡滨江视线的南北向体量转化为连接城市和江岸的桥梁，采用框架结构解决规划道路穿越既有建筑的问题。

Green Hill was originally a six-storey tobacco warehouse with a reinforced concrete frame. The design takes advantage of the waterfront location to create an urban riverside complex integrating municipal infrastructure, green space and supporting services, giving new life to the industrial building. The huge north-south structure that blocks the riverview is transformed into a bridge connecting the riverbank and the hinterland. The use of a frame structure has solved the problem of road planning through existing buildings.

顶视图
Top view

蜿蜒状的漫游路径
Curling promenade

中庭双螺旋楼梯
Spiral staircase in atrium

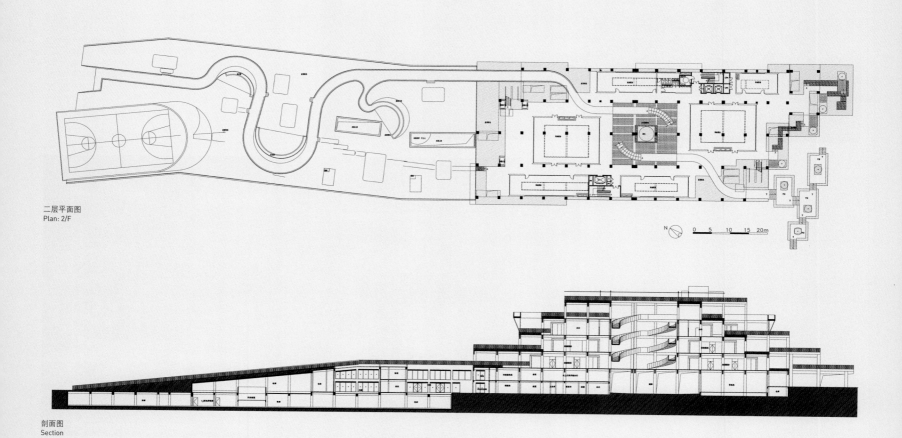

二层平面图
Plan: 2/F

剖面图
Section

五层钢结构环形游廊
Annular verandah of five-storey steel corridor

世界技能博物馆
WORLD SKILLS MUSEUM

整体鸟瞰
Bird's eye view

项目地点：杨浦区杨树浦路1578号永安栈房旧址西楼
设计单位：同济大学建筑设计研究院（集团）有限公司
建设单位：上海杨浦滨江投资开发有限公司
设计时间：2018.5
建成时间：在建
建筑面积：10172m²
用地面积：10920m²

Location: West Building of Yong'an Textile Company's Warehouse, No.1578, Yangshupu Road, Yangpu District
Designer: Tongji Archtectural Design (Group) Co., Ltd.
Constructor: Shanghai Yangpu Riverside Investment and Development Co., Ltd.
Design Period: May 2018
Date of Completion: Under Construction
Gross Floor Area: 10,172m²
Site Area: 10,920m²

安浦路南立面
Southern façade of Anpu Road

建于1930年的永安栈房，现存2座仓库，系现代风格的无梁楼盖结构，为上海市文物保护建筑，经改造建成世界技能博物馆。设计坚持历史信息清晰、结构安全合理的原则，结合规划定位和功能需求，留存场地和历史记忆，焕发近代工业遗产的新生。在保证对结构体系和原修缮成果较小干预的前提下，通过挖掘结构潜力，建构层层递进的共享中庭，营造独特的空间体验。充分保留建筑无梁楼盖的特点，以反梁进行隐性结构加固，以构造空腔作为空调送风的管廊。在首层设置咖啡、文创商店等公共空间，形成良好的互动，让博物馆融入城市日常生活。打破西侧原有耳房对建筑空间的束缚，实现内部空间与城市景观、滨江景观的充分融合，创造独特的滨江观景体验平台。

沿江立面
Façade

The World Skills Museum is formerly known as Yong'an Textile Company's warehouse. It was built in 1930, and the two existing warehouses are modern beamless floor structures, which are under the protection of Shanghai Cultural Relics.
The reconstruction design of the World Skills Museum adheres to the principles of clear historical information, safe and reasonable structure, combines with the functional requirements and planning positioning of the World Skills Museum, and makes it rejuvenate on the basis of preserving the site and historical memory. By tapping the structural potential and ensuring less influence on the system and the original repair results, a progressive shared atrium is carefully constructed to release the unique space experience. The structure of the museum adopts hidden reinforcement - counter beam, and the structural cavity formed by the counter beam is used as the pipe gallery of air conditioning, which fully retains the characteristics of the original building. The open coffee, cultural and creative shops and other public spaces on the first floor integrate the museum into the daily life of the city, creating an interactive space with good atmosphere. The west side of the building breaks the shackles of the original side rooms on the building space, realizes the full integration of the internal space, urban landscape and riverside landscape, and creates a unique riverside viewing experience platform.

剖面图
Section

明华糖厂
MING HUA SUGAR REFINERY

整体鸟瞰
Bird's eye view

项目地点：杨浦区杨树浦路1578号
设计单位：同济大学建筑设计研究院（集团）有限公司原作设计工作室
建设单位：上海杨浦滨江投资开发有限公司
设计时间：2018.2-10
建成时间：2019.11
建筑面积：1400m²
用地面积：800m²

Location: No. 1578, Yangshupu Road, Yangpu District
Designer: Original Design Studio (A subsidiary of Tongji Architectural Design (Group) Co., Ltd.).
Constructor: Shanghai Yangpu Riverside Investment and Development Co., Ltd.
Design Period: February 2018 - October 2018
Date of Completion: November 2019
Gross Floor Area: 1,400m²
Site Area: 800m²

北侧鸟瞰
Bird's eye view from the north

前身为日商开办的明华糖厂，后为上海化工厂的一部分，经历多次修建、改造，直至2010年搬离。设计保留精糖仓库南侧加建的2层建筑，拆除其余历史价值不高且形制混乱的部分。明华糖厂所在的区域，原有防汛墙横亘在码头和后方厂区之间，以高大厚实的姿态强硬地阻隔了城市空间与水岸的连接。为将滨水空间开放给城市居民，设计将二级防汛墙后撤，并藏匿在连续的草坡之中。

建筑东侧界面的一层标高较低，设置弧形挡土墙与草坡衔接，形成地文重塑后室内外空间积极相容的组织模式。作为"呼吸气口"的东侧二层与北侧界面均设置落地玻璃窗，充分展现建筑内部柱脚加腋、轻钢桁架等特征构件，优化建筑与周边环境的交互关系。在屋顶平台"第五开放界面"上，可以沐浴着阳光，眺望波光粼粼的江面，见证老建筑以新姿态守望在水岸与城市之间。

西侧立面保留原有的外墙形制
The original western façade remained

The plant of the Shanghai Chemical Plant, formerly home to the Ming Hua Sugar Refinery, was built by the Japanese in 1924. It underwent multiple constructions and renovations before it was moved away in 2010. During the redevelopment, the two-storey building added to the south of the refined sugar warehouse was preserved, while other buildings demolished because of its low historical value and disordered shape. In the area where Ming Hua Sugar Refinery was located, the original tall and thick flood control wall between the wharf and the factory area stood in the way between the urban space and the waterfront. In order to open the waterfront to urban residents, the secondary flood control wall was set back and hidden in the continuous grass-covered slope.
As the elevation of the ground floor on the east side of the building is low, the designer set up a curving retaining wall to connect with the grass slope. The landscape remodeling connects indoor and outdoor spaces. While as the "breathing outlets", the east side and north side of the two floors were equipped with floor-to-ceiling glass windows, which can fully display the building's original elements, e.g., column bases, light steel truss, and forges close bond between the building and the surrounding environment. The rooftop terrace serves as the "the fifth open space", allows people to enjoy abundant sunshine, to enjoy the scene of the sparkling waves and appreciate the new appearance of the old building.

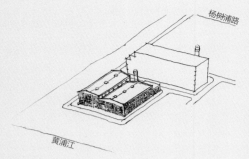

1924年原有厂房。
Original plant built in 1924.

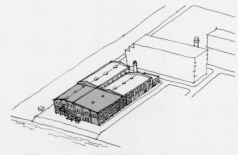

1927—1939年二次贴建，建筑形制较为简单，结构构件较有特色。
Extension in 1927-1939; simple structure and distinctive components.

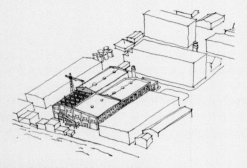

1980年原有屋面拆除，加建上部两层及新屋顶。
In 1980, the original roof was demolished and two upper floors and a new roof were built.

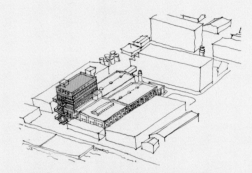

2010年现存东、北立面为原有建筑间的分隔内墙，历史风貌价值较弱。
The existing eastern and northern facades are the interior partition walls originally built in 2010, with medium historical value.

建筑西侧与南侧的界面采用"修旧如旧"的保护方式。保留20世纪30年代钢筋混凝土厚实墙体，精心修复极具历史特征的南侧门窗木框和西侧排风扇。建筑北侧界面在拆除原隔墙后形成悬挑灰空间，落地玻璃窗轻盈低调地藏于后方，显现出独特的空间断面。为强化其特征，在内部楼梯的位置新置入轻盈的悬索楼梯，一路延伸至屋顶人行廊道，从而形成公共开放的望江平台。

新建屋面采用缎面阳极氧化铝，浮于结构加固后的屋面板上。结合原屋面通风井设置采光井道，改善底层采光。新旧对比之下，建筑形态充满张力，不只是单一的历史回溯，而是很多年代之间"旧"与"新"的并置。

The warehouse's western and southern elevations are restored to their original state, and the thick reinforced concrete walls from the 1930s remain intact. The wooden door and window frames on the southern elevation and the ventilation fans on the western elevation have antilevered gray space after the removal of the original partition wall, and the floor-to-ceiling glass windows are hidden in the rear to show a unique space. A light hanging cable staircase is newly placed at the position of the original internal staircase, which extends to the "floating" pedestrian corridor on the roof, forming an excellent open terrace that overlooks the river.

The new roof, made of satin anodized alumina panels, is detached from the wall top of the original building, and "floats" on the existing roof after the later was reinforced. The lighting shaft is set in combination with the original ventilation shaft on the roof to improve the lighting on the ground floor. A fusion of the old and the new, the architecture is rooted in the past but celebrates the new.

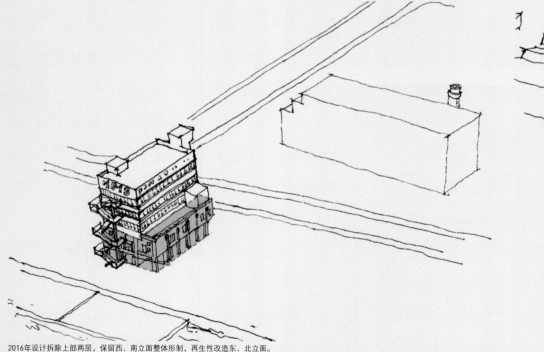

2016年设计拆除上部两层，保留西、南立面整体形制，再生性改造东、北立面。
In 2016, the upper two floors were dismantled, the western and southern façades are retained overall, and the northern and eastern façades were rebuilt.

历史变迁分析图
Diagram of evolution

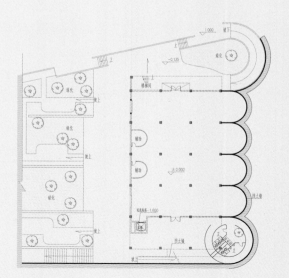

一层平面图
Plan: 1/F

北侧开敞界面
Semi-open space in the north

世界会客厅
The Grand Halls

东南角夜景
Aerial view from the southeast

项目地点：虹口区黄浦路118号
设计单位：华建集团华东建筑设计研究院有限公司
建设单位：上海久事北外滩建设发展有限公司
设计时间：2018.5-2021.2
建成时间：2021.9
总建筑面积：99000m²
用地面积：19562m²

Location: No. 118, Huangpu Road, Hongkou District
Designer: East China Architectural Design & Research Institute Co., Ltd., Arcplus Group PLC
Constructor: Shanghai Jiushi North Bund Development Co., Ltd.
Design Period: May 2018 - February 2021
Date of Completion: September 2021
Gross Floor Area: 99,000m²
Site Area: 19,562m²

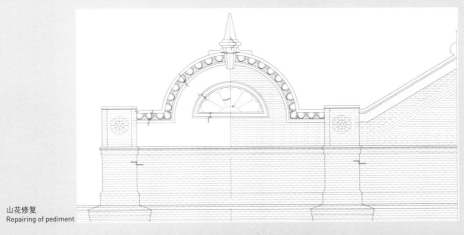

山花修复
Repairing of pediment

整体鸟瞰
Bird's eye view

北外滩作为外滩的延伸，拥有丰富的文化资源和深厚的历史积淀。扬子江码头原为上海日本邮船株式会社码头，又称三菱码头，主要建筑有日本邮船株式会社办公楼（现港务局办公楼位置）、1号仓库、2号仓库和3号仓库等建筑，其中2号仓库和3号仓库两座建筑始建于1902-1903年，以清水青砖与清水红砖为特征的外立面，体现了特有的历史风貌。项目按照"中国故事、上海表达、世界客厅、共筑辉煌"的设计目标，在满足国际级重大会议文化中心的功能基础上，挖掘和传承项目独有的历史底蕴，以"新老融合共生"作为设计理念，将"世界会客厅"作为外滩的延伸。通过拆除原港务办公楼及1号库，保留并改造作为历史建筑的原2号库、3号库，三栋楼在整体更新改造后成为具有国际重大会议接待功能的会议中心，并在形式上与外滩万国建筑群保持协调统一，同时运用了新理念、新模式和新技术，满足核心使用功能，展现庄重的国家形象和上海独特的城市魅力。

As an extension of the historic Bund, the North Bund has rich cultural resources and profound historical legacy. The Yangtze River Wharf was originally the wharf of Shanghai Nippon Shipping Co., Ltd., also known as Mitsubishi wharf. The main buildings include the office building of Nippon Shipping Co., Ltd. (now the office building of the port administration), warehouses Nos. 1, 2 and 3—the latter two dating from 1902 -1903. The green and red brick walls are distinctive features that reflect a unique historical style. To perform the function of an international conference and cultural center, the project explores the unique historical heritage. It reflects the "symbiosis of the old and the new" as the design concept, making "The Grand Halls" an extension of the Bund. The former Port Office Building and Warehouse No. 1 are demolished while the historical buildings of the original warehouses No. 2 and 3 are preserved and renovated into a conference center. It can accommodate international conferences and meetings and is compatible to the architecture of the Bund although it embraces brand-new concepts and designs. It is an architectural treasure that can show case China's image and Shanghai's unique glitz and glamour.

入口大厅
Lobby

东南角鸟瞰
Aerial view from the southeast

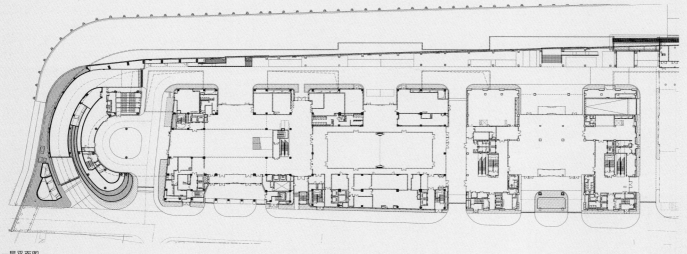
一层平面图
Plan: 1/F

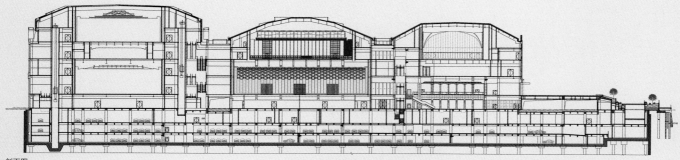
剖面图
Section

沿江立面
Façade

外滩公共服务中心
BUND PUBLIC SERVICE CENTER

建筑街景
Façade

建筑街景
Façade

沿江东立面
Eastern façade

项目地点：黄浦区中山东一路15号甲
设计单位：同济大学建筑与城市规划学院郑时龄工作室
建设单位：上海市申江两岸开发建设投资（集团）有限公司
设计时间：2003-2006
建成时间：2012
建筑面积：11723m²
用地面积：2118.1m²

Location: No. 15 A, East Zhongshan No. 1 Road, Huangpu District
Constructor: Zheng Shiling Studio, CAUP of Tongji University
Design Period: 2004-2006
Constructor: Shanghai Shenjiang Liang'an Development Construction Investment (Group) Co., Ltd.
Design Period: 2003-2006
Date of Completion: 2012
Gross Floor Area: 11,723m²
Site Area: 2,118.1m²

项目是位于中山东一路14号上海市总工会大楼与15号外汇交易中心大楼之间一灰白色大楼。为修复外滩第一界面景观，整合优化滨江历史风貌界面，完善地区公共功能定位，开展外滩风貌补全工程，并进行外滩公共服务中心的设计。设计在外形上参照外滩20世纪初期的建筑风格，采用较为传统的立面分段手法，沿用历史规律，兼具一定现代建筑的特色。外墙材料以石料为主，整体风格平整、简洁。

设计主要协调三方面的关系：一是与整个外滩全景立面的关系；二是与相邻建筑的功能关系；三是与陆家嘴城市标志的关系。为此在各个方向上设置景框，在被视者和观察者之间形成良好的视线关系。
立面在相对严谨的横竖三段分割法、对称轴线、比例关系、虚实对比和天际线等传统手段基础上，融入现代建筑设计手法的简化处理，在门额、内庭等部位局部变化，缩小现代痕迹对历史风貌立面的影响，材质质感、体量感和立面肌理与相邻建筑适度呼应。
通过内庭的组织，在平面上开辟出串联中山东一路、九江路和街坊内部通道的公共通道，向市民敞开，与城市空间形成渗透关系。项目较好地填补外滩连续界面的空缺，形成在历史环境、时代理念、技术演化方面的统一协调。

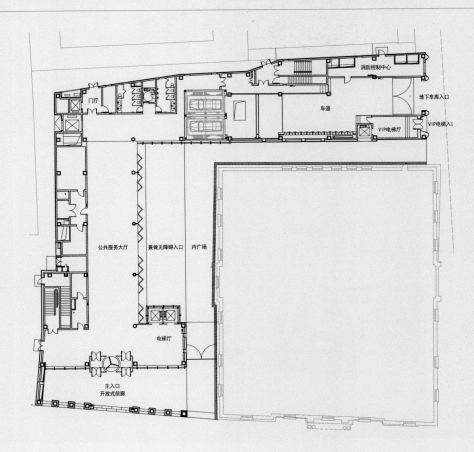
一层平面图
Plan: 1/F

The project is a grayish structure tucked between the buildings of the Shanghai General Labor Union and of the Shanghai Foreign Exchange, at Nos. 14 and 15, East Zhongshan Road No. 1 respectively. In order to fill in the gap between the two historical buildings, and improve the function of this area, a project was launched to construct this building at the otherwise empty ground plot. Referring to the architectural contour of the Bund in the early 20th century, the design adopts traditional façade segmentation while introducing modern architectural features. The exterior wall, greyish and mostly made of stones, shows a pristine style.
The design has to handle three relationships: a visual consistency with the facades of all other buildings on the Bund, a functional coordination with adjacent buildings; and the harmony with iconic landmarks in Lujiazui across the river. For this purpose, view frames are set in all directions to build a harmonious visual relationship between the viewees and the viewers.
The façade follows traditional approaches: three horizontal levels compounded with three vertical sections, symmetrical axis, contrast between the omission and addition, and a harmonious skyline. Except some elaborate details at the gate top and inner courtyard, it shows a minimalist touch of modern architecture to minimize modern traces. The material, volume and architectural profile of the building are consistent with those of neighboring historical grande dames on the Bund.
Internal passages open up the building, linking it to East Zhongshan Road No. 1 and Jiujiang Road as well as the neighbourhood. These public linkages effectively embed the building within the urban fabric. The project fills the gap to form a continuous interface on the Bund, being a perfect piece of engineering that honors the historical environment, the contemporary concept and technical progress.

上海世博会城市最佳实践区
URBAN BEST PRACTICES AREA OF SHANGHAI WORLD EXPO

世博创意秀场
Expo I-Pavilion

入口
Entrance

建筑地点：黄浦区北至中山南路，西至保屯路、望达路，南至苗江路，东至花园港路、南车站路
设计单位：同济大学建筑设计研究院（集团）有限公司、华建集团华东建筑设计研究院有限公司、北京建筑设计研究院有限公司
建设单位：上海世博发展（集团）有限公司
设计时间：2009.1~2016.9
竣工日期：2017.9
总建筑面积：250000m²
用地面积：15.08hm²

Location: Block surrounded by Zhongshan South, Baotun, Wangda, Miaojiang, Huayuangang and South Chezhan Roads
Designers: Tongji Architectural Design (Group) Co., Ltd., East China Architectural Design & Research Institute Co., Ltd., Arcplus Group PLC, Beijing Institute of Architectural Design
Partner: Shanghai World Expo Development (Group) Co., Ltd.
Constructor: Shanghai World Expo (Group) Co., Ltd.
Design Period: January 2009 - September 2016
Date of Completion: September 2017
Gross Floor Area: 250,000m²
Site Area: 15.08 hm²

活水公园中心景观
Middle of Water Theme Park

城市最佳实践区是2010年上海世博会的一大创新项目，集中展示了全球具有代表性的城市为提高城市生活质量所进行的各种具有创新意义和示范价值的最佳实践。当世博会降下帷幕，城市最佳实践区作为街区改造范例的实践过程也悄然开始。该区域延续了世博会期间的建筑较为完整地保留了世博会期间的展览场馆，继承了上海世博会的无形遗产，继续成为交流、分享、推广城市最佳实践的全球平台，不断演绎"城市，让生活更美好"的世博主题；依托存量建筑和设施进行功能转换，打造成为集商务办公、创意设计、文化休闲、会议展览等为一体，具有世博特征和工业遗产特色的文化创意产业集聚区；在建筑、开放空间、基础设施、慢行交通等方面，继续体现街区层面的低碳生态发展理念，成为低碳生态街区。

汉堡馆
Hamburg Pavilion

The Urban Best Practices Area (UBPA) is one of the innovative projects of Expo 2010 Shanghai. It highlights innovative and exemplary best practices in cities around the world to improve the quality of urban life. When the World Expo ended, the UBPA discreetly started neighborhood transformation. This area is a continuation of the construction during the World Expo; it retains the exhibition venues during the World Expo in a relatively complete way and inherits the intangible legacy of the Shanghai World Expo. It continues to become a global platform for exchange and sharing urban best practices; constantly demonstrating the Expo motto, "Better City, Better Life." Relying on existing buildings and facilities, the remodeling has transformed their functions, making the site a cultural and creative industry hub that includes offices, design facilities, leisure facilities, and conference and exhibition venues. The spirit of the Expo and the industrial heritage continues on. In terms of architecture, open space, infrastructure, and non-motorized traffic, it achieves low-carbon ecological development.

城市最佳实践区由南北两个街坊组成，一条连续的步行轴线将两个街坊的广场、绿地、建筑院落等串联起来，形成收放自如的连续开放空间体系和完整的街区步行网络。各具特征的九个建筑组团围绕南北两个开放空间核心布置，北街坊以商务办公为主、商业服务和文化休闲为辅，南街坊以商业服务和文化休闲为主、商务办公为辅，形成复合互补、动静相宜的功能布局。开放空间和建筑组群之间相互渗透，使整个街区具有丰富的层次感和有趣的序列感。

The Urban Best Practice Area is composed of two neighborhoods, one in the north and one in the south. A continuous walking axis connects the squares, green spaces, and architectural yards of the two neighborhoods, forming a continuous open space and a complete neighborhood pedestrian network. The nine architecture groups with different characteristics are arranged around the core of the two open spaces in the north and south. The northern neighborhood is dominated by offices, supplemented by commercial services and cultural and leisure facilities. The southern one is dominated by commercial services and cultural leisure facilities supplemented by offices, forming a composite and complementary functional layout suitable for dynamic events and static display. The integration of open spaces and building clusters delivers a strong sense of layers and order to the block.

沪上生态家
EcoHome

园区鸟瞰
Aerial view

UPBA 行政楼
UBPA administrative center

整体鸟瞰
Aerial view

上海当代艺术博物馆
POWER STATION OF ART

入口大厅
Lobby

项目地点：黄浦区花园港路 200号
设计单位：同济大学建筑设计研究院（集团）有限公司原作设计工作室
设计时间：2011.4-2011.9
建成时间：2012.9
建筑面积：41000m²
用地面积：19103m²

Location: No. 200, Huayuangang Road, Huangpu District
Designer: Original Design Studio (A subsidiary of Tongji Architectural Design (Group) Co., Ltd.)
Design Period: April - September 2011
Date of Completion: September 2012
Gross Floor Area: 41,000m²
Site Area: 19,103m²

烟囱内螺旋展廊
Spiral gallery inside chimney

七层展厅
Exhibition hall: 7/F

这是一个触手可及的艺术馆，一个公平分享艺术感受的精神家园，更是一个充满人文关怀的城市公共生活平台。它以一种历史叙事的方式结束了辉煌的工业时代的使命，经历了从上海南市发电厂主厂房到2010年上海世博会城市未来馆的转变，继而蜕变为上海当代艺术博物馆。六年的时间见证了一个昔日能源输出的庞大机器如何转变为推动文化与艺术发展的强大引擎。

In the renovation project, the design has applied limited intervention to the architecture of the former Nanshi Power Plant; rather, it has given prominence to the industrial heritage, maximizing the benefit of its structure and internal spaces. At the same time, it has also maintained the visible traces of the passage of time—from its original identity of a power plant to the Urban Future Pavilion of the 2010 Expo and then to a brand new treasure

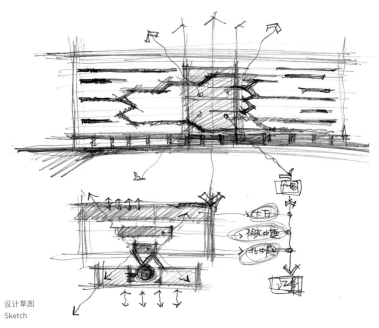
设计草图
Sketch

烟囱及辅助建筑
Chimney and auxiliary buildings

设计对原有南市电厂进行有限干预，最大限度地让厂房外部形态与内部空间的原有秩序和工业遗迹特征得以体现，同时又刻意保留了时空跨度上的痕迹，体现新旧共存的建筑特征。

建筑以开放性与日常性的积极姿态融于城市公共文化生活，以空间的延展蓄意模糊了公共空间与展陈空间的边界，不仅给颠覆人与展品的传统关系创造诸多机会，更为日常性的引入提供最大可能。上海当代艺术博物馆以多样性与复合性的文化表达诠释着人与艺术的深层关系，以漫游的方式打开了以往展览建筑的封闭路径，开拓出充满变数的探索氛围。

---

trove of art. Architectural features old and new are allowed to complement each other. This design enabled the revival of the once Expo pavilion, reflecting the designer's awareness of the natural and cultural environments of the property.
The Shanghai Power Station of Art seeks to establish itself as a powerhouse of the city's cultural life. It deliberately obscures boundary of the public spaces and exhibition spaces. This approach creates a new relationship between viewers and exhibits, reminding the public that art is an essential part of life. The property, with the cultural diversity it accommodates, proves to be a physical reminder of the close relationship between people and art, revolutionize traditional route designed for visitors byusual exhibition spaces, creating an atmosphere awaiting exploration. The Shanghai Power Station of Art has emerged as a gallery for everyone, a home of art and a platform of enjoyable urban life.

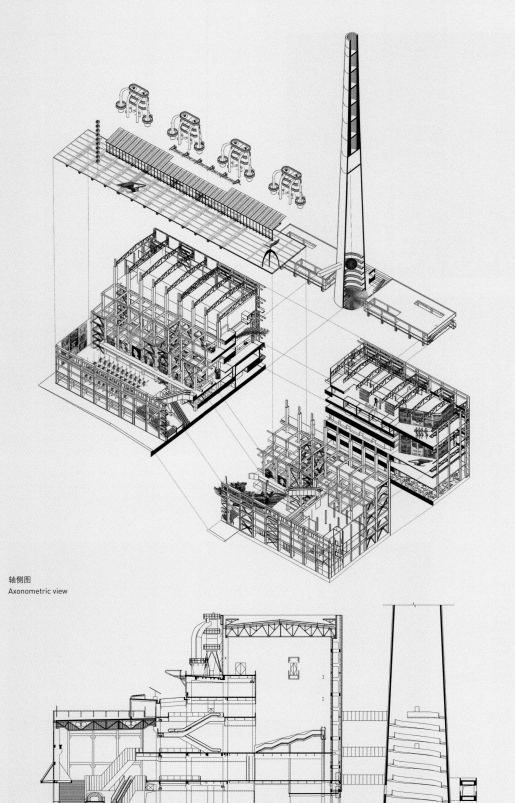

轴侧图
Axonometric view

剖面图
Section

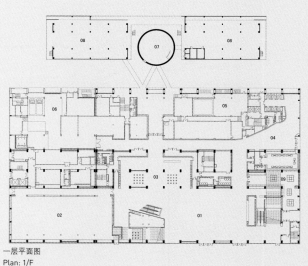

一层平面图
Plan: 1/F

沿江全景
Panoramic view

中庭
Atrium

北中庭
Northern atrium

北立面细部
Northern façade (details)

余德耀美术馆
YUZ MUSEUM

美术馆主入口
Main entrance

项目地点：徐汇区丰谷路35号
设计单位：藤本壮介建筑事务所
合作设计单位：上海都市设计有限公司
建设单位：上海徐汇滨江开发投资建设有限公司
设计时间：2013
建成时间：2014.5
建筑面积：9500m²
用地面积：5442m²

Location: No. 35, Fenggu Road, Xuhui District
Designer: Sou Fujimoto Architects
Partner: Shanghai Urban Architectural Design Co., Ltd.
Constructor: Shanghai Xuhui Waterfront Development Investment and Construction Co., Ltd.
Design Period: 2013
Date of Completion: May, 2014
Gross Floor Area: 9,500m²
Site Area: 5,442m²

立面细部
Façade (details)

东北角外景
Panoramic view from the northeast

原为生产麦道公司MD-82飞机机库，于1950年代由苏联援建。设计对保存完好的机库建筑进行改扩建，保留建筑主体结构，以及坡屋顶、钢结构桁架、钢制大门等精细构件，再现民航历史记忆。同时拓展网架结构的大跨度空间，雕琢局部构造的精细做法，设计继承历史风貌。

针对具有鲜明形体特征的老机库，充分利用大空间，将其设计成一个大空间的艺术展示厅。在尊重历史的原则下，尝试新老空间的融合，采用平屋顶的玻璃大厅与坡屋顶形成相互对比和呼应，再通过营造树木葱郁、明亮的开放式入口大厅，让新老建筑交相辉映。

The Yuz Museum is built on the premises of the former hangar of the Shanghai Aircraft Factory, a facility built with Russian assistance in the 1950s. It was also the place where MD-82/83 aircrafts were assembled in the 1980s under an agreement with the then McDonald Douglas Company. It is the cradle for the Chinese to start its dream of building airliners. The main structure of the building and the original style of the hangar have both been preserved, while a glass hall has been added to serve as its porch. Delicate components such as curved trusses and steel gates remain to honor the memory of its past endeavor in aircraft building.

The large space of the former hangar ideally suits the need for large-scale exhibitions. With due respect for the building's historical architectural contour, scarlet is chosen to be the color of the exterior, evoking a striking visual effect and a historical sense of

红墙与绿树
Green meets red

在色彩上，尊重老建筑的风貌，整体采用"深红色"，富有视觉冲击力与历史的沧桑感。玻璃材质的介入使老建筑的"深红色"砖墙与周边景观相融合，不仅使历史气氛得到继承，而且还营造了新的城市形象。

---

the former hanger. The newly added glass hall ushers in abundant sunshine, adding a welcoming touch to the old building. Thanks to the leafy landscape and genius addition of the glass hall, new and old architectural features coexist harmoniously with each other.

入口大厅
Lobby

展厅
Exhibition Hall

剖面图
Section

北侧外景
Aerial view from the north

改造前的飞机库
The hangar before renovation

星美术馆
START MUSEUM

沿江侧鸟瞰
Bird's eye view

西侧立面
Western facade

项目地点：徐汇区瑞宁路111号
设计单位：让·努维尔建筑事务所
合作设计单位：同济大学建筑设计研究院（集团）有限公司
建设单位：上海徐汇滨江开发投资建设有限公司
设计时间：2016
建成时间：2021
建筑面积：1777m²

Location: No. 111, Ruining Road, Xuhui District
Designer: Ateliers Jean Nouvel
Partner: Tongji Architectural Design (Group) Co., Ltd
Constructor: Shanghai Xuhui Waterfront Development Investment and Construction Co., Ltd.
Design Period: 2016
Date of Completion: 2021
Gross Floor Area: 1,777m²

西侧鸟瞰
Bird's eye view from the west

星美术馆由原铁路南浦站十八线仓库改造而成。始建于1907年的南浦站，为当时上海地区唯一自备专用码头的铁路车站，第一个水陆联运码头。

建筑改造充分尊重历史建筑，以文化展示功能为基础，重塑场地特征，形成一处全新的文化休闲场所，成为上海西岸"美术馆大道"中独特的文化亮点。

美术馆为单层建筑，保持原建筑轮廓，强化与开放空间的联系，构建了一个独特的、与公众互动的展览空间。保留的坡屋顶形式，形成室外公共区域遮风避雨的缓冲空间，屋顶的金属材料和玻璃天窗完美结合，幕墙和玻璃随着光线的变化，呈现微妙而自然的景观效果，为艺术与市民生活之间创造最直接的对话。

The Start Museum is converted from the old warehouses of the now-defunct Nanpu Railway Station. Starting its service in 1907, the former railway station was the only one in Shanghai that had its own wharf and it was also the first railway station in Shanghai that could handle land-and-water coordinated transport.
The renovation effort has remodeled historical buildings to accommodate its new cultural functions. The Start Museum has become a unique cultural landmark in the Museum Mile of Shanghai West Bund.
The art museum is a single-story building that preserves its original structure. After renovation, its architectural features are closely related to open spaces, making it more accessible. The reserved sloped roof provides shelter from wind and rain. The metal structures of the roof and the glass sunroof are perfectly fused together, producing an amazing visual effect with the change of light. The museum building and the adjacent Xuhui Riverfront Community Service Center forge a close bond with the Huangpu River, adding an artistic touch to public life.

西侧人视
Eye-level view from the west

龙美术馆西岸馆
LONG MUSEUM WEST BUND

东侧外观
Eastern square

展厅空间
Interior

项目地点：徐汇区龙腾大道 3398 号
设计单位：大舍建筑设计事务所
合作设计单位：同济大学建筑设计研究院（集团）有限公司（结构机电）
建设单位：上海徐汇滨江开发投资建设有限公司
设计时间：2011.11-2012.7
建成时间：2014.3
建筑面积：33007m²
用地面积：19337m²

Location: No. 3398, Longteng Avenue, Xuhui District
Designer: Atelier Deshaus
Partner: Tongji Architectural Design (Group) Co., Ltd. (Mechanical and electrical design)
Constructor: Shanghai Xuhui Waterfront Development Investment and Construction Co., Ltd.
Design Period: November 2011 - July 2012
Date of Completion: March 2014
Gross Floor Area: 33,007m²
Site Area: 19,337m²

外部空间
Exterior

这里曾是黄浦江煤运码头，设计保留了20世纪50年代建造的煤料斗卸载桥，以及两年前建成的两层地下车库。采用独立墙体的"伞拱"悬挑结构，自由布局的剪力墙插入地下室与原有框架结构柱浇筑在一起。地下一层的车库空间由于剪力墙体的介入转换为展览空间，地面以上的空间由于"伞拱"在不同方向的相对连接形成多重意义的空间指向。机电系统都整合在"伞拱"结构的空腔里，地面以上的"伞拱"覆盖空间。这样的结构性空间，在形态上不仅对人形成庇护感，也与保留的煤料斗产生视觉呼应。
墙体和天花采用清水混凝土，形成模糊的几何分界。建筑内部空间呈现一种原始的野性魅力，而大小不一的空间以及留有模板拼缝和螺栓孔的清水混凝土表面又带来一种现实感。这种"直白"的结构、材料、空间形成的直接性与朴素性，加上大尺度出挑产生的力量感或轻盈感，使整个建筑与原有场地的工业特质间取得一种时间与空间的接续关系。螺旋回转、层层跌落的阶梯空间，连接地面以上的清水混凝土、

The Long Museum West Bund is converted from Coal Wharf, reserved for loading and unloading of cargo. The design retains the coal hopper bridge built in 1950s, and the two-story underground garage built two years ago.
The four-storey main building features a unique "vault-umbrella" structure. The peculiar arched spaces of the first and second floors above ground level are built by smooth-textured, fair-faced concrete. It visually echoes the "Hopper Corridor" converted from the former coal hopper of the wharf. In addition to a sense of tranquility and primitivism pertaining to industrial heritage, it shows a sharp contrast between strength and lightness, and a sense of modernity and creativity ideal of a museum. The exhibition hall on the basement floor is a "white cube" in the traditional rectangular shape. Functionally, it provides public spaces with enhanced openness to engage public involvement: Art is not inaccessible but relevant to daily public lives.

建筑局部
Details

建筑局部
Details

建筑局部
Details

"伞拱"下的流动展览空间和地下一层传统"白盒子"式的展览空间。既原始又现实的空间，与古代、近代、现代以及当代艺术的展览陈列，带来并置的张力，呈现出一种具有时间性的展览空间。

The building is surrounded by approximately 160 cherry blossom trees, which are a fitting decoration of the architecture. The Long Museum is a must-see place to enjoy cherry blossoms in the spring.

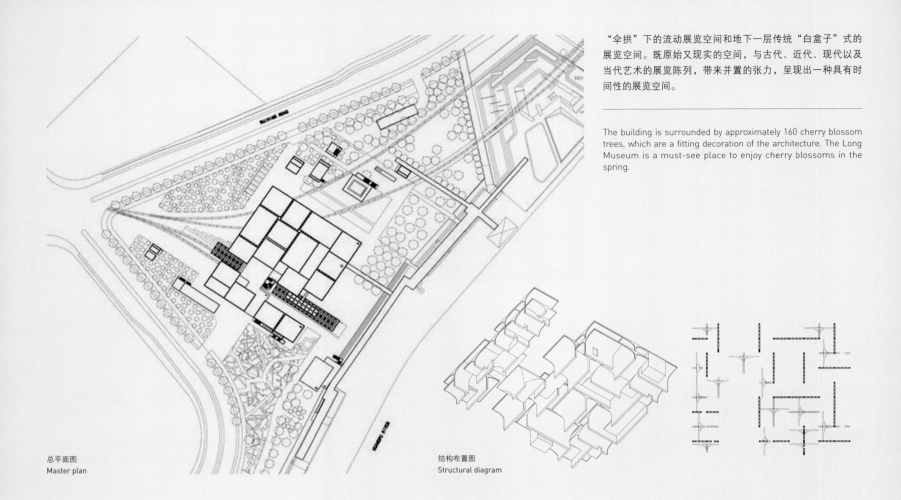

总平面图
Master plan

结构布置图
Structural diagram

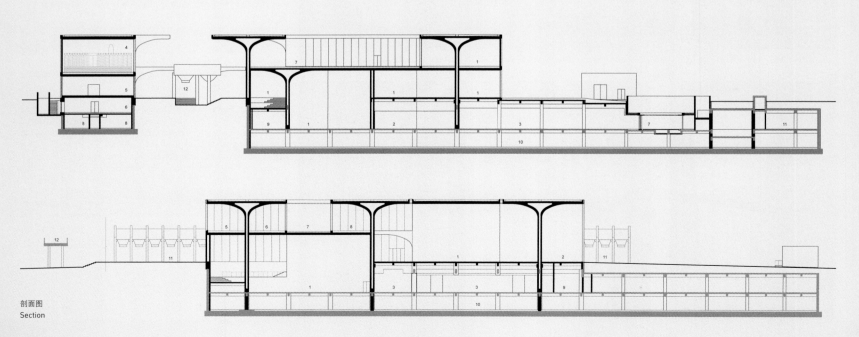

剖面图
Section

美术馆与煤料斗卸载桥
The museum and the neighboring coal-unloading bridge

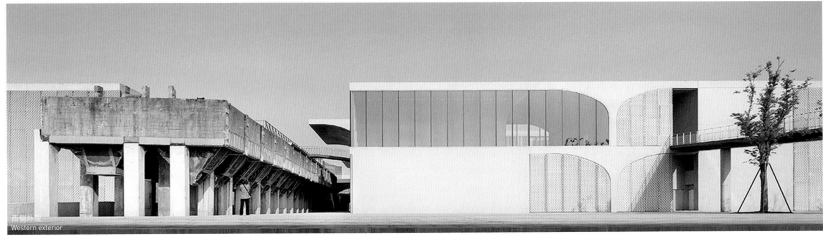
西侧外观
Western exterior

油罐艺术公园
TANK SHANGHAI

鸟瞰图
Aerial view

项目地点：徐汇区龙腾大道 2530号
设计单位：Open建筑事务所
合作设计单位：同济大学建筑设计研究院（集团）有限公司
建设单位：上海徐汇滨江开发投资建设有限公司
设计时间：2014
建成时间：2019
建筑面积：10845m²

Location: No. 2530, Longteng avenue, Xuhui District
Designer: Open Architecture
Partner: Tongji Architectural Design (Group) Co., Ltd.
Constructor: Shanghai Xuhui Waterfront Development Investment and Construction Co., Ltd.
Design Period: 2014
Date of Completion: 2019
Gross Floor Area: 10,845m²

改造前的中航油储油罐
The tanks before renovation

基地原有7个储油罐体，是中航油为虹桥机场提供燃油补给的。设计通过选择性地保留其中5个，将其改造成为展览和活动空间，同时与室外场地紧密联系，营造可生长的"大地艺术"公园的核心理念。这是全球为数不多的油罐空间改造案例之一。

油罐罐体最大限度地保持工业痕迹和原始的美感，只新增一些圆形、胶囊形的舷窗和开洞，形成外立面独特的表情，同时给罐内营造了朝向公园和黄浦江的优美框景。覆土绿化的新地形——"超级地面"串联五个油罐，形成一个高低起伏的公园。邻近公路的1、2号罐相对独立，位于超级地面之上；3、4、5号罐则有一半位于超级地面之下，三个罐之间形成开阔的半地下公共空间，包含艺术中心门厅、展览空间、报告厅、咖啡厅和艺术商店等功能，隔着通透的落地玻璃面向下沉式广场。公共空间吊顶设有大量天窗，可引入自然光。三个罐的入口护坡由10～25毫米厚的钢板围合，层叠呼应着原始罐体的曲面结构，部分台阶则改造为咖啡厅的座椅。

改造后的油罐
The tanks after renovation

The original seven oil storage tanks were provided by CNOOC for fuel supply to Hongqiao Airport. The designer selected 5 of them and transformed into exhibition and activity space. At the same time, the Tank Art Park is connected with the outdoor space, to create a viable "land art" park. This is one of the few oil tank space transformation cases in the world.

Conceived of as both an art center and an open park simultaneously, the project not only pays tribute to the site's industrial past, but also seeks to dissolve the conventional perceptions of art institutions with formidable walls and definitive boundaries that separate the museum goers from those who are not. TANK Shanghai sets out to be an art center for all, and a museum without boundaries. Central to the project's design is the merging of architecture and landscape through a Z-shaped "Super-Surface"—a five-hectare landscaped swath of trees and grasses which connects the five tanks and weaves different elements of the site together. The Super-Surface brings a number of both aesthetic and practical benefits to its riverside site, which shares a 115 meter-stretch of shoreline with the Huangpu River. The lush, thickly planted landscape creates beautiful and urgently needed parkland in a city with just 17.56% green space, contributing to the rejuvenation of the area and the return of animal life in the city. Underneath the Super-Surface is the new construction of a large

城市广场
Square

城市广场
Square

从西侧的龙腾大道到东侧的黄浦江边，室外公园拥有多条参观路径，人们可自由漫步其间。沿着逐层下降的"阶梯水景"进入"城市广场"，凉爽的喷雾使它成为夏季最受欢迎的乐园，而喷雾的平面形态与一个已拆除的油罐呼应。贯穿基地南侧的是一大片"都市森林"，是城市中的自然；东侧为一片开阔的"草坪广场"，为人们提供奔跑、休憩的空间，也可以作为室外音乐节等活动的观众席。

场地中散落着艺术装置，以及两个用于举办艺术活动和小型展览的独立小建筑。一座为镜面不锈钢包裹的"无影画廊"，掩映在都市森林之中，若隐若现；另一座为拥有连续锯齿形天窗的"项目空间"，靠近黄浦江边，以其鲜明的几何形态与白色的油罐群形成强烈反差。

设计将城市公园与艺术展览、大地景观与建筑空间、工业遗存与创新未来，天衣无缝地融合起来，并以开放友善、谦逊包容的姿态融入城市生活中。一个无边界的美术馆发挥着它的无限可能，同时潜移默化地塑造着一种平等、共享的人文精神。

free-flowing open space that connects three out of five tanks from the underground level.
Visitors arriving from the street would descend a pebble path accompanied by a terraced waterscape toward the plaza, and then to the main entry, hardly realizing that now they are at underground level. Hidden misting devices create a cooling fog in the center of the plaza during summer times. At once poetic and practical, this circle of mist was designed to evoke the memory of a sixth tank which was torn down for the construction of the neighboring heliport, while at the same time improving the microclimate of the Plaza. Flanking the south side of the plaza is an "Urban Forest" which provide much desired greenery and shades to the urban residents. Covering the eastern portion of the site, a second grassy plaza provides open space for outdoor events and leisure, and doubles as a standing room for audiences during music festivals. The Super-Surface is dotted with a collection of freely accessible public art installations.

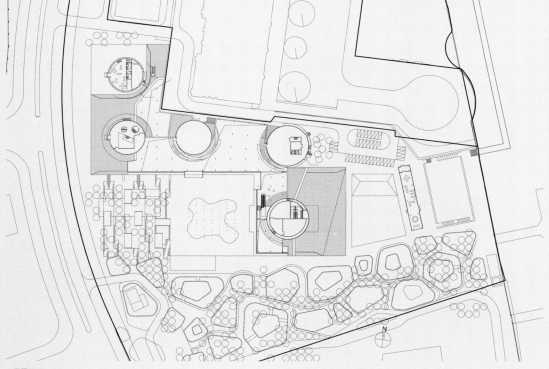

一层平面图
Plan: 1/F

剖面图
Section

阶梯水景
Water area

民生码头八万吨筒仓
80,000-TON SILOS AT MINSHENG WHARF

整体鸟瞰
Panoramic view

筒仓内部的结构得以保留
The interior remained

外挂扶梯处的景观视野
Broad vision

西侧立面
Western façade

项目地点：浦东新区民生路3号
设计单位：大舍建筑设计事务所
合作设计单位：同济大学建筑设计研究院（集团）有限公司（结构机电）
建设单位：上海东岸投资（集团）有限公司
设计时间：2016.12-2017.10
建成时间：2017.10
建筑面积：16322m²
用地面积：20000m²

Location: No. 3, Minsheng Road, Pudong New Area
Designer: Atelier Deshaus
Partner: Tongji Architectural Design (Group) Co., Ltd (M&E)
Constructor: Shanghai East Bund Investment (Group) Co., Ltd
Design Period: December 2016 - October 2017
Date of Completion: October 2017
Gross Floor Area: 16,322m²
Site Area: 20,000m²

"适应性再利用"原则下的筒仓改造

八万吨筒仓是浦东民生码头保留工业建筑之一，曾是亚洲最大容量的散粮筒仓，已列入上海市文物保护建筑。考虑到筒仓建筑相对封闭的空间适合作为展览空间，将其定位为多功能艺术展览空间。方案设计依据"适应性再利用"的原则，以满足区域发展和适应建筑空间特点为出发点。

外挂楼梯重塑空间流线

为满足2017年上海城市空间艺术季的展览计划，筒仓空间的改造需要在半年内快速完成，并要兼顾后续改造需求。空间艺术季主要使用筒仓建筑的底层和顶层作为展览空间。筒仓建筑高达48米，改造需要整合底层和顶层形成顺畅的观展流线，同时增加必要的消防疏散等设施。其难点在于解决筒仓原本生产功能带来的封闭性与新的公共文化功能带来的开放性之间的矛盾。

方案在筒仓北侧外挂一组直接引至顶层展厅的三层自动扶梯，既解决了交通问题，又为观展的人们提供了欣赏黄浦江以及整个民生码头壮丽景观的视角。筒仓本身几乎不做任何改动，最大限度保留原本风貌，并通过重新利用注入新能量。

The 80,000-ton Silo is one of the conserved industrial buildings located at the Minsheng Wharf in Pudong, its heritage value coming from its unique architectural form that is associated with the era when it was built. As a case of adaptive reuse of a historic building, planning studies and research by the architects were conducted to facilitate the development, where a new program of multi-functional art exhibitions was eventually decided on, which would be suitable to both the neighborhood's development and the unique spatial characteristics of the existing building. Culture and art programs have become the main proponents in the public space productions of Shanghai's recent urban regeneration projects, in which art exhibition spaces are a good fit for the enclosed spaces

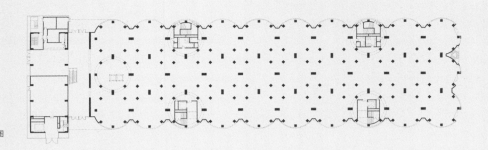

一层平面图
Plan: 1/F

螺旋形坡道
Spiral ramp

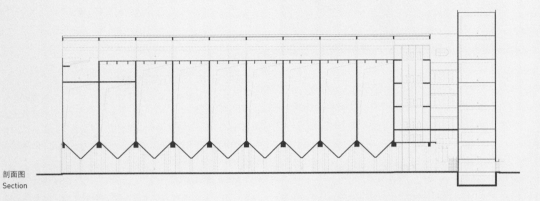

剖面图
Section

外挂扶梯将八万吨筒仓与滨江公共空间、浦江景色联系起来，增加了建筑的公共性。同时，在筒仓中部预留了接口，便于在后续改造设计中利用这30个直径约12.5米的仓筒组合，通过内部分隔、连通，形成立体展览空间。
未来，从江边直上筒仓三层的粮食传送带将被改造为自动人行坡道，形成从江边直接上至筒仓顶层的公共空间。

---

of the existing Silos. The 2017 Shanghai Urban Space Art Season (SUSAS) held a three-month exhibition in the Silo. To prepare for the SUSAS, the Silos rapidly were renovated within half a year, providing a foundation for future renovation efforts.

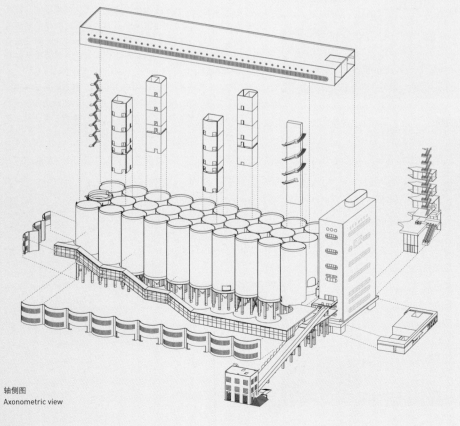

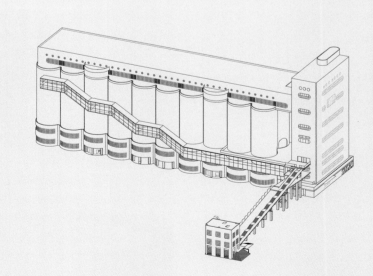

轴侧图
Axonometric view

北侧建筑近景
Close-up view from the north

上海浦东船厂 1862
SHANGHAI PUDONG SHIPYARD 1862

建筑外观
Exterior

对原有结构的巧妙利用
The original structure remained

项目地点：浦东新区滨江大道1777号
设计单位：隈研吾建筑都市设计事务所、华建集团上海建筑设计研究院有限公司
建设单位：中船置业有限公司
设计时间：2012.6-2015.7
建成时间：2016.12
建筑面积：31626m²
用地面积：11340m²

Location: No. 1777 Binjiang Avenue, Pudong New Area
Designer: Kengo Kuma and Associates, Shanghai Architectural Design & Research Institute Co., Ltd., Arcplus Group PLC
Constructor: CSSC Real Estate Co., Ltd.
Design Period: June 2012 - July 2015
Date of Completion: December 2016
Gross Floor Area: 31,626m²
Site Area: 11,340m²

南侧立面
Southern facade

上海船厂前身为招商局造船厂和英联船厂，具有悠久历史，是中国船舶工业集团公司下属的五大造修船基地之一。为传承船厂悠久的工业文明，设计保留船台及造机车间，充分利用老厂房原有的层高和外立面结构及滨江优势，业态定位以娱乐休闲、餐饮、演艺为主，包含剧院、多功能厅、商业店铺。

The history of the Shanghai Shipyard can be traced to the Shanghai Merchants Shipyard and the British-run Shanghai Dockyards founded in 1862. It is also one of the five major ship building and repairing bases of the China State Shipbuilding Corporation. The renovation of the preserved workshops and shipways on Plot 2E1-1 takes full advantage of their immense heights and proximity to the river. It has converted the buildings into an array of facilities, ranging from entertainment, catering, performing art and retail premises.

To design the layout of the plan, the designer set up commercial facilities at the main entrance on the west side and an 800-seat theater on the east side where the river can be seen. Inside the building, an atrium connecting horizontally and vertically is set up to give a sense of the overall scale of the building. At the same time, the rusted steel staircase and the slogans on the columns are preserved as much as possible so that the historical atmosphere can be felt. Materials with industrial characteristics such as oxidized rusted steel plates, aerated concrete blocks and stainless steel metal mesh are mainly used in the new construction part. In

保留的混凝土柱与鱼腹梁
The retained concrete columns with fish bally beams

新旧材料交映
The new meets the old

平面布局上，在西侧主入口方向设置商业设施，在可以观江的东侧设置800座剧院。在建筑内部，设置连接横向和纵向的贯通中庭，使人感受到建筑的整体尺度。同时，尽可能地保留锈迹斑斑的钢楼梯与柱子上的标语，可感受到历史氛围。新建部分主要采用氧化锈迹钢板、加气混凝土砖、不锈钢金属网等具有工业特征的材料。在山墙墙体中，使用四种不同颜色的陶砖，以直径8毫米的不锈钢拉索连接吊挂，形成砖幕墙，与历史风貌相协调。为衔接南侧已拆除的砖墙，整面砖块以水平向渐变的方式随机排布，形成随机性与混杂感。

the wall of the hill wall, four different colors of ceramic bricks are used and hung with stainless steel cable connection of 8 mm in diameter to form a brick curtain wall, which is in harmony with the historical style. In order to articulate the demolished brick wall on the south side, the whole brick is randomly arranged in a horizontal to gradual manner, creating a sense of randomness and mixture.

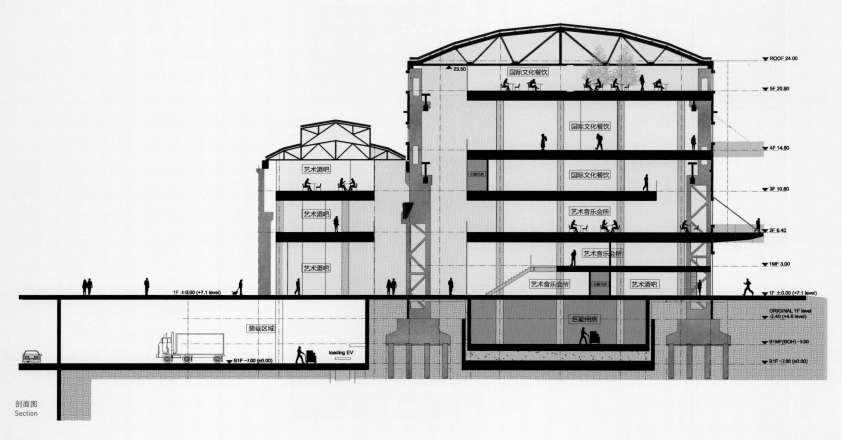

剖面图
Section

陶砖与室内光影
Pottery bricks andshadows

上海艺仓美术馆及其长廊
SHANGHAI MODERN ART MUSEUM AND ITS WALKWAYS

东南角外观
Bird's eye view from the southeast

长廊下的咖啡馆
Café in corridor

项目地点：浦东新区滨江大道4777号
设计单位：大舍建筑设计事务所
合作设计单位：同济大学建筑设计研究院（集团）有限公司（结构机电）
建设单位：上海浦东滨江开发建设投资有限公司
设计时间：2016.2-2016.10
建成时间：2016.12
建筑面积：9180m²

Location: No. 1777 Binjiang Avenue, Pudong New Area
Designer: Atelier Deshaus
Partner: Tongji Architectural Design (Group) Co., Ltd. (M&E)
Constructor: Shanghai Pudong Binjiang Development Construction Investment Co., Ltd
Design Period: February - October 2016
Date of Completion: December 2016
Gross Floor Area: 9,180m²

长廊与滨江空间
Corridor along riverside

艺仓美术馆的旧址是位于上海浦东的老白渡煤炭码头，码头上既有的煤仓建筑险些被拆除，建筑师通过在既有的煤仓废墟中成功地进行了一次临时展览，说服了业主将原有建筑保留了下来。为更好地组织空间，并尽可能减小对现有煤仓结构的破坏，设计采用悬吊结构，利用屋顶已经拆除后留下的顶层框架柱支撑一组巨型桁架，并层层下挂，在完成流线组织的同时也构建了原本封闭的仓储建筑所缺乏的与黄浦江景观之间的公共性连接。

建筑首层是有着八个煤仓漏斗天花的多功能大厅，旧有的结构直接暴露在室内，与室外精致的阳极氧化铝板的外立面形成了鲜明的对比。二层是低矮的由煤料斗的斜面间隔的多媒体展厅，三层是美术馆的主展厅，它利用原来储煤的八个方形煤斗，相互打通形成独一无二的带着原本粗粝质感的混凝土墙体展厅。四层则又是一个高大空间的展览厅，原本斜向运煤的通道被改造为钢结构大楼梯可以由室外直达这一层，从而也为美术馆的多种运营和展览方式提供了多重的人行流线。旧有结构并未暴露在外，而是深藏在看上去是一个崭新建筑的核心之中。

长廊二层的观景平台
Platform on the corridor 2/F

Shanghai Modern Art Museum's site was the Laobaidu Coal Wharf in Shanghai's Pudong. The coal storage infrastructure that remained had nearly been demolished, but through some of the temporary exhibitions that the architect organized amongst the ruins, the client was persuaded to preserve the remaining infrastructures. To better organize the spaces, while minimizing further damage to the coal infrastructures, a suspension structure was designed, with a large set of trusses hung from the upper most columns that remain after the roof was removed. Layering the trusses downwards, the floor slabs were connected to the suspended hooks on one side and to the original structure of the coal bunker on the other side. The design both completes the circulation organization and also creates an open connection to the scenery of the Huangpu River that had been closed in the former warehouse.

On the ground level, a multi-functional lobby with eight original coal hoppers forming its ceiling, contrasting the rough exposed infrastructure inside to the polished anodized aluminum cladding on the outside. On the second floor is a compressed multi-media exhibition space that sits between the diagonal walls formed by coal hoppers. The main exhibition space of the art museum is on the third floor, where the original walls separating the eight coal hoppers opened and enclosed by the original coarse concrete walls. On the fourth floor, there is another full-height exhibition space that is directly accessible from the exterior by a grand steel stair, which is converted from the former coal-loading channel, and which offers an alternative circulation to accommodate the operations and exhibitions of the museum. Rather than exposing the historical structure, its concealment is at the core of the new architecture.

The elevated coal-loading channel of the Laobaidu Coal Warehouse is approximately 250 meters long. Before renovation, only rows of parallel concrete frames remained and they were connected by perpendicular beams at intervals. The newly-built walkway on the coal-loading channel both belongs to the Modern Art Museum, and is part of the reconstructed urban waterfront public space along the Huangpu River. The upper level is the elevated walkway for pedestrians, while the ground level is occupied by art shops and cafés, which both contribute to the waterfront public space, as well as offering potential uses for the Museum.

The elevated walkway brings together a new beam string and suspended steel structural system with the existing concrete frames, which has weathered on the surface, forming a new

煤仓北侧的高架运煤廊道长约250米，改造前只留下一排排平行阵列的混凝土排架，间隔有一些垂直方向的横梁连接。这些运煤的廊道既属于艺仓美术馆，也是沿黄浦江重新构建的城市公共空间的一部分。长廊上部是高架的人行通道，下部则布置了艺术品商店和各有特色的咖啡馆，为江边的公共空间服务，也为美术馆的运营带来更多的可能性。

长廊的设计采用张弦梁加悬吊钢结构系统与表面已经有一定风化腐蚀的混凝土排架结构相结合，形成一个新的整体结构，也持了各自的视觉完整性。它既承担了上部廊道的地面结构，也向下挂着玻璃盒体商店的屋顶部分。既使商店的盒体建筑从构架中游离出来，又仍然是结构整体的一部分。这种做法并没有着意去强调既有的结构体，而是使已然废墟般的结构再次融入新的建造，它使曾经的工业碎片悄然进入上海的日常生活环境。

改造前的煤斗
Coal hoppers before renovation

改造后的多功能大厅
Multi-functional hall after renovation

改造前的混凝土排架
Parallel concrete frames before renovation

structural entity while still maintaining their own material integrity. This structural entity both supports the elevated walkway's structure above, and provides a roof for the glass-enclosed shops on the ground. This structural plinth both floats the shop volumes within the concrete frames, while still being part of the entire structure. Rather than underlining the existing structures on the site, it allows the historic ruin to become meld into new construction. In this way, the industrial fragments of the past could remain subtly part of Shanghai's everyday life.

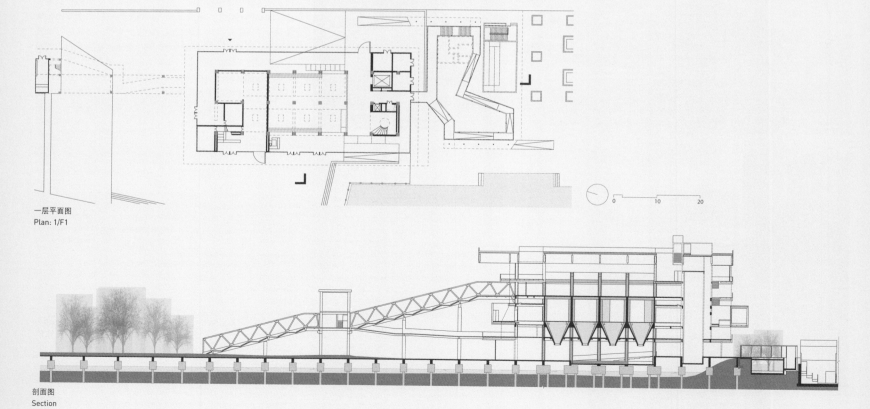

一层平面图
Plan: 1/F1

剖面图
Section

北侧外观
Northern exterior

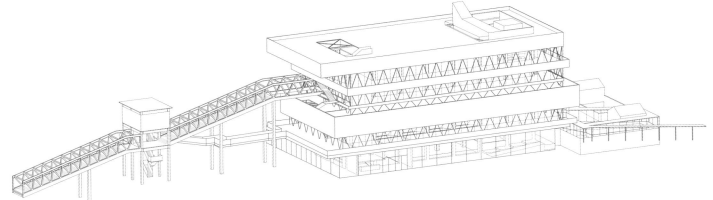

轴侧图
Axonometric view

163

上海宋城·世博大舞台
SHANGHAI SONGCHENG . WORLD EXPO GRAND STAGE

东侧立面
Eastern façade

项目地点：浦东新区世博大道 1750 号
设计单位：华建集团华东建筑设计研究院有限公司
建设单位：上海宋城世博演艺发展有限公司
设计时间：2016-2020
建成时间：2021.4
建筑面积：42571m²
用地面积：18039m²

Location: No. 1750, Shibo Avenue, Pudong New Area
Designer: East China Architectural Design & Research Institute Co., Ltd., Arcplus Group PLC
Constructor: Shanghai Songcheng World Expo Performance Development Co., Ltd.
Design Period: 2016 - 2020
Date of Completion: April 2021
Gross Floor Area: 42,571 m²
Site Area: 18,039 m²

顶视图
Top view

建筑前身是上钢三厂的特钢车间，世博会期间改造成宝钢大舞台，是各省市馆举办馆日庆典以及群众性综艺演出的场所。将其设计为演艺公园——上海宋城.世博大舞台，形成演艺剧场集群。采用超大型剧场上下重叠的设计方式，与紧邻的世博文化公园共同构成"一江一河"城市会客厅中的人文生态空间。为了构建"多功能、全天候的旅游文化活动中心"，设计既保持工业元素，延续场馆记忆，创造文化艺术空间，确保与城市文脉的良好关系，同时又满足演艺技术和舒适观演的需求，达到演艺需求与公众效益的协调统一。设计在原有厂房的结构框架内，置入新的使用功能，保留原厂房的主要结构构件，复原年久失修的连铸车间屋顶。

传统文化类建筑强调自身的标志性，空间也较为内向封闭，但上海宋城的建筑体量区别于周边其他场馆，且位于世博公园内，从空间属性上更应融入滨江文化休闲带。因此设计通过局部架空的方式连接地块南侧总部央企办公区和北侧滨江绿地，形成"商企-人文-自然"的完美过渡。

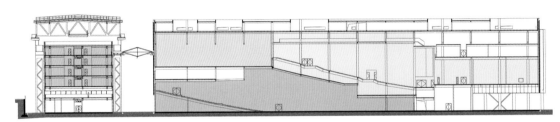
剖面图
Section

The building was formerly the special steel workshop of the Shanghai No. 3 Steel Plant, which was transformed into the Baosteel Stage during the World Expo, a venue for different provinces and cities to hold celebrations and gala performances. It is designed as a performing arts park - Shanghai Songcheng that emcompasses a cluster of performing arts theaters.
In order to build a multi-functional, 24/7 tourism and cultural venue, the design preserves the industrial elements to recall the past of the venue while forging bonds with the cultural life of the city. It also satisfies the needs for performing arts technology and comfortable viewing, and achieving the balance between performing arts needs and public benefits. The design has assigned new functions to the structural framework of the original plant, preserving the main structural elements of the original plant and restoring the roof of the dilapidated continuous casting workshop.
Traditional cultural buildings tend to emphasize their nature as landmark and mostly assume enclosed spaces. But the building volume of Shanghai Songcheng is different. As it is located in the Expo Park, it is compatible to the riverfront cultural and recreation zone in terms of spatial properties. The design connects the office area of enterprises directly under the central government on the south side and the riverfront green area on the north side by means of elevation passage, forming a perfect integration of "business - humanities - nature".

北侧鸟瞰
Bird's eye view from the north

设计·师说
MASTERS' WORDS

人类与水的共存

隈研吾

新冠疫情将成为城市历史乃至人类历史的重大转折点。准确地说，我认为这是一个具有决定性的转折点，历史将发生逆转。在这一逆转中，亲水空间将起到很大的作用。

在发生新冠疫情以前，人类的历史是一个简单的上坡道，如果用一句话概括，就是朝着向城市、人类聚集的方向前进的。首先，从狩猎采集时代到农业时代作为爬坡的开端，接着出现了街道、城市，最终以超高层建筑的形式，加速了人类的聚集。

但是，通过这次新冠疫情带来的城市封锁，我注意到以人类聚集为目标的上坡道已经超越了人类这种生物的极限。新冠疫情警告着我们，如果就这样放任一切继续"跑"下去的话，我们的身体和精神都将无法承受。

重要的是全世界的人们都有同样的感受。甚至全人类都感到，如果不转换方向，人类将无法继续生存下去。这一点是通过亲身体验，跨越了国境，全球所有人都同时感受到的。新冠疫情之后的世界，将会朝着自然的方向发生巨大转变，我们无法阻挡这股潮流。

在这一逆转中，城市中的水体将发挥巨大的作用。因为人类从狩猎采集时代开始，离开水就无法生存。因此水才是人类最需要的自然资源。靠近水源附近，会给人类带来无穷的安全感。

回顾历史就会发现，在通往人类聚集的坡道的顶点，也是摩天大楼最鼎盛的20世纪，人类彻底远离了水。水域周边被认为是最适合布局工厂和仓库的位置，全世界的城市水域周边都被工业设施占据，人们从水边被赶了出来。人类不仅因为超高层大厦而与大地分离，更因为工业化社会使人类越来越远离水域。

出于对这一悲剧状况的危机感，其实早在新冠疫情发生之前，世界各地已经开始试图让城市恢复水源，让人与水重新紧密结合。其实，人们在新冠疫情之前就已经感受到不想再继续攀爬在以人类聚集为目标的坡道上，不想再远离自然。上海黄浦江两岸的滨水空间就是其中最具代表性的项目。

从这个意义上来说，设计黄浦江沿岸的船厂1862项目的重要性远远超过单纯的建筑设计。通过船厂1862这栋建筑，我们在不知不觉中参与到恢复世界各地水源，重建人类和水域关系的大项目当中。

在设计类似的滨水空间项目的时候，我最重视的是建筑能够尊重自然，使其自由流动。自然绝对是片刻不停的。它是各种流动的集合体，有水的流动，有风的流动，也有光的流动。在避免阻挡这些流动的前提下，我们设计了船厂1862。

首先，在建筑中打通几个连接河流和街道的孔。在船厂1862中也开了个孔，同一时期我设计的苏格兰海滨项目（维多利亚和艾伯特博物馆邓迪分馆）中，也将面向河流的孔作为建筑物的主角。

虽然船厂1862的主要材料是砖块，但通过在砖块与砖块之间留出的小小缝隙，我们用崭新的做法向带有厚重感的砖块发起了挑战。利用不去阻碍流动性的细节使砖块和水一同奏响出最和谐的音符。无论是在规划上，还是在细节上，水边的建筑物都需要开孔。

这样一来，城市里的水域就会和森林里的水域一样，变成可以透风、透光的舒适空间。从狩猎采集时代开始，水域周边原本就是这样的地方——人们聚集于水域，与水共存。想要恢复到这种关系绝不只是梦。走在黄浦江岸边，眺望着波光粼粼的水面，渐渐可以看到未来人与水的关系。

Coexistence Of Human and Water
Kengo Kuma

The Covid-19 pandemic will be a major turning point in the history of the city and in human history. To be precise, I think it is a significant turning point in which history can be reversed. In this reversal, the space by water bodies will play a major role.

Prior to the pandemic, human history was simply an "uphill ramp" that prompts human beings to gather in cities. First, when the climb began from the hunter-gatherer era to the agricultural era, streets and cities emerged, eventually accelerating human aggregation with the invention of super tall buildings.

However, through the lockdown caused by Covid-19, I have noticed that the uphill ramp of the human gathering is well beyond the limits of human's capability. The pandemic has warned us that if we continue to live like this, we will not be able to afford the price either physically or mentally.

The important thing is that people all over the world feel the same way. The entire human race feels the need to change the way we live, otherwise we will not survive. Through personal experience, people regardless of their nationality felt this simultaneously. The post-pandemic world will shift dramatically in favor of nature, and we cannot stop the trend.

During this reversal, water bodies in cities play a critical role. This is because humans have not been able to survive without water since the hunter-gatherer era; water is the nearest and most needed natural resource. Proximity to water sources gives human beings enormous sense of security.

A review of history will show that at the top of the ramp leading to the human gathering, which is also the 20th century, known as the age of skyscrapers, has forced human beings to completely move away from water. The perimeter of water bodies was considered to be the most suitable locations for plants, factories and warehouses, and the perimeter of urban waters around the world was occupied by these inhuman facilities where people have been elbowed out. Humans have been separated not only from the ground by high-rise buildings, but also from the water by the industrialization of the society.

Out of a sense of anxiety over this tragic situation, attempts have been made universally to bring water back to the cities and reconnect people with water long before the beginning of the pandemic. In fact, before the pandemic, people already felt that they did not want to continue climbing up the ramp where humans gathered and did not want to be separated from nature any further. Waterfront spaces on both sides of the Huangpu River in Shanghai are the most representative project of such kind.

In this sense, the design of the Shipyard 1862 project along the Huangpu River is much more important than a mere practice of architectural design. Through the Shipyard 1862 building, we have unknowingly participated in a grand undertaking of restoring water sources worldwide and re-establishing the tie between people and water.

One thing I valued the most while participating in designing such waterfront project is that architecture should not stop the flow of nature. Nature is a composite of never-ending flows: the flow of water, of wind, and of light. On the basis of not stopping the flow, we designed the Shipyard 1862 project.

Firstly, a few openings connecting the river and the streets will have to be made in the architecture. Shipyard 1862 was opened up through a void, and meanwhile in the Scottish waterfront project (V&A Museum Dundee Branch) that I designed, the cavity facing the riverfront also became the key feature of the building.

In Shipyard 1862, despite the main material being brick, by leaving small gaps between them, the material usually correlated with heaviness and bulkiness was once challenged by such a new way of exploitation. Through such facade detail that doesn't block the flow, bricks and water can meet harmoniously.

In this way, the waters in the city would become comfortable spaces as those in forests, both of which are permeable to wind and light. Since the hunter-gatherer era, waterfront has been a place like this - people gather here, and coexist with water. Restoring this relationship is by no means a dream. While strolling along the Huangpu River and looking at rippling surface of the river, I begin to see the future relationship between humans and water.

HUANGPU RIVER SUZHOU CREEK

THE WATERFRONT SPACE & ARCHITECTURE IN SHANGHAI

多元融合创新
INNOVATION OF NEW BUILDINGS

上海白玉兰广场 Sinar Mas Plaza

外滩 SOHO Bund SOHO

外滩金融中心 The Bund Finance Center

董家渡金融商业中心 Dongjiadu Financial Center

世博会博物馆 World Expo Museum

西岸美术馆 West Bund Museum

西岸智慧谷 West Bund AI Valley

西岸传媒港 West Bund Media Port

浦东美术馆 Museum of Art Pudong

上海中心大厦 Shanghai Tower

上海 JW 万豪侯爵酒店 Shanghai JW Marriott Marquis Hotel

新开发银行总部大楼 New Development Bank Headquarters

上海大歌剧院 Shanghai Grand Opera House

上海久事国际马术中心 Shanghai Juss International Equestrian Center

前滩国际商务区 New Bund International Business Distric

上海长滩 Shanghai Long Beach

设计·师说
MASTERS' WORDS

新建筑开发赋能

沈 迪

黄浦江两岸的建筑如同一幅打开的历史长卷，完整而真实地记录着上海城市发展走过的历程。它又如同一扇窗口，向世人展示在当下城市更新阶段，上海城市建设中的新理念、新探索带来的成果。

黄浦江两岸作为上海城市更新的重点地带，沿江滨水区域的开发建设成为上海产业结构调整、经济结构转型，以及城市能级提升的重要发展引擎，带动两岸腹地的城市经济和社会全方位发展。从本书所列举的建成或在建的新建筑案例中可以发现，黄浦江两岸的新建筑基本是围绕上海金融和贸易中心、科技创新产业，以及高等级文化体育设施等建设主题开展的。这不但丰富了滨江地区的功能复合度，大大提升了滨江区域的土地价值，对激发黄浦江两岸滨水区域的城市活力也起到积极作用。总之，黄浦江两岸的新建筑在与沿岸历史建筑的对话与融合中，成为上海城市建设的新成就和城市发展的新形象。

今天，仔细观察黄浦江两岸的建设更新，可以看到其开发模式也在悄悄地发生着变化。区域性整体开发更新成为主要趋势，如董家渡金融中心、西岸传媒港等项目都是区域性的整体开发更新的典型案例。这些项目将区域综合性交通组织、提升公共服务设施、打造公共开放空间，以及保护与保留老建筑等各方面的规划要素纳入区域性的开发更新工作中，以实现全方位提高城市能级和环境风貌品质。本书案例中介绍的几个新建筑单体，同样也是在区域性开发的背景下，具有区域代表性或标志性的建筑。如新开发银行，它作为代表性案例向我们呈现2010年上海世博会A、B片区在后世博这轮城市更新发展中整体开发建设的状况与面貌。

黄浦江两岸区域性的整体开发在规划和设计理念上的改变，为我们带来不一样的建设成果。在公共开放空间的打造上，规划和设计更突出它对城市整体环境和风貌的提升度，更关注它与市民日常工作和生活的结合度，更加强调它对绿化和生态环境修复的贡献度。在总体交通方面，规划设计将地下空间开发与城市立体交通组织结合在一起，努力化解高密度开发给城市交通带来的挑战，并以轨道交通和公共交通优先为设计理念。与此同时，对慢行交通的重视也提到新的高度，将以人为本的开发更新思想落实到建筑总体规划设计中。

强调整体性的规划和设计理念为打造沿黄浦江世界一流的滨水建筑和空间环境提供了新的思路和建设成果，初步形成一套可复制、可借鉴的上海经验和模式，为将上海建设成为全球卓越城市树立标杆性的样板。

从黄浦江两岸的城市风貌与建筑环境的角度来看，本书呈现的新建筑案例的开发模式和理念的转变，给建筑设计带来了新的变化。设计从对单个建筑形象的追求转向对建筑群整体形象的关注，并自觉地将建筑放在滨江的整体环境中加以考虑和表达，注重建筑的尺度、肌理，以及与周边风貌环境的协调成为普遍的设计共识。对于开发更新区域内的保护、保留建筑，规划设计不仅要满足政策规范要求，更成为了发掘项目的历史意义，赋予项目文化内涵的重要手段。在新与老的有机融合中，保护、保留建筑获得新的生命意义，新建筑则在以保护为前提的设计理念中找到自己的历史位置。本书中的外滩SOHO就是典型的案例。

当然，黄浦江两岸的新建筑在实现与老建筑对话的同时，没有忽略建筑本身对新技术、新材料的探索与创新性运用。从最为直观的建筑外立面到建筑内部机电设备，建筑的智能化与绿色生态都成为新建筑设计的标配。如新开发银行的设计，建筑全面对接国际标准，在强调建筑高效节能的同时，全面引入绿色健康建筑的理念和标准，将建筑造型、空间布局、可再生材料的应用等各方面元素综合地融入建筑设计之中，全方位满足高标准的节能环保与绿色健康建设目标。尤其是建筑外幕墙的设计，将国际先进技术与国内施工技术相结合，实现建筑外观效果整体性与室内自然通风要求完美结合，并首次由国内幕墙专业单位承担超高层建筑外幕墙的制作与施工。本书的新建筑案例，从内到外都在以各自的设计理念和手法充分地表达黄浦江两岸新建筑在技术创新和绿色环保实践等方面的探索。它们也从一个重要的侧面反映了上海建筑的当代性思考。

其实，今天黄浦江两岸的新建筑只是上海城市发展历史过程的一个片断，所反映的是当代人对一江一河与上海城市发展关系的认识；对建筑与城市关系，尤其是与城市经济、文化、社会，包括生态环境等多种要素高度聚集的城市轴带相互关系的认识。在城市更新的大背景下，让黄浦江两岸的新建筑成为新时期上海城市的新名片。

FORGING A NEW TRAIL IN ARCHITECTURAL DEVELOPMENT
Shen Di

The buildings on both sides of the Huangpu River are like a long scroll of history, which preserves a complete, true record of Shanghai's urban evolution. It also showcases the results of new ideas and new adventures in the explorations of urban development and construction in Shanghai.

As the key area of Shanghai's urban regeneration, the development of the Huangpu River waterfront has become an important engine for industrial structure adjustment, economic structure transformation and city level upgrade in Shanghai. It also drives the all-round development of the urban ecosystem. As illustrated in this book the examples of new buildings completed or under construction, those buildings on both sides of the River emphasize Shanghai's key functions as the financial and trade center, technology innovation hub and the host city of major cultural and sporting events. This not only enriches the functions of the waterfront and greatly enhances the land value, but also plays a positive role in stimulating the urban vitality of the waterfront. In short, the fusion of the new and historic building alongside the riverbanks has become the new achievement in urban development of Shanghai.

Today, a closer look at the regeneration of both sides of the Huangpu River reveals a gradual changing the development pattern. Regional holistic development has become a mainstream, such as the Dongjiadu Financial Center and the West Bund Media Port projects. These projects have taken into account organizing regional transport, upgrading public service facilities, creating public open spaces, and protecting and preserving old buildings in the regional development and regeneration work. As a result, they achieve all-round improvement in urban city level and landscape quality. The individual new buildings presented in this book are also iconic in the context of development in their respective areas. The New Development Bank, for example, is a prime example of the former World Expo site sections A and B redevelopment in this round of urban regeneration.

The change in design ideology of the holistic development on both sides of the Huangpu River has yielded different accomplishment. In the creation of public open space, the design has focused on the enhancement of the overall environment and appearance of the city, the integration of citizens daily work and life, and the contribution to the greening and ecological restoration of the environment. In terms of general transportation, the design combines underground space development with three-dimensional urban transportation organization, which successfully resolved the challenges posed by high-density development and gave priority to rail transportation and public transport. At the same time, serious attention has been given to slow traffic, with the renewed idea of human-centered development implemented in the overall design of the building.

The emphasis on holistic planning and design concepts provides new ideas and accomplishments for creating a world-class waterfront architecture and space alongside the Huangpu River, presenting classic references and models and setting a benchmark for building Shanghai into a world class metropolitan .

From the urban landscape and built environment perspectives, this book presents new architectural examples where changes in development patterns and concepts have brought new changes in architectural design. The focus on the overall image of the building group replaces the pursuit of individual building images, and consciously places the building in the overall environment of the riverfront for design consideration and architectural expression, and the scale and texture of the building, as well as its harmony with the surrounding landscape environment, becomes a general design consensus. For the conservation and preservation of buildings in the developmental and regeneration area, the planning and design not only observe the policies and regulations, but also explore the historical significance of the developmental and regeneration project and give it a cultural connotation. In the combination of the new and the old, the protected and preserved buildings gain new meaning, while the new buildings find their own historical place in the design concept based on the premise of preservation. The Bunds SOHO in this book is a typical case in point.

Of course, the new buildings on both sides of the Huangpu River do not neglect the exploration and innovative use of new technologies and materials while the old and the new are engaged in a dialogue. From the easily perceivable building facades to the energy-efficient, environmentally friendly mechanical and electrical equipment inside, building intelligence and green ecology have become the standard for new buildings. For example, in the design of the New Development Bank, international standards are applied. While emphasizing high performance in energy efficiency, the concept and standards of green and healthy buildings are introduced, incorporating elements such as building shape and spatial layout, and the application of renewable materials into the building design to meet the high standard of energy efficiency and environmental protection and green and healthy construction goals. In particular, the design of the building's external curtain wall combines advanced international technology with domestic construction techniques to achieve a perfect combination between the integrity of the building's exterior and the ventilation of the interior, and for the first time, a domestic curtain wall specialist has undertaken the production and installation work of the external curtain wall on a super high-rise building. Therefore, the new buildings in this book represent the exploration of technological innovation and green practices in the new buildings with their respective design concepts and methods. They also highlight our reflection on architectural contemporaneity in today's Shanghai.

In fact, the new buildings along the Huangpu River are only a temporary phenomenon in the process of Shanghai's urban development, reflecting the contemporary understanding of the relationship between the river and Shanghai's urban development; the relationship between architecture and the city, especially the relationship with the city's economic, cultural, social, including ecological environment and other elements of the highly concentrated urban axis. In the context of urban renewal, the new buildings on both sides of the Huangpu River are made to become a calling card of the city of Shanghai in the new era.

上海白玉兰广场
Sinar Mas Plaza

塔楼顶部
Top view from tower

项目地点：虹口区东长治路588号
设计单位：华建集团华东建筑设计研究院有限公司
合作设计单位：美国SOM公司（方案）
建设单位：上海金港北外滩置业有限公司
设计时间：2007.1-2009.3
建成时间：2016.12
总建筑面积：414798m²
用地面积：56670m²

Location: No. 588, East Changzhi Road, Hongkou District
Designer: East China Architectural Design & Research Institute Co., Ltd., Arcplus Group PLC
Partners: Skidmore, Owing & Merrill LLP (Scheme design)
Constructor: Shanghai Jingang North Bund Realty Co., Ltd.
Design Period: January 2007 - March 2009
Date of Completion: December 2016
Gross Floor Area: 414,798m²
Site Area: 56,670m²

塔楼外观
Exterior

上海白玉兰广场的设计概念源自花与江河。白玉兰是上海市的市花，象征上海开路先锋和奋发向上的城市精神特征。主塔楼高320米，其平面与立面来自对白玉兰花朵几何图案的抽象化处理，顶部设有当前上海最高的直升机停机坪。裙房建筑借鉴蜿蜒的河谷，西侧下沉式广场和东侧中庭空间形成两条曲线，一内一外，阴阳互补，寓意上海的黄浦江和苏州河。在江河交汇处，主塔楼犹如玉兰花绽放，表达了对城市精神的尊重和对本土文化的传承。上海白玉兰广场已成为环

The overall design concept of the Sinar Mas Plaza is inspired by a blossoming magnolia, Shanghai's city flower that celebrates a rousing, pioneering spirit. The plan and elevation of the 320-meter main building is based on an abstraction of geometric patterns of magnolia, with the steel structure at the top shaped like petals. The podium looks like a meandering river. The sunken plaza on the left and the atrium on the right are like two mutually complementary curves that represent the Huangpu River and the Suzhou Creek nearby. The exterior and interior design of the building reflect local elements, expressing a respect for Shanghai spirit and a commitment to local culture. The Magnolia Plaza has become a fine example of environment-friendly architecture. The

境友好设计的典范。设计一体化考虑多层次立体绿化与滨江景观，改善区域小气候，为人们提供丰富且高品质的公共活动场所，带动整个地区的城市活力，努力实现"城市，让生活更美好"的愿景。

---

multi-level vertical green space design is ideally incorporated into the waterfront landscape along the Huangpu River, providing the community with intriguing spaces for various activities and fueling the revitalization of the entire neighborhood. It is a statement of the vision of the Shanghai World Expo: Better City, Better Life.

中庭空间
Atrium

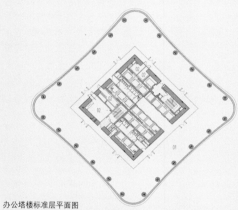

办公塔楼标准层平面图
Plan of typical floors of office tower

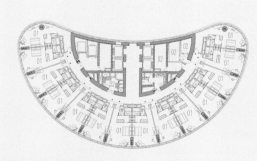

酒店塔楼客房标准层平面图
Plan of typical floors of hotel tower

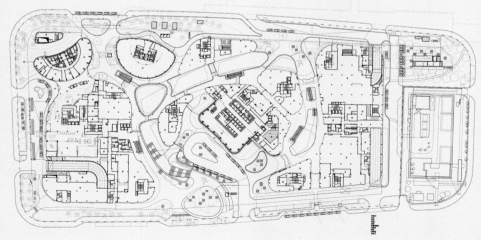

裙房首层平面图
Plan of podium 1/F

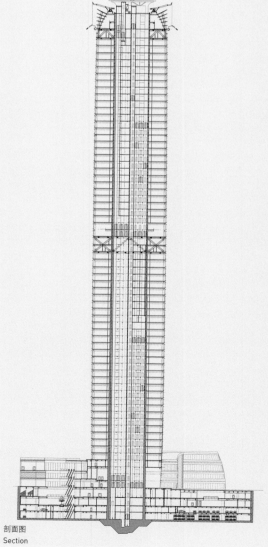

剖面图
Section

北侧鸟瞰
Bird's eye view from the north

外滩 SOHO
Bund SOHO

北侧外观
Northern view

立面的竖向线条
Stretching toward the sky

项目地点：黄浦区新开河北路10号
设计单位：华建集团华东建筑设计研究院有限公司
合作设计单位：德国gmp 国际建筑设计有限公司（概念方案）
建设单位：鼎鼎房地产开发有限公司
设计时间：2012
建成时间：2015
总建筑面积：189909m²
用地面积：22462m²

Location: No. 10, North Xinkaihe Road, Huangpu District
Designer: East China Architectural Design & Research Institute Co., Ltd., Arcplus Group PLC
Partners: GMP International Architectural Design Co., Ltd. (Concept)
Constructor: Shanghai Dingding Real Estate Development Co., Ltd.
Design Period: 2012
Date of Completion: 2015
Gross Floor Area: 189,909m²
Site Area: 22,462m²

沿江
Panoramic view

外滩SOHO与城隍庙商圈近在咫尺，北侧连接外滩万国建筑群，东面紧邻黄浦江，远眺东方明珠、金茂大厦、上海中心等标志性建筑。项目由多栋错落排列的高层办公楼和商业裙楼围合而成，巧妙地形成楼宇中央独特的小型广场。贯穿东西向的户外步行街一直延伸至外滩，不同形态的公共休闲空间沿街分布，塑造活泼生动的商业步行街区。设计从狭长的"里弄"、星罗的街道网和竖向石材的老建筑中获得灵感，既有对传统文化及历史建筑的深刻理解与继承，也展现"国际化街区"的未来感、前瞻性、开放和包容。

The Bund SOHO, on the corner of Renmin and XinYong'an roads, is at the southern end of the Bund. Richly endowed by nature, the sprawling complex borders on the Ancient City Park and the Town God Temple Trading Area in the south, and close to the legendary Bund thronged with historical Western-style buildings in the north. It faces the Huangpu River on the east, vying for attention with landmark buildings in Pudong, such as the Oriental Pearl TV Tower, Jinmao Tower and the Shanghai Center.
The property is composed of a number of high-rise office buildings and commercial podium buildings, forming a unique open space at the center. The outdoor pedestrian street running through the east and west of the property extends to the Bund, along which various forms of public leisure space are distributed, creating a lively commercial pedestrian street. Inspired by the long and narrow "lane", the starry street network and the old vertical stone buildings, you can enjoy the profound understanding and inheritance of traditional culture and historical buildings by walking along the Bund SOHO. At the same time, you can clearly feel the openness and inclusiveness of the "international block" with a strong sense of future and foresight. In addition to the identification degree in modeling, the block space design inside the plot is also very bright. In the limited land use, it not only meets the design requirements of high plot ratio, but also combines underground space and landscape elements as far as possible to create a fashionable, high-end and comfortable atmosphere.

楼宇中央的小型广场
Lovely square

宜人的街区尺度
Well-designed spatial distribution

通向陆家嘴的视觉通廊
Lujiazui within the vision

入口和雨篷
Entrance and rain awning

在有限的用地上，方案既满足高容积率的设计要求，又尽可能地结合地下空间、景观元素，营造时尚、高端、舒适的街区氛围。围合形成的内部空间，仿佛一条蜿蜒、极具动感的"水溪"。楼宇间的缝隙朝向陆家嘴方向，形成一条条敞亮的视觉通廊，这些景观"窗口"提供了欣赏黄浦滨江的独特视角。立面的竖向线条巧妙地"分解"了建筑体量，与周边环境形成较好的比例关系。行人可以感受宜人的街区尺度，而不是摩天大楼挺立的压迫感。

在2016年竣工后，项目先后荣获地产设计大奖金奖与首届中国高层建筑奖。如今，外滩SOHO为老外滩优雅起伏的天际轮廓线添写了出色的一笔。

Walking in the Bund SOHO has a different feeling from walking on the Bund Street: firstly, the space has been enclosed several times to form a unique small square in the center of the building, and a meandering "water stream" connects the building space, which is very dynamic; secondly, the space between the buildings faces Lujiazui, forming a bright visual corridor, which is opposite to the view of Lujiazui buildings; third, the "split" volume of the Bund SOHO also has a good proportion with the surrounding environment. Walking in it, you can deeply feel the pleasant scale of the block, but not the pressure of skyscrapers.

Since its completion in 2016, the Bund SOHO has won the Gold Award of Real Estate Design Prize and the first China High Rise Building Award, each a prestigious honor. The property has extended the fascinating skyline of the century-old Bund.

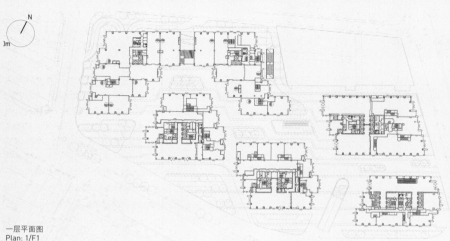
一层平面图
Plan: 1/F1

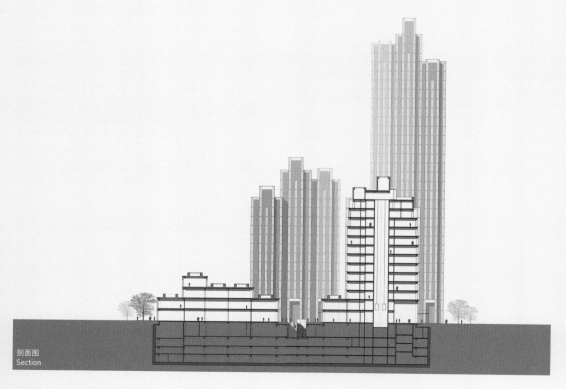
剖面图
Section

外滩侧日景
View from the Bund side

外滩金融中心
THE BUND FINANCE CENTER

南侧鸟瞰
Bird's eye view from the south

回望陆家嘴
Lujiazui within the vision

项目地点：黄浦区中山东二路600号
设计单位：华建集团华东建筑设计研究院有限公司
合作设计单位：福斯特建筑事务所、海德威克设计工作室
建设单位：上海证大外滩国际金融服务中心置业有限公司
设计时间：2011.4-2012.12
建成时间：2016.5
总建筑面积：428332m²
用地面积：45472m²

Location: No. 600, East Zhongshan No. 2 Road, Huangpu District
Designer: East China Architectural Design & Research Institute Co., Ltd., Arcplus Group PLC
Partners: FOSTER+PARTNERS, Heatherwick Studio
Constructor: Shanghai Zengda Bund International Financial Service Center Real Estate Co., Ltd.
Design Period: April 2011 - December 2012
Date of Completion: May 2016
Gross Floor Area: 428,332m²
Site Area: 45,472m²

Panoramic view

外滩金融中心毗邻上海外滩，地处传统老城边界，与浦东隔江相望，与河滨绿地傍路相依。设计以黄浦江为主要景观朝向，临江布置带有层层退台的办公楼及多功能艺术中心。在建筑体量的处理中，北区以小型办公楼延续东侧老城厢的城市肌理，南区以双子办公塔楼形成与浦江对岸陆家嘴金融区的遥相呼应，重塑城市天际线。多功能艺术中心采用全球首创的三层流苏帘幕系统，以浦东天际线作为文化舞台背景，重新诠释浦江地标，为基地注入活力。设计运用玻璃与石材

The development project is adjacent to the Bund of Shanghai, at the boundary of the traditional old city, overlooking Pudong across the river, close to the Bund riverside green space. In the design, the Huangpu River is the main landscape orientation, and office building and multi-functional Art Center are designed along the river. In terms of architectural volume, the small office buildings continue the urban fabric of the old Shanghai town on the east side, while the twin towers echo Lujiazui Financial District on the other side of the Huangpu River. The multifunctional art center is the world's first three-tier curtain system, which takes Lujiazui

外滩侧街景
View from the Bund side

黄浦江篇 · 多元融合创新　HUANGPU RIVER · INNOVATIVE OF NEW BUILDINGS

的有机组合，立面形成从上到下由虚到实的渐变效果。在浦江景观立面延展出高度不一的矩形，不仅主体大楼的轮廓突出，也回应了周围建筑物的立面语言。从黄浦江远眺，建筑立面光影与粼粼波光上下呼应，相映成景，形成协调而富有特色的城市景观。

scenery as the background. In architectural design, the organic combination of glass and stone is used to form a gradual elevation effect. The facade of the building shows rectangular shapes at different heights when viewed from theHuangpu River. The rippling reflections of the facade in the Huangpu River makes for a very impressive sight.

人民路街景
View from Renmin Road

1. 北区N1
2. 北区N2
3. 北区N3
4. 北区N4
5. 北区N5
6. 南区S1
7. 南区S2
8. 南区S3

1:500

总平面图
Master plan

东南角鸟瞰
Bird's eye view from the southeast

董家渡金融商业中心
Dongjiadu Financial Center

整体鸟瞰
Panoramic view

塔楼间空间
Space between towers

项目地点：黄浦区中山南路998号
设计单位：华建集团华东建筑设计研究院有限公司、华建集团建筑装饰环境设计研究院、上海申元岩土工程有限公司
合作设计单位：KPF建筑师事务所（建筑方案设计、办公幕墙）、大象建筑设计有限公司（住宅地块设计）、Aedas凯达环球有限公司（商业设计、商业幕墙）、DLC地茂景观设计（上海）有限公司（景观设计）、章明建筑事务所等
建设单位：中民外滩房地产开发有限公司
设计时间：2015.5-2018.6
建成时间：2022
总建筑面积：1140000m²
用地面积：127hm²

Location:No. 998, South Zhongshan Road, Huangpu District
Designers: East China Architectural Design & Research Institute Co., Ltd., Shanghai Xian Dai Architectural Decoration & Landscape Design Research Institute Co., Ltd., Arcplus Group PLC; Shanghai Shenyuan Rocky & Civil Engineering Co., Ltd.
Partners: Kohn Pederson Fox Associates PC (Scheme design, official curtainwall design), Group of Architects (Residential district design), Aedas Architects (Shanghai) Co., Ltd. (Commercial design, commercial curtainwall design), Design Land Collaborative (Landscape design), ZhangMing Architects, ETC
Constructor: Zhongmin Bund Real Estate Development Co., Ltd.
Design Period: May 2015 - June 2018
Date of Completion: 2022
Gross Floor Area: 1,140,000m²
Site Area: 127hm²

沿江边跌落的建筑体量
Panoramic view

该地块是上海市"一城一带"中外滩金融集聚带南外滩核心区内最大的开发用地，地处黄浦江沿岸核心区段，与陆家嘴地区隔江相望，是未来完善外滩金融功能和优化浦江沿岸整体风貌的重点发展区域。

由于地处传统浦西外滩的南延伸段，在建筑设计上充分考虑与外滩历史风貌以及老城厢历史保护建筑形态的协调。建筑体量呈阶梯状，塔楼与塔楼、塔楼与裙房之间，在高度上形成过渡，同时在沿江立面上呈现出连续的天际线。商业裙房通过退台形成不同标高的室外平台，延伸和丰富了室内空间。北侧的室外平台与室内商业动线结合，形成连续活泼的室外动线。不同标高的平台之间以楼梯和自动扶梯串联，并设有绿化景观和露台，首层的公园绿化得以延展到裙房各层，实现"公园式商业"的设计意向。以流畅开敞的形态实

整体鸟瞰
Panoramic view

Standing on the biggest developing plot in the core sector of the South Bund Financial Zone, the property is opposite to the Lujiazui Financial Hub, being the key development area to improve the financial function of the Bund and to redevelop the Huangpu waterfront.

Given its location on the southward extention of the historical Bund, the property fully considers the compatibility with the historical features on the Bund and the old Shanghai town. This follows the ladder shape. It can produce a continuous gradation between towers and between towers and podiums, meanwhile, a transitional skyline above the Huangpu River is formed. To build interaction between the new and old, the tower façade empathies vertical fins and the transition of glass curtain texture. The width of the aluminum frame changes following the building height. Architects hope the architectural typology can organically compact with the old urban context. The commercial podium designed as the retreating volume forms lots of balconies on the diverse levels. Those balconies effectively extend space from indoor to outdoor. Meanwhile, the northern exterior platform integrates with interior circulation, which can produce an organic relationship between

现"回到江边"的构思，自然地将陆地的城市肌理过渡到有机的水岸形态，有效联系来自不同方向的人群，形成立体高效的空间纽带，营造绿色自然的空间体验。建筑形态整体统一，将建筑、景观、交通等多种城市要素融为一体。

inside and outside. The concept of Park Business means those platforms connecting by staircases and escalators can adequately extend green structures to every podium level. The establishment of this project will make significant progress and bring a wide vision to the South Bund Financial District. Moreover, we hope this project can be a successful case and landmark to the rest of the Shanghai regions.

流畅开敞的水岸空间
Broad and friendly waterfront park

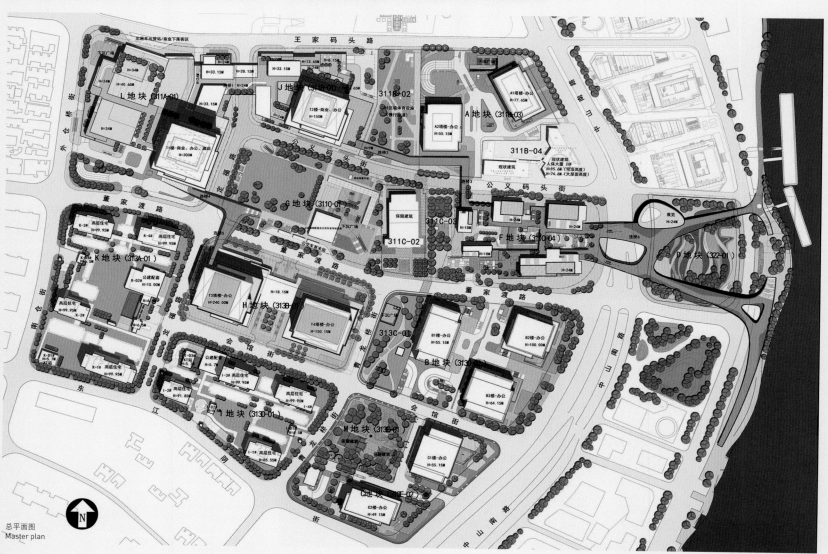

总平面图
Master plan

塔楼夜景
Towers

世博会博物馆
WORLD EXPO MUSEUM

整体鸟瞰
Bird's eye view

入口广场
Entrance

项目地点：黄浦区蒙自路818号
设计单位：华建集团华东建筑设计研究院有限公司
建设单位：世博会博物馆
设计时间：2012-2017.5
建成时间：2017.5
总建筑面积：46550m²
用地面积：40000m²

Location: No. 818, Mengzi Road, Huangpu District
Designer: East China Architectural Design & Research Institute Co., Ltd., Arcplus Group PLC
Constructor: The World Expo Museum
Design Period: 2012 - May 2017
Date of Completion: May 2017
Gross Floor Area: 46,550m²
Site Area: 40,000m²

历史河谷
Valley of history

世博会博物馆由上海市政府与国际展览局合作建设，作为世博会专题博物馆和国际展览局的文献中心，也是这座城市最激动人心和别具想象力的文化建筑之一。设计以"永恒的瞬间"为主题，呼应世博会160年的辉煌历史。将博物馆构建为"承载人类欢乐记忆的时间容器"，收纳所有美好而短暂的世博会回忆。世博会博物馆的主体就像从巨大、永恒的岩石中剖开的"历史河谷"，象征时间的"永恒"。一座特殊的自支撑钢结构玻璃体空中展厅形成"欢庆之云"，高度致敬了第一届世博会的主展馆——水晶宫，代表着人类文明的传承和未来。两者有机组合，形成丰富的室内外互动空间，融合了世博记忆与城市生活。

项目位于2010年上海世博会的黄浦江北岸园区中，是"后世博"再开发的第一座重要的公共文化建筑，承担着延续实践"绿色世博、生态世博"理念的责任，深度参与城市生活，通过提供陈列展览、收藏保护、科学研究、文献服务、学术交流、社会教育、社区公益等七项主要的公共服务功能，成为上海的"城市客厅"。博物馆的建设带动了周边城市功能的重新规划和公共交通提升，优化了滨江地区的公共环境。

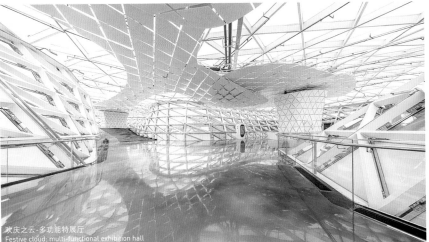

欢庆之云-多功能特展厅
Festive cloud: multi-functional exhibition hall

The World Expo Museum, jointly built by the Shanghai municipal government and the International Exhibitions Bureau, is the only museum of this kind in the world. The main structure of the museum forms a "valley of history" that is carved out from a giant rock. A unique midair exhibition hall with the glazing facades takes the shape of "festive cloud." It recalls the Crystal Palace, the seat of the first World Expo, to suggest the perpetuation of human civilization.

The museum is located in the northern section of the Expo Park where the 2010 World Expo was held, being the first major public facilty developed in the post-Shanghai Expo era. It demonstrates the commitment to making World Expo a green, eco-friendly event, and an effort of turning former Expo sites into spaces for exhibitions, collection and protection, research, documentation, academic exchange, education and community welfare. The construction of the museum has fueled the replanning of the city functions and improved the waterfront environment.

南侧外景
Exterior from the south

"云柱"室内
"Cloud column" interior

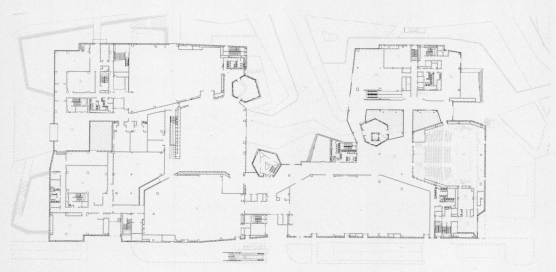
一层平面图
Plan: 1/F

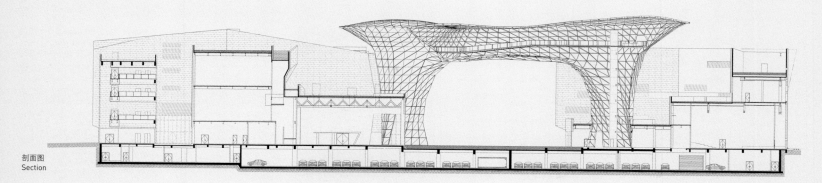
剖面图
Section

"河谷"空间
Valley of history

西岸美术馆
WEST BUND MUSEUM

北侧视角
The northern space

二层展厅 Exhibition hall: 2/F

项目地点：徐汇区龙腾大道2600号
设计单位：大卫·奇普菲尔德建筑事务所
合作设计单位：上海都市建筑设计院
建设单位：上海徐汇滨江开发投资建设有限公司
设计时间：2013-2018
建成时间：2019
总建筑面积：22355m²
用地面积：12591m²

Location: No. 2600, Longteng Avenue, Xuhui District
Designer: David Chipperfield Architects
Partner: Shanghai Urban Architectural Design Co., Ltd.
Constructor: Shanghai Xuhui Waterfront Development Investment and Construction Co., Ltd.
Design Period: 2013-2018
Date of Completion: 2019
Gross Floor Area: 22,355m²
Site Area: 12,591m²

东立面 Eastern façade

西岸美术馆位于上海黄浦江西北岸，是西岸总体规划——将工业区转变为城市的新兴文化区的重要组成。美术馆所在地块呈三角形，位于宽阔的滨江绿化带的最北端，连接了徐汇区与外滩历史文化风貌区。

项目扮演了优化基地与黄浦江、公园之间联系的公共角色，人们可通过西侧龙腾大道和东侧滨江步道进入美术馆门厅，书店以及咖啡厅与龙腾大道直接联系。美术馆建筑的地面广场抬升至洪泛区之上，让黄浦江景致一览无余。广场东边的台阶连接邻近河岸的休息平台。在布局上，三座承载着展览功能的建筑呈风车状旋转排布，通过中央双层通高的门厅联系，带来各自独立运营的可能性。

The West Bund Museum is located on the northwest bank of the Huangpu River in Shanghai. It is an important part of the West Bund master plan -- transforming the zone from an industrial into a cultural one.
The museum occupies a triangular plot at the northernmost tip of a new public park, at the point where Longteng Avenue and the river converge. A raised public esplanade above the flood plain surrounds the building, offering views to the river. The edge of the esplanade on the east side is delineated by a continuous series of steps with landing stages leading to the riverbank. The site offered the opportunity to create a completely freestanding structure and its location allowed for improved access to both the river and the park. The building consists of three main gallery volumes placed in a pin-wheel formation around a central lobby with a double-height atrium.
Each of the three main volumes is 17-metre high, with an aboveground floor and a semi-underground one. The aboveground floor of each volume contains a top-lit gallery space, while the spaces on the semi-underground floor vary in function, housing a multipurpose hall, an art studio and education spaces. These are partially sunken, and lit by clerestory windows. A low pavilion housing the café sits at the river's edge at the level of the esplanade. Its elongated form is intended to maximize river views while its roof serves as a generous terrace fronting the upper entrance. The roof of the atrium remains below the roofline of the major volumes and cantilevers far out beyond the building towards the river and the road. At either end, the roof is supported by a colossal tapering column that draws attention to the entrances. The three dominant volumes are clad with translucent recycled glass, lending the complex an opaline quality. These facades, appearing iridescent during the day and prismatic at night, stand in contrast with the smooth brightness of the plaster clad hovering roof

沿街入口 Entrance

三座主要建筑高均为17米，分为地上和半地下两层。地上层为顶部采光的展览空间，半地下层利用侧高窗采光，包含多功能厅、艺术工坊和教育空间等丰富功能。咖啡厅位于滨江侧首层的低矮体量内，修长的造型试图将更多的江景带给参观者，同时也为二层入口创造一个宽敞的露台。在三座建筑之间，一个较低矮的屋顶覆盖整个中庭并向外延伸，形成面向黄浦江和道路的悬挑结构。在东西两侧，由巨大的锥形柱支撑屋顶，下方为建筑入口。三个主要体量由半透明、玉石般的玻璃包裹，赋予建筑乳白色的质感。这种立面材质将斑斓的日光带入室内，入夜透出绚丽的灯光，与悬挑屋顶光洁的质感涂料饰面形成鲜明对比。

canopies. The pin-wheel configuration of the galleries is reinforced with large windows at their outer ends offering panoramic views over the park, the river and the city.

This configuration allows for different components of the museum to operate independently. The lobby itself is accessed from Longteng Avenue to the west and from the riverside to the east. The latter offers two entrance options: the visitor can either descend into a sunken courtyard at a lower level or climb a set of stairs from the esplanade to arrive in the central lobby at a higher level. Both of these levels give access to the main gallery spaces. The book shop and café can be found at an intermediate level directly connected to Longteng Avenue.

黄浦江景
Accessible Huangpu River

体块分析图
Segmentation diagram

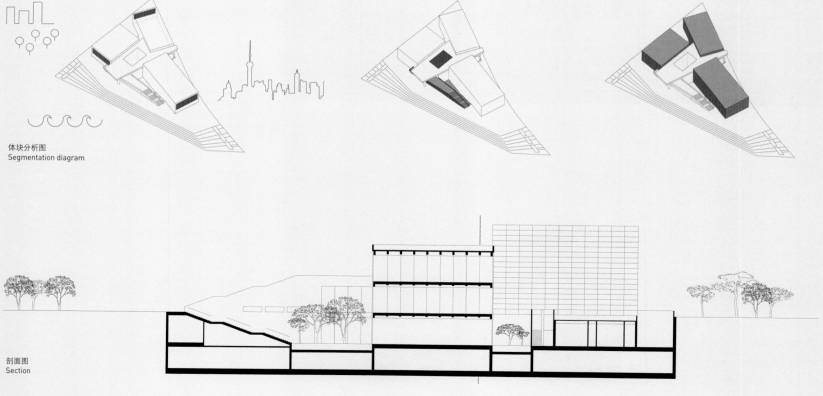

剖面图
Section

门厅
Lobby

西岸智慧谷
WEST BUND AI VALLY

整体鸟瞰
Bird's eye view

项目地点：徐汇滨江龙耀路以北
设计单位：日建设计、SANAA、骏地设计、KPF、福斯特建筑事务所、SOM、Gensler、OPEN、创盟等
合作设计单位：华建集团上海建筑设计研究院有限公司、同济大学建筑设计研究院（集团）有限公司等
建设单位：西岸集团、城开寰宇、隧道股份、梦响强音、阿里巴巴等
设计时间：2013年-今
建成时间：在建
总建筑面积：1200000m²
用地面积：34hm²

Location: Xuhui section of Huangpu waterfront, to the north of Longyao Road
Designers: Nikken, SANAA, JWDA, KPF, Foster, SOM, Gensler, OPEN, Archi-Union etc.
Partners: Shanghai Architectural Design & Research Institute Co., Ltd., Arcplus Group PLC, Tongji University Architectural Design Research Institute (Group) Co., Ltd.
Constructors: West Bund Development Group, Huanyu Company of Shanghai Urban Development Co., Ltd., Shanghai Tunnel Engineering Co., Ltd., Voice of Dream, Alibaba, etc.
Design Period: 2013 - now
Date of Completion: Under construction
Gross Floor Area: 1,200,000m²
Site Area: 34 hm²

西岸智慧谷位于全市人工智能产业"人"字形布局交汇点，是徐汇区对接国家和上海发展战略、推动上海科创中心建设、部署"一核一极一带"人工智能产业空间格局的重要项目载体。
智慧谷以西岸智塔为核心，构筑集总部办公、国际交流、应用展示、研发转化于一体的综合型地标，以上海期智研究院、树图区块链研究院、微软亚洲研究院、阿里巴巴、华为等企业机构为引领，打造具有全球影响力的人工智能产业集聚区。
项目追求土地价值的最大化，靠近云锦路为高密度塔楼带，跌落的塔楼高度营造良好的沿江景观视野；以沿江博物馆等文化建筑为锚点，联动沿江公共开放空间，形成富有创造力的绿地系统。开发地块内的开放式街坊最具多样性和灵活性，形成连续的商业界面，也为步行者提供多种路径选择，同时激活多样的首层城市界面，增加与行人的互动。

The West Bund AI Valley is a key project of Shanghai's artificial intelligence industry, at the intersection of " 人 " shape, reflecting Xuhui District's commitment in establishing Shanghai as a science and technology innovation center—a national and municipal strategy reflecting the city's development. By presenting the West Bund AI Tower as an architectural icon, it seeks to develop the 1.2 million-square meter space to make it a hub of the artificial intelligence industry, so that it can bear the same mandate in this sector as the Zhangjiang Science City in Pudong and the National Science Center, becoming a new landmark of China's AI endevours.

从西岸智塔望向黄浦江（西侧鸟瞰）
Bird's eye view from the west

西岸智塔45F
45/F of AI Tower

流线形裙房
Streamline podium

场地景观
Greenbelts

西岸智塔
West Bund Ai Tower

项目地点：徐汇区云锦路701号
设计单位：日建设计
合作设计单位：华建集团上海建筑设计研究院有限公司
建设单位：西岸开发集团有限公司
设计时间：2014
建成时间：2021
总建筑面积：299726m²
用地面积：72210m²

Location: No.701, Yunjin Road, Xuhui District
Designer: Nikken
Partner: Shanghai Architectural Design & Research Institute Co., Ltd., Arcplus Group PLC
Constructor: Shanghai West Bund Development Group Co., Ltd.
Design Period: 2014
Date of Completion: 2021
Gross Floor Area: 299,726m²
Site Area: 72,210m²

建筑造型设计：双子塔的主要设计思路是把徐汇滨江沿江的景观、自然风、绿色生态带到"内陆"，为滨江腹地创造发展条件。在建筑形体方面，设计的主要目标是沿江建筑尽可能地向滨江腹地延伸，不阻挡江风，缓解市中心普遍存在的热岛现象。通过专家对黄浦江风向的流体模拟进行研究，明确了对江风阻碍最小的双子塔的朝向、造型和流线弧度。流线形的建筑裙房，不仅像凤凰展翅一般"托举"起双子塔，也起到集结、疏导自然风的作用。
场地景观设计从黄浦江滨江公园沿着街道伸展出一条绿化轴，使道路充满绿色，为街区抵御炎热。从江边到腹地尽量种植本地植物，实现地区生物多样化，从而抵御太阳高温并抑制热岛效应。

双子塔与景观绿轴
Twin towers and greenbelts

The architectural design: The design of the twin-tower seeks to extend the landscape, natural wind and green ecology of Xuhui section of the waterfront to the hinterland to create development opportunities for the latter. While considering the architectural form, the main concern was to prevent natural wind from Huangpu River from being blocked by riverside buildings, so that the southeast wind from the river can penetrate to alleviate the common heat island effect of the city center. Through experts' research on the fluid simulation of the wind direction of Huangpu River, the orientation, shape and streamline radian of the twin-tower are designed to minimize the obstruction to the wind from the river. Visually, the streamlined podium not only "holds up" the twintowers like a flying phoenix, but also guides the natural wind.
The landscape design: A green axis is formed by extending the riverside park along the streets, so the whole road is full of green, which can cool the heat of the neighborhood. From the riverside to the hinterland, local plants are planted as far as possible to achieve a biodiversity, so as to inhibit the solar heat and prevent the generation of heat island effect.

西岸艺岛
West Bund Art Tower

项目地点：徐汇区龙耀路8号
设计单位：SANNA建筑事务所
合作设计单位：同济大学建筑设计研究院
建设单位：西岸开发集团有限公司、华鑫置业（集团）有限公司
设计时间：2016-2018
建成时间：2020
总建筑面积：90185m²
用地面积：14690m²

Location: No.8, Lonyao Road, Xuhui District
Designers: Kazuyo Sejima + Ryue Nishizawa, SANAA
Partner: Tongji Architectural Design Group Co., Ltd.
Constructor: Shanghai West Bund Development Group Co., Ltd, China Fortune Properties
Design Period: 2016-2018
Date of Completion: 2020
Gross Floor Area: 90,185m²
Site Area: 14,690m²

鸟瞰
Bird's eye view

项目东邻黄浦江，西邻城市高层建筑。为了在两者之间形成空间过渡，更好地融入周边环境，设计以"伫立于公园之中的建筑"为理念，将较大的建筑体块拆解为几个小体块，体块高度自西向东逐渐降低，留出可以远眺黄浦江美景。

首层室外庭院与周边绿地连通，形成开放的商业空间，人们可以在其间自由地穿行和漫步。各建筑体块在首层时分时合，四个面均可用作零售，同时在龙启路一侧形成连续的商业界面。商业区设置多个中庭，并通过上下层的立体连接实现商业价值最大化。一层设有两层通高的独立办公入口，人们通过门厅内的自动扶梯到达二层的办公大堂。二层的空中平台将两个办公大堂与一层人行道及庭院空间连接起来，并通过楼梯和扶梯的设置使一、二层形成立体环路。

地下一层的设计希望建立起地铁人流与商业的连接，设置商业中庭、集会空间，以及与一层庭院连通的下沉花园，满足消费者、办公人群及其他人群的需求。在商铺上部，竖井和中央回廊位置布置了和集会空间以及地上中庭相连续的下沉式花园，用与一层部分立体性连接的方式创造出各种各样的设施关系。享受购物乐趣的人群，上下班的职员等，都可以利用这些开放空间。

总平面图
Master plan

The project borders on the Huangpu River in the east and the high-rise buildings in the west. In order to provide a spatial transition between the two and better integrate itself into the surrounding environment, the design takes the concept of "a building standing in the middle of a park", breakinga huge volume into several smaller ones, each lower than the one on the west, leaving gaps for appreciating the beautiful view of the Huangpu River.

The perforated ground level allows visitors to stroll around and through the courtyard, taking various routes to neighbouring parks, which makes the retail area an open space. The building blocks are partially overlapped and separated on ground levels, and this allows the building to have retail on four sides and creates the linear retail facade on Longqi Road. The office lobby with direct access from the porch leads to the main office entrance located on the second floor. Visitors go up the escalators within the double-height atrium and are welcomed by the spacious reception. Several atriums with retail entities around maximize the value of retail space by having three-dimensional connection between different floors. The "Sky deck" along the west connects to the second floor, allowing access to and from the pedestrian walkway and the office lobby, furthermore linked with atrium spaces and stairs that connect to the courtyard to create 3-dimensional circulations on the first two floors.

The basement allows for the free flow of people between the metro station and the retail shops. The atrium in the shopping mall and the sunken garden connected to the first-floor courtyard can create positive experience for shopping mall visitors and office workers alike.

屋顶花园
Roof garden

阿里巴巴徐汇滨江 T 地块项目
Ailbaba Xuhui Riverside Plot T Project

项目地点：徐汇区龙腾大道、云启路交叉口
设计单位：福斯特建筑事务所
合作设计单位：华建集团上海建筑设计研究院有限公司
建设单位：传橙科技（上海）有限公司
设计时间：2020.5-2020.12
建成时间：2023.12
总建筑面积：61442m²
用地面积：11907m²

Location: Intersection of Longteng Avenue and Yunqi Road, Xuhui District
Designer: Foster and Partners Limited
Partner: Shanghai Architectural Design & Research Institute Co., Ltd., Arcplus Group PLC
Constructor: Chuancheng Holding Limited
Design Period: May 2020 - December 2020
Date of Completion: December 2023
Gross Floor Area: 61,442m²
Site Area: 11,907m²

人视效果
Eyelevel perspective drawing

阿里巴巴徐汇滨江T地块项目设计灵感源于阿里巴巴的创新多样性，意图通过建筑体现阿里巴巴的企业文化和精神。设计积极回应地方文化和气候，将成为黄浦江沿岸新文化科技区不可或缺的一部分。

项目坐落在徐汇江畔的活力新区，拥有黄浦江和陆家嘴中央商务区的壮丽景色。方案创造性地在场地中心创建一个大型的公共城市广场，形成活跃的社区核心空间。位于不同标高和朝向的观景平台可俯瞰中央广场。

独特的建筑形式是创新设计过程的结果。设计使用遗传算法形成最佳体量，通过算法创建最合适的形式，实现对环境条件高度的响应能力、最大化的外部视野，以及针对不同功能的特定面积要求。该设计通过在中心广场提供大尺度的遮蔽空间，使内部免受夏季的日晒以及冬季强风的侵袭。

建筑的外观极为通透，人们能够看见阿里巴巴的办公世界，同时为企业员工提供了绝佳的景观视野，实现"看与被看"的协同和对话。同时为阿里巴巴的各个部门创建量身定制的工作区解决方案，优化用户的舒适度。

办公区、休息区和会议室都经过认真布置，通过视觉和物理连接以鼓励团队协作。建筑采用自然采光，联通外部空间，为工作空间赋予生气，同时创造了更加专注和积极的工作氛围。设计将利用工厂预制进行质量控制，采用模块化建造减少浪费和现场操作，提高施工效率。

Inspired by the incredible diversity and creativity of Alibaba, China's foremost platform of e-commerce, the design seeks to reflect the company's corporate culture. It has taken into account the cultural elements and the climate of Shanghai, making it a sparkling jewel of this high technology park along the Huangpu River.

Situated at a new, vibrant Xuhui section of Huangpu waterfront, it commands stunning views of the Huangpu River and the Lujiazui CBD. The masterplan innovatively carves out a large plaza at the center of the property to make it the essential space for public activities. Viewing terraces at different levels and facing different directions can overlook the central plaza.

The unique architecture results from an innovative design process: used a genetic algorithm to calculate the most ideal volume. The algorithm works out several aspects crucial to the property, making it highly responsive to the environment conditions. Meanwhile, it can provide the maximum view and satisfy requirements on areas to accommodate different functions, thus creating the best possible form. The design also optimises user comfort at the heart of the plaza by providing a large shaded space in summer and shelter from wind in winter.

The exterior of the building is totally transparent, showing the internal spaces of Alibaba offices and enabling the staff to have a broad view. This is a magic dialogue of "to see and to be seen."

The office spaces, break-out spaces and meeting rooms are all carefully arranged to encourage collaboration and teamwork. Visual and physical connectivity are designed to encourage interactions. Natural light and external spaces animate the workspaces and create a more focussed and engaged workforce. Following a modular approach, the design will utilise off-site production for quality control, reducing wastage, and minimising on-site operations to create an efficient construction programme.

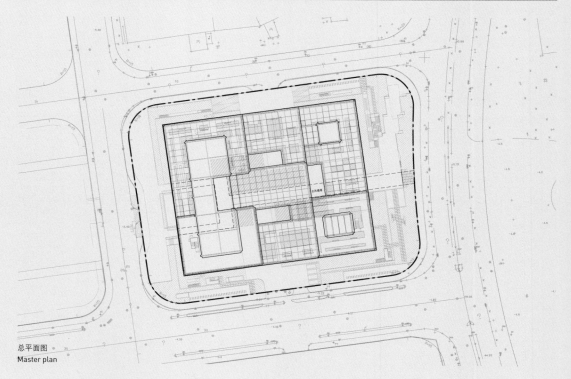

总平面图
Master plan

影音传媒综合商办新建项目（梦响强音）
Dream Sound

项目地点：徐汇区云锦路、龙启路、云谣路、龙华机场路围合地块
设计单位：上海江欢成建筑设计有限公司
建设单位：梦响启岸文化发展（上海）有限公司
设计时间：2017
建成时间：在建
总建筑面积：66791m²
用地面积：14690m²

Location: The plot surrounded by Yunjin, Longqi, and Yunyao Roads and Longhua Airport
Designer: JAE
Constructor: Shanghai Mengxiangqi'an Culturd and Medlia Co., Ltd.
Design Period: 2017
Date of Completion: Under construction
Gross Floor Area: 66,791m²
Site Area: 14,690m²

梦响强音位于上海市徐汇区，包含商办地块和代建公共绿地。项目主楼为20层办公楼，裙楼为2层音乐业态商业，北侧沿规划道路布置透明演播厅音乐盒子。

为了打造具有文化气质和品位的滨水区域和一流的文化产业聚集地，设计利用滨江公共绿地和开放空间建设滨江功能发展带，围绕多样化、差异化的音乐主题相关业态，依托品牌效应，最大限度地利用景观资源打造音乐综合体。开放式景观及公共系统结合建筑布局，在地块中设置多样化的开放空间，作为音乐主题及相应业态的载体，容纳场地内不同活动的需求，为整个社区提供一个既安全友善、富有活力、利于交流的环境。

The Dream Sound project includes a commercial building and a green space. The former is a 20-storey office building whose two-storey podium is reserved for organizations in the music industry; a transparent "music box" studio is located along a road that has been planned in the north.

In order to create tasteful waterfront and first-class cultural cluster, the design makes use of the riverfront green space and open space to build a functional development zone, relying on the branding to maximize the use of landscape resources to create a music-themed complex. Referring to the building blueprint the design has set up diversified open spaces in the site to accommodate music themed events. The needs of different events are considered to make it a safe, friendly, energetic and communication-friendly environment for the whole community.

鸟瞰效果
Bird's eye view

西岸传媒港
West Bund Media Port

整体鸟瞰
Aerial view

Eyelevel view

项目地点：徐汇滨江龙耀路以南
设计单位：Nikken、KPF、MVRDV、BENOY、Gensler等
合作设计单位：华建集团上海建筑设计研究院有限公司、华建集团华东建筑设计研究院有限公司
建设单位：西岸集团、腾讯、诺布、湘芒果、游族、国盛、大众交通
设计时间：2012年起
建成时间：在建
总建筑面积：1100000m²
用地面积：55hm²

Location: Xuhui section of Huangpu waterfront, to the south of Longyao Road, Xuhui District
Designer: Nikken, KPF, MVRDV, BENOY, Gensler etc.
Partner: Shanghai Architectural Design & Research Institute Co., Ltd., East China Architectural Design & Research Institute Co., Ltd., Arcplus Group PLC
Constructor: West Bund Group, Tencent, Nuobu, Mango-Xiang, Yoozoo, Guosheng, Dazhong Transportation
Design Period: Since 2012
Date of Completion: Under construction
Gross Floor Area: 1,100,000m²
Site Area: 55 hm²

西岸传媒港是徐汇滨江建设重要的先导项目，以中央广播电视总台长三角总部、上海总站为旗舰，聚集腾讯、湘芒果、游族等文化传媒类企业，打造国际级的城市地标、文化地标、传媒地标。

核心区9个街区，采用创新开发模式，即"三带"（带地下空间、地上方案、绿色建筑标准）土地出让方式和"四统一"（统一规划、设计、建设、运营）开发模式，统筹100万平方米建筑体量（地上53万平方米，地下47万平方米），协调9家业主、15个项目、50余家设计单位和3家总包单位整体推进，实现区域组团式开发和地下空间统一建设，是对提升城市综合体开发水平与土地利用价值的有力探索。

统一开发的设计亮点

城市核心：连通B3-2F，集合了标识、交通、信息、采光、通风等多项功能，成为地区形象和使用功能的显著特点；

公共环通道：位于B2层，规划4进3出出入口，由市政道路直接进入，连通9个地块，共享5200个停车位；

二层平台：近5万平方米平台建立第二地面，缝合9个地块，通过结构优化和种植技术，实现环境优美、人车分流的共享空间；

敞开式地下商业街：连接西侧地铁11号线龙耀路和沿江西岸梦中心，优化业态布局和动线组织，提高地下商业品质；

能源中心：采用溴化锂+水蓄冷技术，提供约50万平方米商办的空调和冷热水。

项目采用地下、地上分别出让的方式，地下由西岸集团统一实施，地上分块出让后，由各主体单位予以自行建设。遵循"三带""四统一"的要求和整体开发规则，最大限度确保各地块在空间与功能上的完美衔接和建设品质上的高度统一。

The West Bund Media Port is a major pilot project of the Xuhui Waterfront: To this goal, it has attracted two flagship enterprises, i.e., the CCTV Yangtze River Delta Headquarters and the Shanghai General Station. It has become home to an array of top-notch cultural and media companies, notably Tencent, Manguohub and Yoozoo, aiming to create acultural and mediaenterprise hub of international statue.

The nine blocks in the core area have been developed in a brand-new way: lease model (with underground space, with above-ground planning, with green building standards) and the "4-prong unified" development model (unified planning, unified design, unified construction, unified operation). In order to promote the comprehensive development of the area, it coordinates the development of 1 million square meters of building volume (530,000 square meters above ground and 470,000 square meters underground), involving the cooperation of 9 owners, 15 projects, more than 50 design units and 3 general contractors. It can be regarded as a powerful endeavor to improve the development level of urban complexes and the land usage value.

Design highlights of integrated development

The Urban Core enables interconnectivity throughout all the floors from B3 to 2F, performing multiple functions such as signs, visitor traffic guidance, information, lighting, ventilation, etc. This is a prominent feature of the complex and of its functions;

Public ring road: located on the B2 floor, with 4 entrances and 3 exits linking to main roads, it connects 9 plots; offering 5200 parking lots;

The second-floor platform provides a "additional loft level" that knits together nine plots spanning a total of 50,000 square meters. By means of structural optimization and planting technologies, the shared space offers a beautiful environment and effectively separates vehicles from pedestrians.

The underground pedestrian pass extends to Longyao Road Station of Metro Line 11 and the West Bund Dream Centre, offering optimized commercial space to improve the shopping experience.

The Energy Center uses lithium bromide fused water cold storage technologies, providing air-conditioning, hot and cold water availability in the 500,000-square-meter commercial space.

The underground spaces and the ground level up spaces are leased separately in this project. The underground spaces are managed and constructed by West Bund Group. The spaces above ground level are developed by different contracted organizations. The requirements of "3-with" and "4-united" and the general principles of development are followed to ensure the perfect connection of spaces, functions and high unity in construction quality of each plot.

人视效果
Eyelevel perspective drawing

西侧鸟瞰
Aerial view from the west

腾讯华东总部大厦
Tencent East China Headquarter Mansion

项目地点：徐汇区黄浦江南延伸段WS5单元188S-E-2地块
设计单位：吕元祥建筑师事务所
建设单位：腾讯控股有限公司
设计时间：2017.1
建成时间：2021.6
总建筑面积：77163m²
用地面积：13000m²

Location: Plot 188S-E-2, WS5 Unit, Huangpu River's Southward Extension Section, Xuhui District
Designer: RLP
Constructor: Tencent
Design Period: January 2017
Date of Completion: June 2021
Gross Floor Area: 77,163m²
Site Area: 13,000m²

腾讯华东总部大厦是一个富有原创性和企业个性的城市新地标，由高150米的办公大楼和高24米的裙楼组成。项目作为腾讯位于上海的企业园区，备有办公区、共享空间、协作空间、企业展厅、生活馆、游戏试玩区、员工餐厅等设施。
建筑立面的设计灵感来自"万家灯火"，由不规则的窗户组成，错落有致，象征秩序、韵律及原则，数千个独特的窗户反射，有如黄浦江的粼光闪闪。这与腾讯的前瞻愿景巧妙呼应——数千个和而不同的窗户，寓意腾讯作为科技和数码领域的领导者，正积极面向全球，致力提供多元产品和服务，以迎合千千万万个家庭的独特需求，也代表腾讯团队"兼容并蓄，厚积薄发"的多元企业文化。

Tencent East China Headquarters building is an innovative landmark that echoes the corporate culture of the owner. It consists of a 150-meter-tall office building and a 24-meter-tall podium. As Tencent's property in Shanghai, the project includes office areas, sharing space, collaboration space, an exhibition hall, a game trial area, and a staff canteen.
The facade of the buildings is diversified and changeable, and composed of well-arranged irregular windows, symbolizing order, rhythm and principle. The design is inspired by "lights of myriads of families", the glass reflection of thousands of unique windows of the buildings, like sparkling of Huangpu River, looks harmonious yet different. The design also ingeniously echoes Tencent's forward-looking vision: these thousands of harmonious but different windows mean that Tencent, as a leader in the field of technology and digitization, is actively committed to providing diversified products and services to meet the unique needs of millions of families all over the world. At the same time, it also represents the Tencent team's iconic diversified corporate culture, inclusive and open mindset, and well-grounded development.

游族大厦
Yoozoo

项目地点：徐汇区云锦路、黄石路路口
设计单位：Nikken
合作设计单位：华建集团上海建筑设计研究院有限公司
建设单位：游族网络股份有限公司
设计时间：2012
建成时间：在建
总建筑面积：77000m²
用地面积：13000m²

Location: Intersection of Yunjin Road and Huangshi Road
Designer: Nikken
Partner: Shanghai Architectural Design & Research Institute Co., Ltd., Arcplus Group PLC
Constructor: Yoozoo Games
Design Period: 2012
Date of Completion: Under construction
Gross Floor Area: 77,000m²
Site Area: 13,000m²

游族大厦由一栋高157米的办公塔楼和一栋附属商业裙房组成。建筑塔冠设计形如引擎，寓意游族大厦是一座产生驱动力的活力发生器，能激发出更多创造力。塔楼和裙房虽高度有异，但形成连续的整体感，塔楼面朝东北方向传媒港的其他地块，形成积极的展示面。办公塔楼内每三层设置挑空中庭，创造人与人沟通交流的空间，柔和的曲面包裹在建筑外立面上，并按照1.5米间隔排列幕墙的竖向铝制装饰条，其角度逐步调整，形成色彩和反射的细微变化，在表达柔和的建筑表情的同时，削弱了体量的压迫感。

The Yoozoo project is composed of a 157-meter-tall office tower and a commercial podium. The tower crown is designed in the shape of an engine, symbolizing that the game company's building is a driving force for creativity. The tower and podium are congenial in their architectural contour. The tower faces the other ground plots on the Media Port in the northeast, showing a fascinating façade. Inside the office tower, an atrium is set every three floors, which creates spaces for people-to-people communication. On the gently-curved facade of the building, vertical decorative aluminum strips of the curtain wall are set at an interval of 1500 mm. The strips can turn to different directions and this brings to changes of colors and reflection. Such a design leads to a subtle architectural expression, offseting the oppressive feeling caused by the collosal volume of the building.

西北侧人视
Eyelevelview from the northwest

东北侧鸟瞰
Bird's eye view from the northeast

挑空中庭
Atrium

大众交通
Dazhong Transportation Towers

项目地点：徐汇区云锦路、龙台路、黄石路、瑞江路围合地块
设计单位：MVRDV
建设单位：大众交通
设计时间：2012
建成时间：在建
总建筑面积：35000m²
用地面积：10000m²

Location: The block surrounded by Yunjin, Longtai, Huangshi and Ruijiang Roads, Xuhui District
Designer: MVRDV
Constructor: Dazhong Transportation
Design Period: 2012
Date of Completion: Under construction
Gross Floor Area: 35,000m²
Site Area: 10,000m²

大众交通为高40米和60米的两栋塔楼。设计将塔楼横向剖景观面，各个盒子通过平面上的位移创造出许多露台，也切成多个盒子，通过盒子体量的缩放增强辨识度，丰富活跃了内部空间。每个叠加的盒子可以在彼此间有联系的徐汇滨江天际线。同时，为了创造更多可遥望浦东的滨水条件下承载不同的办公功能需求。

The Dazhong Transportation Towers are twin high-rise buildings of 40 and 60 meters respectively, which can highlight the changes of the overall skyline in the master plan of Xuhui section of Huangpu waterfront river surface. In addition, in order to shape different recognizable features, the tower is horizontally separated into several boxes, which are enlarged or reduced in volume to enhance recognizability. At the same time, to create more waterfront landscape surfaces that can look into Pudong in the distance, multiple balconies are created and active internal space changes are provided on the plane of each box through displacement. Each stacked box can meet different functional needs.

人视效果
Eyelevel perspective drawing

央视上海总站
CCTV Shanghai Headquarters

项目地点：徐汇区龙文路、云锦路、龙台路、瑞江路围合地块
设计单位：KPF
建设单位：上海徐汇滨江开发投资建设有限公司
设计时间：2012
建成时间：在建
总建筑面积：202000m²
用地面积：46000m²

Location: the block surrounded by Longwen, Yunjin, Longtai and Ruijiang Roads, Xuhui District
Designer: KPF
Constructor: Shanghai Xuhui Waterfront Development Investment and Construction Co., Ltd.
Design Period: 2012
Date of Completion: Under construction
Gross Floor Area: 202,000m²
Site Area: 46,000m²

内街
Interior structure

项目总平面以一条东西向的步行贯穿轴线为核心，交织出活跃在三个不同高度上的商业和餐饮动线，并以各类公共广场点缀其中。主要内街形成一个由办公塔楼和各类文化、商业功能围合的"峡谷"空间；塔楼沿主要街面向上逐渐后缩，给其下的步行空间带来自然采光，同时也有效地建立起地铁站和滨江区域的视觉通廊。塔楼的外侧幕墙在南北两侧构筑出连续的沿街面，朝向内侧的幕墙则在形式上更为多变，由金属和玻璃交织成独特的跌级幕墙表皮，在垂直方向上逐渐向上收进。

In the master plan of the project, the core is a pedestrian axis running from east to west, while the commercial and catering lines are interwoven and run on three floors at different heights, interspersed with various squares. A canyon-like space is formed by the office tower and various cultural and commercial facilities along the main inner streets. The tower gradually shrinks backward along the main street, bringing sunshine and natural light to the pedestrian space and forming a visual corridor between the metro station and waterfront. The outer curtain wall of the tower forms a continuous surface along the street on the northern and southern sides, while the inward curtain wall presents a changing form. Metal and glass form a unique curtain wall facade, which gradually narrows as the height goes up.

灯光秀
Lighting show

浦东美术馆
Museum of Art Pudong

入口大厅
Lobby

镜厅
Glass hall

项目地点：浦东新区滨江大道2777号
设计单位：让·努维尔建筑事务所
合作设计单位：同济大学建筑设计研究院（集团）有限公司
建设单位：上海陆家嘴（集团）有限公司
设计时间：2016
建成时间：2020.11
总建筑面积：40487m²
用地面积：13000m²

Location: No.2777, Binjiang Avenue, Pudong New Area
Designer: Ateliers Jean Nouvel
Partner: Tongji Architectural Design (Group) Co., Ltd.
Constructor: Shanghai Lujiazui (Group) Co., Ltd.
Design Period: 2016
Date of Completion: November 2021
Gross Floor Area: 40,487m²
Site Area: 13,000m²

朝向花园的立面
Façade facing the garden

浦东美术馆位处小陆家嘴滨江核心区，在视觉上同时属于外滩、黄浦江和浦东。建筑本身就应是这个时代的艺术品，既能满足人们的生活，又能为城市和文化带来吸引力。设计以第四维度为切入点，即将艺术符号化并赋予特色，将其和周围环境区分开；打造一个承接艺术作品展示的场地，一个向外部展示的场所；采用白色的石材，厚实的基座，雕塑般起伏的体量；两侧环抱的景观让美术馆变得显而易见。

浦东美术馆共设13个展厅，创造了一系列"呈现的排列"，为艺术创造安置的空间。设计中采用"领地"的概念，希望观众的参观之旅始自踏入建筑场景边界的那一刻。

美术馆室内采用自然光与人工光相结合的照明方式。人工光源的布置以现代艺术杰出代表马列维奇（1878—1935）的经典几何线条为灵感，在满足照明的功能需求的同时营造几何阵列的美感。自然光线呈现三种体验模式：全暗息、全天光和部分遮挡滤光，可根据展览的不同要求设定。在自然光源的营造上，设计还大量采用"框景"手法。

通高中央展厅
Central exhibition hall

夕阳中的大厅
Lobby in the sunset

Located at the heart of the Little Lujiazui stretch of the Huangpu waterfront, the Museum of Art Pudong has a visual bond with the Bund, the Huangpu River and Pudong at the same time. Aiming to be a work of art on its own, the building enriches the public's cultural life while adding cultural weight to the city. The design introduces the "fourth dimension", i.e. to symbolize art and give character to it; distinguishing it from its surroundings. It is a venue to display artworks and also a venue to showcase its own charm to the world out there. The white stones, thick base, and a sculpture-like undulating contour' fused with the surrounding landscape make the museum stand out from all other buildings.

The Museum of Art Pudong has 13 exhibition halls, creating spaces for the display of artworks. Based on the concept of "Domain", it hopes that one can begin his artistic journey the moment he steps past the building's gates.

The interior of the museum is lit by both natural and artificial light. The placement of artificial lighting fixtures is inspired by the classic geometric lines of Kazimir Malevich (1878-1935), a representative of modern art, to create the aesthetics of a geometric array while meeting its functional needs. Natural light is presented at three levels of brightness: full darkness, full daylight and partially obscured filtered light. The mode can be chosen according to the requirement of each exhibition. The design also has natural light penetrating through a range of "frames".

大厅
Lobby

除常规展厅以外，浦东美术馆还设有数个独一无二的特殊展厅。位于美术馆中央区域的中央展厅，贯通地下一层至地上四层。观众可获得对同一件艺术作品不同的视角和观赏体验。而在建筑朝向外滩的一面，则是美术馆标志性的镜厅（Glass Hall）——安置了整面高反光的LED屏幕，设计理念亦取自艺术家杜尚在其作品《大玻璃》中所提出的"第四维度"。

In addition to the traditional-style exhibition halls, the museum also has several unique and special ones. The Central Gallery, located in the central area of the museum, runs from the first floor to the fourth floor above ground. Visitors can feel an artwork in different ways. On the Bund-facing side of the building is the Museum's iconic Glass Hall. The design is also based on the "fourth dimension" proposed by Duchamp in his work, "The Big Glass Piece".

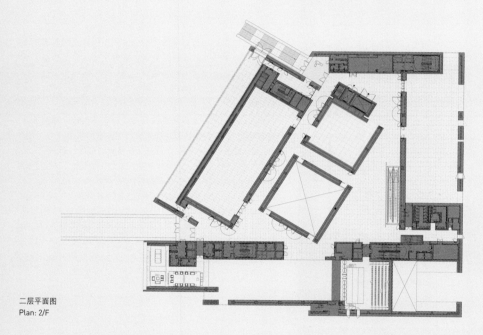

二层平面图
Plan: 2/F

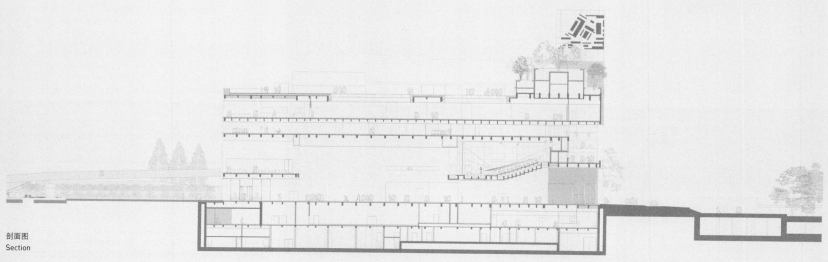

剖面图
Section

沿江立面
Façade facing the river

上海中心大厦
SHANGHAI TOWER

城市中的上海中心大厦
Shanghai Tower and its neighbors

上海中心大厦日景
Shanghai tower under blue sky and sunshine

项目地点：浦东新区陆家嘴银城中路501号
设计单位：Gensler（方案），同济大学建筑设计研究院（集团）有限公司（施工图）
建设单位：上海中心大厦建设发展有限公司
设计时间：2008.11-2010.10
建成时间：2015
总建筑面积：574158m²
用地面积：30368m²

Location: No. 501, Middle Yincheng Road, Lujiazui Pudong New Area
Designer: Gensler (Concept), Tongji University Architectural Design Research Institute (Group) Co., Ltd. (Construction drawing)
Constructor: Shanghai Tower Construction and Development Co., Ltd.
Design Period: November 2008 - October 2010
Date of Completion: 2015
Gross Floor Area: 571,458m²
Site Area: 30,368m²

螺旋上升的体量
Spiral form

上海中心大厦位于陆家嘴金融贸易区，与上海环球金融中心、金茂大厦呈品字形布置。主体建筑高度632米，共127层。竖向分为九个功能区，一区为大堂、商业、会议、餐饮区，二区至八区为办公区，七区、八区为酒店和精品办公区，九区为观光区，九区以上为屋顶皇冠。一区至八区顶部为设备避难层。大厦拥有全球最高的建筑内观景台，同时配有全球最快的电梯，最高速度可达18米/秒。

建筑外观呈螺旋式上升，宛如一条盘旋升腾的巨龙。设计摆脱传统的外部结构框架，以旋转120°且建筑截面自下朝上收分（缩小）、不对称的外立面，有效减少了大楼结构的风力荷载。

建筑采用双层玻璃幕墙设计，改善大厦内的空气质量，减少建筑的能耗。在内外幕墙之间形成9个垂直中庭空间，这一系列垂直社区用于不同用途，每个垂直社区设计为独立的生物气候区进行调节和使用，创造出宜人的休息环境，营造花

观光层
Observation level

内外层玻璃幕墙
Interior and exterior curtain walls

Located in the Lujiazui Financial and Trade Zone, the Shanghai Tower and the Shanghai World Financial Center and Jinmao Tower are placed in a triangular formation. The height of the main structure of the Shanghai Tower rises to 632 meters with 127 floors. It is vertically divided into nine functional sections. The lobby, commercial, conference and catering facilities are in Section 1, while the office section coverssections 2 to 8, the hotel and boutique offices covers sections 7 and 8, and the sightseeing platform in section 9, and a crown on the top of it. Above each of the first eight sections is the equipment and refuge floor. The building boasts the world's highest in-building observation deck and it is also equipped with the world's fastest elevator with a maximum speed of 18 meters per second.

The exterior of the building spirals upward in the shape of a spiraling dragon. The design abandons the traditional external structural framework but presents an asymmetrical façade that turns 120°. The building, which is a giant tapering tower from bottom to top, effectively reduces the impact of wind on its structure.

The building is designed with a double-glazed curtain wall to improve the air quality inside the building and reduce the building's energy consumption. Nine vertical atrium spaces are formed between the interior and exterior curtain walls. This series of vertical communities are used for different purposes, and each vertical community is designed to be regulated and used as a separate bioclimatic zone, creating a pleasant resting environment

园、咖啡馆、餐厅和零售空间。
同时，工程采用分布式能源利用技术、变风量空气调节技术、热回收利用技术、地源热泵技术、涡轮式风力发电技术、雨水收集处理和回用技术等多项最新绿色环保技术，理论计算可节约成本25%以上，是同时符合国家绿色建筑三星认证和美国LEED绿色建筑认证体系的"绿色摩天大楼"。

---

that includes gardens, cafes, restaurants and retail spaces.
The project adopts many latest green technologies such as distributed energy utilization technology, variable air volume air conditioning technology, heat recovery and utilization technology, ground source heat pump technology, turbine wind power generation technology, rainwater collection, treatment and reuse technology, etc. They can save more than 25% of the cost according to theoretical calculation. The Shanghai Tower is a "green skyscraper" that has received both the certification of three-star green building in China and of the LEED green building in the US.

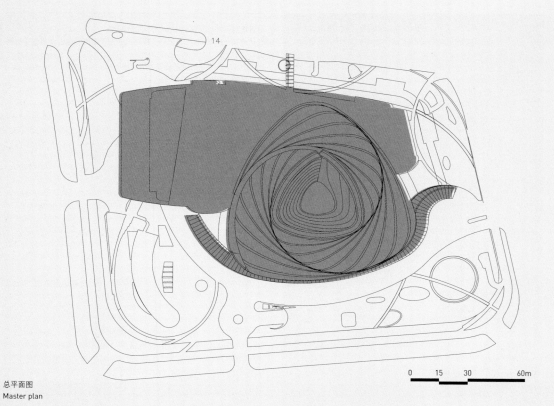

总平面图
Master plan

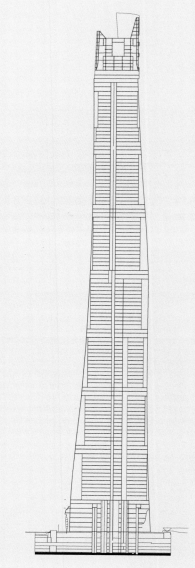

二层平面图
Plan: 2/F

剖面图
Section

沿江立面
Façade facing the river

上海 JW 万豪侯爵酒店
SHANGHAI JW MARRIOTT MARQUIS HOTEL

沿江外景
Façade facing the river

13层通高的A形中庭
13-storey A-shaped atrium

项目地点：浦东新区浦明路988号
设计单位：华建集团华东建筑设计研究院有限公司
合作设计单位：SOM（方案）、思迈建筑咨询(上海)有限公司（机电）
建设单位：上海申电投资有限公司
设计时间：2014.4-2015.4
建成时间：2019.1
总建筑面积：113875m²
用地面积：45667m²

Location: No.988, Puming Road, Pudong New Area
Designer: East China Architectural Design & Research Institute Co., Ltd., Arcplus Group PLC
Partner: SOM (Scheme design), Squire Mech (Shanghai) Co., Ltd. (M&E)
Constructor: Shanghai Shendian Investment Co., Ltd.
Design Period: April 2014- April 2015
Date of Completion: January 2019
Gross Floor Area: 113,875m²
Site Area: 45,667m²

车行入口
Entrance and vehicle passage

上海JW万豪侯爵酒店项目位于黄浦江畔、陆家嘴和世博会之间的浦东滨江核心区中段，与董家渡隔江相望，距陆家嘴中央商务区仅2公里。建筑塔楼采用单廊双面客房，核心筒沿东侧布置使面江客房最大化。酒店一至十三层设有通高的大堂中庭，总高约60米。沿街面设计10米高的半室外入口广场，其上方是位于裙房三层的大型宴会厅。

The Shanghai JW Marriott Marquis Hotel is located in a prominent position along the river in Pudong between Huangpu River, Lujiazui and World Expo, in the middle section of Pudong riverside core area, opposite Dongjiadu that is 2 km away from Lujiazui CBD. The building tower adopts a simple and effective way of single corridor and rooms arranged on double sides, the core tube is arranged along the east side to maximize the number of rooms facing the river. The hotel has a total height of about 60 m, a lobby atrium having a full-height from 1st floor to 13th floor, which runs through the entire podium and reaches the 13th floor of the tower. In the atrium's design of the entrance lobby, the problem where the tower is offset from the axis of the riverside open space is skillfully solved so that the residence can receive their due amount of sunshine. The massive complex banquet hall is arranged on the 3rd floor of the podium, which combines the volume requirement with drop-off demand, with a personalized spacious entrance square formed along the street. The largest banquet hall is located above the semi-outdoor entrance square, which is nearly 10 m high, having a floor area of about 1,700 m² and can accommodate 2,600 people at the same time.

The design team sets a series of gates on the connecting axis of the Shanghai JW Marriott Marquis Hotel, which becomes the picture frames of the lobby, park, river view and distant Bund landscape. Like the gate of the garden, this orderly arranged entrance of the building establishes a striking direct connection between the hotel and the river. The arch-shape design of the hall entrance connecting to the park and river emphasizes the moment guests pass through the walkway, and highlights the importance of the

通透的主入口
Bright and airy main entrance

折叠式幕墙降低光污染
Collapsible curtain wall

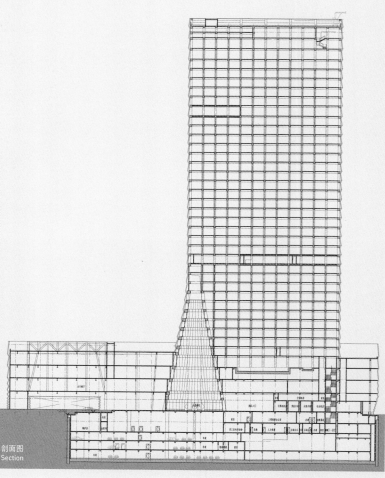

剖面图
Section

建筑轴线上设置一系列大门，成为大堂、公园、江景及远处外滩景观的画框。各建筑入口按次序排列，在酒店与黄浦江之间建立了直接的视觉联系。拱形的大厅入口强化了通过时的空间感受，突显出入口以及内部景观的节奏。入口设计与大堂空间相结合，为进入的宾客带来富有冲击力的体验。裙房结构设计采用桁架从屋顶吊住整个裙房，保证大宴会厅下部半室外落客区、南端餐厅下部与滨江相连的室外景观区的设计。裙房内部通过垂直管井连接屋面室外型设备，减少设备机房、提高室内功能空间利用率。

entrance itself and the interior landscape in the upward direction. The grand entrance design is combined with an extravagant lobby space. This creates an impactful experience when guests enter, the magnificent lobby atrium and plot itself are hence emphasized. In the structural design of the podium, trusses are used to hang the whole podium from the roof, this ensures an outdoor landscape area is formed by connecting the semi-outdoor drop-off area at the lower part of the grand banquet hall at the south to the riverside. In the podium, the pipe shaft is designed for the connection of outdoor equipment on the roofs, greatly reducing the space of equipment rooms and improving the utilization rate of indoor functional space.

简洁鲜明的建筑形象
Concise and distinct design

新开发银行总部大楼
NEW DEVELOPMENT BANK HEADQUARTERS

鸟瞰
Bird's eye view

项目地点：浦东新区国展路1600号
设计单位：华建集团华东建筑设计研究院有限公司
建设单位：上海市机关事务管理局
设计时间：2016.6-2020.8
建成时间：2020.8
总建筑面积：126423m²
用地面积：12067.4m²

Location: No. 1600, Guozhan Road, Pudong New Area
Designer: East China Architectural Design & Research Institute Co., Ltd., Arcplus Group PLC
Constructor: Shanghai Municipal People's Government Affairs Administration
Design Period: June 2016 - August 2020
Date of Completion: August 2020
Gross Floor Area: 126,423m²
Site Area: 12,067m²

局部凹进的立面细部
Concave façade (details)

南立面
Southern façade

塔楼与裙房间的连廊
Corridor between tower and podium

设计诠释了新开发银行"创新、平等、透明、可持续发展"的机制理念，突显新开发银行的国际地位，展现上海国际金融中心风貌。建筑强调独特性与融合性的统一，注重实用性与人性化设计，实现运行经济性与绿色环保。

设计从新开发银行的Logo中获取灵感。三角形象征稳定和平衡，旋转产生动力，代表创新和发展。塔楼平面亦以三角形为母题，通过切削角部，创造更为开阔的转角办公空间；三条边局部凹进，提供更长的室内景观面。结合功能将塔楼分成四段，在竖向上不断旋转上升，形成稳重而富有变化的独特造型。建筑融入世博A片区规划结构，贯通生态绿带，延续空间通廊。形体简洁大气，与世博A片区企业总部的整体建筑风格和谐统一。

标准层由三个矩形办公空间组成，易于分隔与组合，贴近新开发银行的使用需求，平面使用率超过80%；配套服务功能集中设置，便于高效使用；裙房相对独立，利于安保管理；裙房底部设置全天候半室外礼仪广场，便于举办各类高规格

The design interprets the concept of "innovation, equality, transparency and sustainable development" and highlights the international status of the New Development Bank. The unique architectural design is known for its inclusiveness. Focusing on practicality and user-friendliness, it seeks to pursue both environmental protection and economical efficiency in its operation.

The design concept originates from the logo of the bank. The triangle symbolizes stability and balance, while the image of "rotation" generates the power for innovation and development. The floor plan of the building is constantly rotating and rising above the plane, forming a stable and varied unique shape. This is highly recognizable along the Huangpu River. The planning structure of Expo Zone A is integrated into the design scheme, with the ecological green belt running through and the space corridor continued. The architectural form is simple and splendid and is in harmony with the overall architectural style of corporate headquarters in Zone A of Expo.

The building is designed to be in a regular shape, and utilization rate exceeds 80%. The standard floor is composed of three rectangular office spaces, which are easy to be divided and combined, satisfying the needs of the New Development Bank. The supporting service functions are set centrally, which is convenient for efficient use. The podium is relatively independent which is conducive to security management. An all-weather semi-outdoor etiquette square is set at the first floor of the

仪式。所有的办公空间实现三面自然采光和通风，创造健康舒适的室内环境；多层次室外观景平台，使人们能从不同高度，360°观赏城市景观；办公楼层之间设有共享空间，促进不同文化背景员工之间的休憩与交流。

技术运用上，通过建筑自身遮阳、屋顶绿化、多方位自然采光、中庭通风效应等措施，实现建筑"低成本投入、低能耗运营、高舒适度"的目标。采用透水铺装、雨水收集、中水系统、太阳能利用等节能环保技术，实现场地径流雨水零排放、建筑室外零水耗、室内常规水源消耗减少50%以上、车库100%可再生能源照明等。建筑达到我国绿色建筑三星标准及LEED铂金级标准。

podium, which is convenient for the bank to hold various high-standard ceremonies. The design theme implements the concept of people-orientation. All office spaces can realize natural lighting and ventilation on three sides so that a healthy and comfortable indoor environment is created. From the multi-level outdoor viewing platform, people can watch the urban landscape with a 360° lens at different heights. There is a shared space between the office floors, which promotes the relaxation and communication between employees with different cultural backgrounds.

The goal of "low cost investment, low energy consumption operation and high comfort" is realized through building self-shading, roof greening, multi-directional natural lighting, atrium ventilation effect and other measures taken in the green design. The energy-saving and environmental protection technologies such as permeable pavement, rainwater collection, reclaimed water system and solar energy utilization, etc. are used to realize the zero discharge runoff and rainwater on the site, zero water consumption outside the building, more than 50% reduction of indoor conventional water consumption, 100% renewable energy lighting in the garage, etc. The building shall meet the requirements of 3-star green building of China and LEED platinum level building.

旋转上升的塔楼形体
Spiral tower form

南入口与景观水池
South entrance and pool

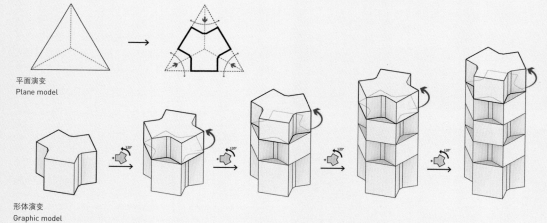

平面演变
Plane model

形体演变
Graphic model

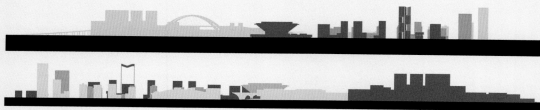

城市天际线
City skyline

夜景鸟瞰
Bird's eye view

上海大歌剧院
SHANGHAI GRAND OPERA HOUSE

西侧外观
View from the west

项目地点：上海浦东新区世博文化公园C02-01地块
设计单位：华建集团华东建筑设计研究院有限公司
合作设计单位：挪威斯诺赫塔建筑设计事务所（方案）、英国剧场顾问公司（剧场工艺）、永田音响（声学）
建设单位：上海大歌剧院
设计时间：2019.1-2020.7
建成时间：在建
总建筑面积：146786m²
用地面积：53023.2m²

Location: Shanghai Expo Cultural Park C02-01, Pudong New Area
Designer: East China Architectural Design & Research Institute Co., Ltd., Arcplus Group PLC
Partner: Snohetta (Scheme design), Theatre Projects Consultants (Theatre Technical), Nagata Acoustics (Acoustics)
Constructor: Shanghai Grand Opera House
Design Period: January 2019 - July 2020
Date of Completion: Under construction
Gross Floor Area: 146,786m²
Site Area: 53,023m²

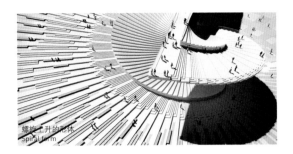
螺旋上升的形体
Spiral form

沿江远景
Distant view

上海大歌剧院旨在打造一个与黄浦江两岸景观和世博文化公园协调、开放共享的城市公共客厅，成为上海城市新地标。上海大歌剧院的主要功能包括2000座的大歌剧厅、1200座的中歌剧厅和1000座的小歌剧厅，同时设有排练、创作、制作、艺术教育、展示和研究等辅助功能，是一座按全产业链要素设计建造的专业歌剧院。

建筑形体取"中国扇"之意，结合歌剧艺术的动态美感，以螺旋上升的形体动态展开，自然形成可上人的屋面和屋面之下的空间。作为整体形态一部分的螺旋楼梯连接地面与天空，创造出面向上海核心城区和黄浦江岸的绝佳观景点。剧院的屋面成为可上人的"舞台"和城市客厅，用于大型庆典活动和日常参观。屋面下的公共大厅是建筑空间的核心，结合建筑形体呈半环形。公共大厅连接西侧主入口和东侧面向世博公园的次入口，并将室内空间划分为南区和北区，南区为大歌剧厅和中歌剧厅，北区布置小歌剧厅和教育培训展览等空间，各功能空间依循中心发散逻辑并随空间高度变化进行布局。

螺旋楼梯
Spiral stairs

The Shanghai Grand Opera House is not only aniconic cultural facility, but also agorgeous landmark of Shanghai by the Huangpu River. The whole complex includes a 2000-seat lyric theater, a 1200-seat opera theater and a 1000-seat studio theater. It can accommodate multiple functions required by opera productions, ranging from rehearsal, creation, production, art education, exhibition to research.

The shape of the building takes the meaning of a "Chinese fan" and combines the dynamic aesthetics of opera art with the dynamic unfolding of the spiral form, which naturally forms the roof and the space underneath the roof that can be occupied. As part of the overall form, the spiral staircase connects the ground to the sky, creating a great viewpoint towards the heart of Shanghai and the Huangpu River. The roof of the theater becomes an accessible "stage" and urban living room for large celebrations and daily visits. The lobby under the roof is the core of the building space and is semi-ring-shaped in conjunction with the building form. The public hall connects the main entrance on the west side and the secondary entrance on the east side facing the Expo Park, and divides the interior space into the south and north areas, with the grand opera hall and the Drama Thearter in the south and the Studio Theater and education and training exhibition in the north.

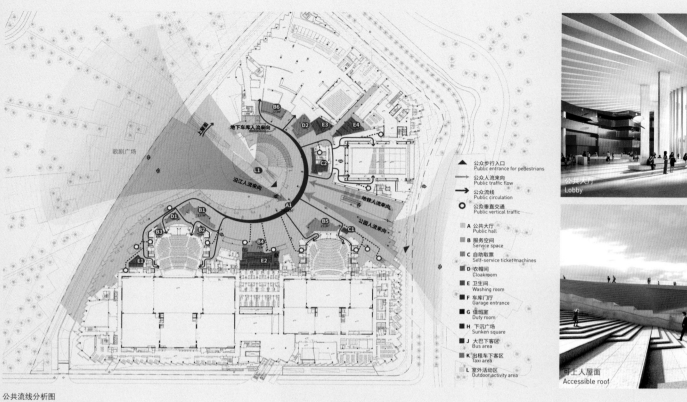

公共流线分析图
Public traffic flow diagram

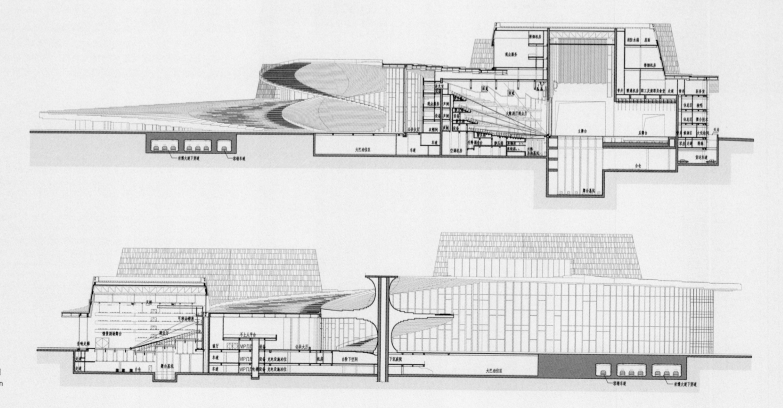

剖面图
Section

整体鸟瞰
Panoramic view

上海久事国际马术中心
SHANGHAI JUSS INTERNATIONAL EQUESTRIAN CENTER

模型照片
Sand table model

项目地点：浦东新区世博文化公园C04-01地块
设计单位：华建集团上海建筑科创中心
建设单位：上海久事体育产业（集团）有限公司
设计时间：2018.2-2020.10
建成时间：在建
总建筑面积：84558m²
用地面积：33217.5m²

Location: Plot C04-01, World Expo Culture Park, Pudong New Area
Designer: Shanghai Archi-Scientific Creation Center, Arcplus Group PLC
Constructor: Shanghai Jiushi Sports Industry (Group) Co., Ltd.
Design Period: February 2018 - October 2020
Date of Completion: Under construction
Gross Floor Area: 84,558m²
Site Area: 33,217.5m²

融入城市景观的马术中心
Shanghai Juss International Equestrian Center: An Eye-catching Landmark

上海久事国际马术中心是中国内地首座符合国际马术运动五星级赛事标准的城市永久性专业赛场，主体包含一个90米×60米的竞赛场地，其场地看台可容纳约5000名观众，同时配套有热身场地、训练场、马厩等功能空间。

Housed in the World Expo Park, the Shanghai Juss International Equestrian Center is the first permanent venue of its kind in mainland China that meets the international equestrian five-star competition standard. It contains a 90m x 60m competition arena with a grandstand that can seat about 5,000 spectators, as well as supporting facilities such as a warm-up area, training ground and stables.

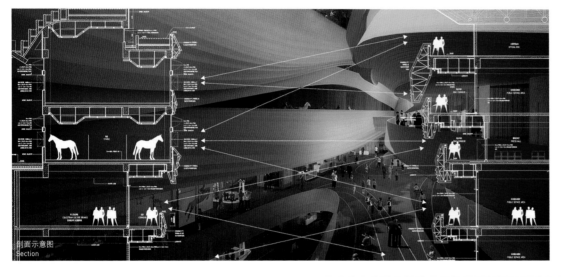

剖面示意图
Section

专业赛场
Race course

连续的步行空间
Continuous pedestrian path

设计理念为"马术谷",概念源自对公园整体起伏的园景特征的回应。建筑延续公园双子山的山势,面园背城,抬起成山,下沉为谷,结合功能置入赛场,文化景观体现马术运动特色。

"谷"在基地西侧的地铁19号线后滩站与东侧的双子山之间建立空间联系,形成面向公众开放的、连续的步行空间,串联不同标高的平台花园,提供公共服务功能,形成可达、可观、可赏的游览路径,成为世博公园的有机组成部分。未来,该马术中心将成为上海马术活动与国际接轨的场所,发展马术体育运动产业,提供文化、旅游与休闲等复合功能。人们在观赛、交流之际,能充分感受到马术文化及其延伸出地进取、乐观、沟通、合作等精神品质。这也正是作为全球城市的上海所传达的精神。

入口广场
Entrance

The concept, an "Equestrian Valley", stems from the undulating topography of the park. With the park in its front and the urban area behind it, it extends the rolling hills of the park, containing both hills and a valley. The property is a foremost destination of equestrian sports.
The design offers a spatial connection between the Houtan Station of the planned Metro Line 19 to the west of the "Valley" and the twin hills on the east, and leaves a continuous pedestrian path open to the public. The terrace gardens at different heights and public service functions are connected with accessible and lovely tour paths, and the latter are an integral part of the Expo Park. In the future, the Shanghai Juss International Equestrian Center will put Shanghai on the world map of equestrian sports, and it will also be a destination of cultural, tourism and leisure events. The public can savor the equestrian culture and be buoyed by its underlying spirit: enterprising, optimism, communication and cooperation. This is also the spirit of Shanghai as an international metropolis.

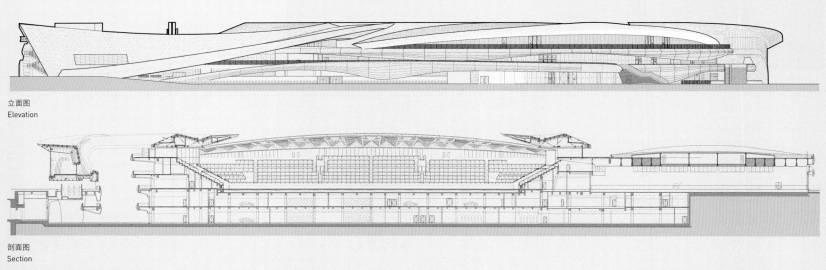

立面图
Elevation

剖面图
Section

城市天际线
City skyline

整体鸟瞰
Bird's eye view

前滩国际商务区
New Bund International Business District

北侧鸟瞰
Bird's eye view from the north

从裙房挑空处仰望塔楼
View up from the podium

前滩中心
New Bund Center

项目地点：浦东新区海阳西路555号
设计单位：同济大学建筑设计研究院（集团）有限公司
合作设计单位：KPF建筑设计事务所（设计顾问、建筑设计）、宋腾添玛沙帝建筑工程设计咨询（上海）有限公司（结构设计）、上海旭密林幕墙有限公司（幕墙设计）、种子设计合作有限公司（景观设计）
建设单位：上海前滩实业发展有限公司
设计时间：2013.10-2015.3
建成时间：2020.8
总建筑面积：191387m²
用地面积：10,216.8m²

Location: No. 555, West Haiyang Road, Pudong New Area
Designer: Tongji Architectural Design (Group) Co., Ltd.
Partners: KPF (Design consultant and architectural design), Thornton Tomasetti Consulting (Shanghai) Co., Ltd. (Structural design), Schmidlin Energy Technologies (Curtain wall design), Seed Design Collaborative Ltd. (Landscape design)
Constructor: Shanghai New Bund Industrial Development Co., Ltd.
Design Period: October 2013 - March 2015
Date of Completion: August 2020
Gross Floor Area: 191387m²
Site Area: 10217m²

前滩中心项目位于前滩新区的中心位置，16万平方米、280米高的办公塔楼是该区域的标志性建筑，成为天际线上的新焦点。从形体上看，核心办公楼可以理解为垂直的船锚，固定和聚集整个区域；悬浮的水平裙房则包括9000平方米的商业，覆盖整个场地，同时为5.1万平方米的酒店塔楼和酒店设施提供支撑。造型设计灵感来自三种肌理，水（流动线条）、木（纹理分叉）和地（石头叠层）。流动的"线条"将三个不同的体块（酒店，办公和裙房）融为一体，根据不同的体块功能，视觉语言各异。

The New Bund Center, located at the heart of the New Bund Business District, is an iconic 280-meter-high office tower that shapes the skyline. The 160,000 m² office building can be understood as a vertical "ship anchor" that solidifies the entire neighborhood. The horizontal podium with a "floating" visual effect includes a 9,000-square meter com mercial space, covering the entire site and supporting the 51,000 –squaremeter hotel tower. The design draws its inspiration from three elements, water (fluid lines), wood (texture with forks) and ground (stacked stones).The fluid "line" runs through the site to unite the three different parts of the building (i.e., the hotel, office and the podium) in a dynamic visual idiom that varies according to the functions of different parts. The office tower's vertical curtain wall, the "wall stack", is aligned at a 3-meter interval and extends from the bottom to the top of the tower. It has minimal impact on the view from the interior. In contrast, the curtain wall "cladding" at the top of the tower begins to branch and twist, responding to the design concept of wood texture, while creating the iconic tower "lantern". The facade of

顶层花园
Roof garden

办公塔楼下部的幕墙竖梃间距为3米，对室内视线影响较小。塔楼顶部的幕墙模拟木纹理的形态，逐渐分叉和扭曲，创造出标志性的塔楼"灯笼"。酒店的立面以层间的水平线条为主，通过装饰条的巧妙设计，创造出立体的波浪效果，同时避免对客房的视线阻挡，酒店顶部逐步演变为双曲面线条，强化了波浪的形态，增加视觉冲击力。简洁的裙房体块以微曲的幕墙竖梃装饰，衬托出两栋塔楼的动感，并通过屋顶花园将两栋塔楼联系起来。

---

the hotel is composed of horizontal decorative strips (upper and lower horizontal lines, with a curved wavy outline in the center). The decorative strips are located on the interstorey walls to avoid obstructing the view of the guest rooms, evolving from a purely parallel relationship at the bottom to a hyperboloid line at the top, while the length of the waves varies according to the height of the tower. The top of the hotel tower is also conceived as a "lantern", while the horizontal contorted lines create an exciting random pattern. The simplified fluid podium volume contrasts and accentuates the dynamic facades of the two towers, while connecting them with a perforated roof garden.

塔元立面细部
Facade of upper tower

裙房和塔立面细部
Facade of podium and tower

顶层花园
Roof garden

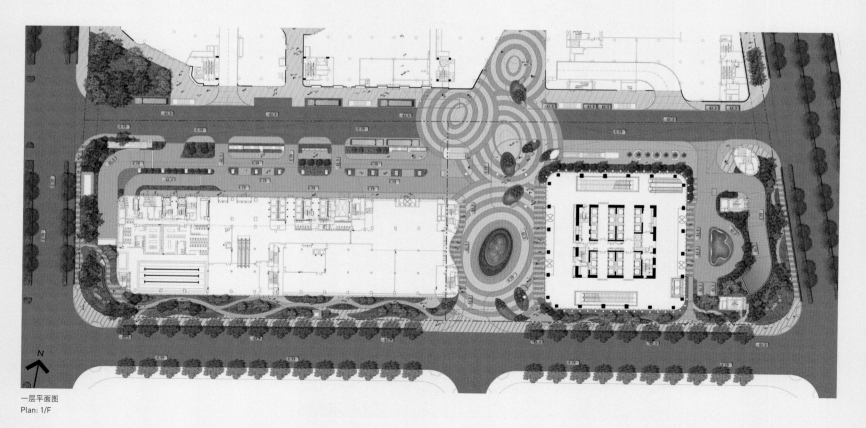
一层平面图
Plan: 1/F

西北角外景
View from the northwest

游泳馆和跳水池
Natatorium and diving pool

西侧全景
Bird's eye view from the west

上海东方体育中心
综合体育馆游泳馆项目
The Natatorium and Indoor Stadium of the Shanghai Oriental Sports Center

项目地点：浦东新区泳耀路300号
设计单位：华建集团上海建筑设计研究院有限公司
合作设计单位：德国gmp国际建筑设计有限公司
建设单位：上海市体育局
设计时间：2008.12-2011.12
建成时间：2012.5
总建筑面积：163841m²
用地面积：347500m²

Location: No. 300, Yongyao Road, Pudong New Area
Designer: Shanghai Architectural Design & Research Institute Co., Ltd., Arcplus Group PLC
Partner: Germany GMP International Architectural Design Co., Ltd.
Constructor: Shanghai Municipal Sports Bureau
Design Period: December 2008 - December 2011
Date of Completion: May 2012
Gross Floor Area: 163,841m²
Site Area: 347,500m²

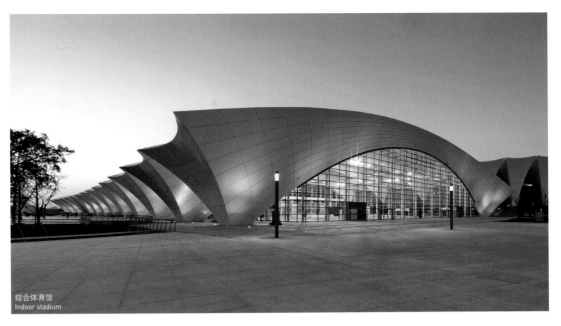
综合体育馆
Indoor stadium

上海东方体育中心紧邻黄浦江和新建世博园区，由一座18000人规模的综合体育馆、5000人规模的游泳跳水馆、1000人规模的室外跳水池和一座新闻中心组成，是2011年世界游泳锦标赛（简称"世游赛"）的主场馆。其中综合体育馆总建筑面积77243平方米，是上海目前座席最多的室内运动场馆，固定和活动座位总数可达18000个。体育馆的使用灵活性高，可进行室内球类、体操、冰上运动等，世游赛期间搭建1个标准游泳池、1个热身池，先后举行游泳和花样游泳比赛。世游赛后，体育中心向社会开放，成为浦东新区居民业余生活的亮点，同时也发展成为上海新的体育运动中心。

The Shanghai Oriental Sports Center is seated to the east of the Huangpu River, south of Chuanyang Creek and west of Jiyang Road in the Pudong New Area, adjacent to the newly-built Expo Park. It consists of an indoor-stadium that seats 18,000 spectators, a natatorium with a seating capacity of 5,000, and an outdoor diving pool that can accommodate 1,000 spectators and a media center. Among them the indoor stadium with a gross floor area of 77,243 m² is the venue of the 2011 FINA World Championships. The indoor stadium can flexibly accommodate ball games, gymnastics and ice sports events. Since the 2011 FINA World Championships, the Sports Center has been open to the public, being a go-to place for sports fans in Pudong.

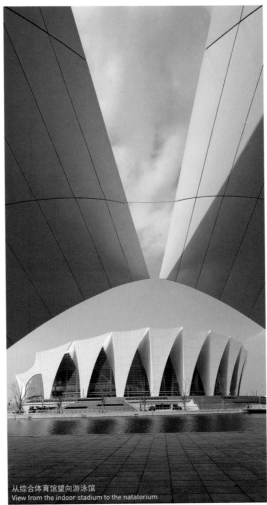
从综合体育馆望向游泳馆
View from the indoor stadium to the natatorium

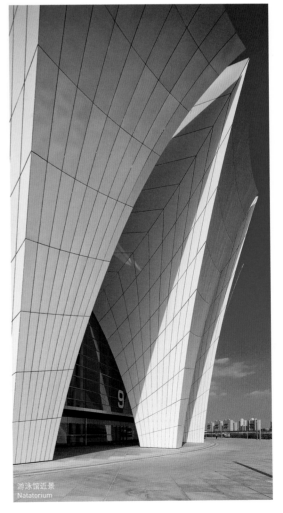
游泳馆近景
Natatorium

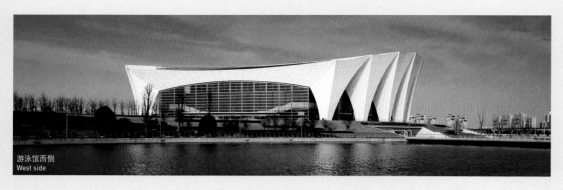

游泳馆西侧
West side

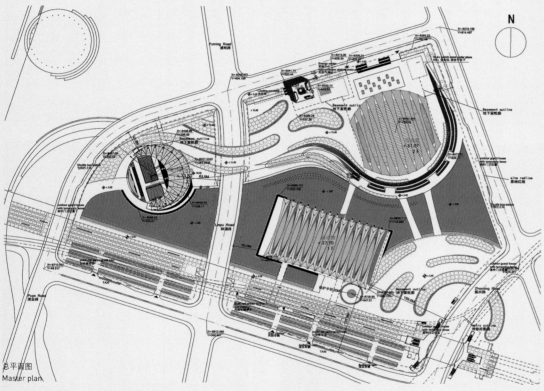

总平面图
Master plan

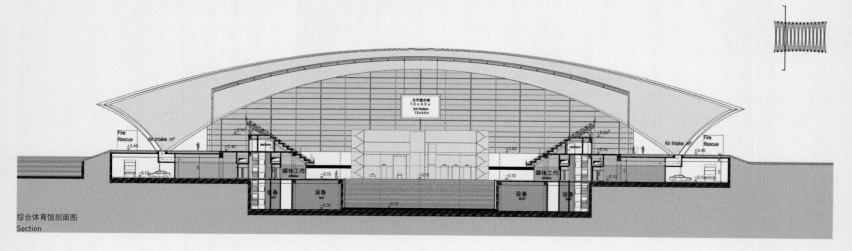

综合体育馆剖面图
Section

东侧全景
Panoramic view from the east

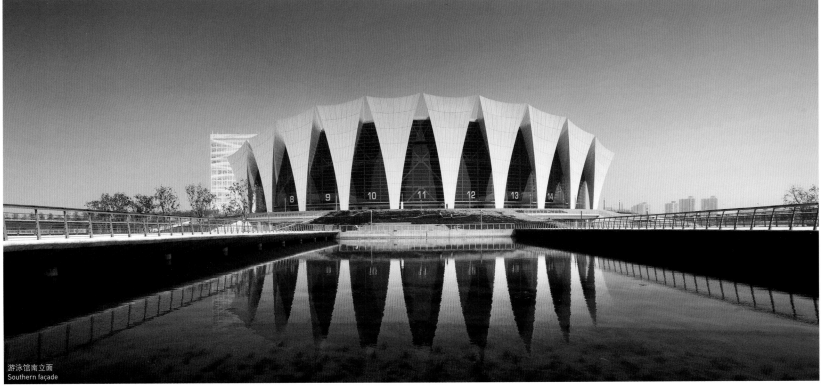

游泳馆南立面
Southern façade

整体鸟瞰
Bird's eye view

华东师范大学第二附属中学前滩学校
The New Bund School, Attached to No.2 High School of East China Normal University

项目地点：浦东新区晨晖路555号
设计单位：上海浦东建筑设计研究院有限公司
合作设计单位：致正建筑工作室（设计顾问）
建设单位：上海前滩国际商务区投资（集团）有限公司
设计时间：2014.4-2016.11
建成时间：2018.4
总建筑面积：53792m²
用地面积：28585m²

Location: No. 555, Chenhui Road, Pudong New Area
Designer: Shanghai Pudong Architecture Design & Research Institute Co., Ltd.
Partner: Atelier Z+(Design Institute)
Constructor: Shanghai New Bund International Business District Investmen (Group) Co., Ltd.
Design Period: April 2014 - November 2016
Date of Completion: April 2018
Gross Floor Area: 53,792m²
Site Area: 28,585m²

校园与城市
Details

华师大二附中前滩学校九年一贯制学制与前滩国际社区定位相辅相成，有效地推动社区的整体开发，增强社区聚合力，为进一步提升城市空间品质提供品质保障。学校设有小学25个班，中学20个班。室外设2片篮球场地，一个250米环形跑道。校区内包含一栋五层教学楼及一栋五层文体楼。总体采用围院式布局，建筑之间通过连廊与屋顶平台相互连接，在营造多层次、流动性、富于变化的室外空间的同时，形成多种类型与个性的景观空间。

建筑设计上强调空间的开放性和互动性，在满足基本教学功能的同时，为丰富多样的教学模式提供更多的空间弹性，让学生能够在多元环境下进行学习和锻炼。教学部分为"8"字形组团，其中南侧为小学普通教室用房，中间的矩形组团容纳专用教室和办公用房，北侧为中学普通教室用房，西侧挑出错落的矩形盒子为图书阅览室联系中小学，提高教学区的高效性，并成为一个更加活跃的场所。四个功能体块通过挑空和搭接的方式联系，高低错落的布置方式形成富有节奏感和韵律感的建筑天际线。

建筑立面处理追求简洁明快现代的效果。教学部分采用竖向彩色百叶立面处理手法，各功能空间则根据不同的使用需求进行开窗处理；面向内部庭院的部位采用落地玻璃，充分建立庭院内外的空间交流；体育活动室考虑通风和采光采用玻璃百叶围合；阅览室挑高天窗的个性化造型处理，使建筑形态更加生动，并且渲染了室内空间的特殊氛围。片状平台削减块的体量感，形成轻盈通透的立面效果。在整体色调统一和谐的前提下，利用活泼明快的色彩作为修饰，增强了功能空间的可识别性。

初中部庭院
Junior high school

The New Bund School Attached to No. 2 High School of East China Normal University is a Nine-year School of the New Bund International Business Zone, Pudong. It is adjacent to the Wellington International School in the north across the river and the New Bund "9-grid" residential area in the west. This public school matches the positioning of the New Bund international community, which can effectively promote the overall development of the community, enhance community cohesion and provide quality assurance for further urban space quality. The school has 25 primary school classes and 20 middle school classes. It has two outdoor basketball courts and a 250-meter-long circular track. On the campus there is a 5-storey teaching building and a 5-storey gym building. The former houses an indoor swimming pool, a basketball gym and a lecture hall that can seat 700 people. The architectural design emphasizes the openness and interaction of spaces. To meet the needs of education, an enhanced space flexibility is provided for students to learn and exercise in a diverse environment.

阶梯平台
Semi-open space

幼儿园中庭
Atrium of kindergarten

行政楼南侧局部外观
Southern side of the administrative building

西南侧局部外观
Southwest façade

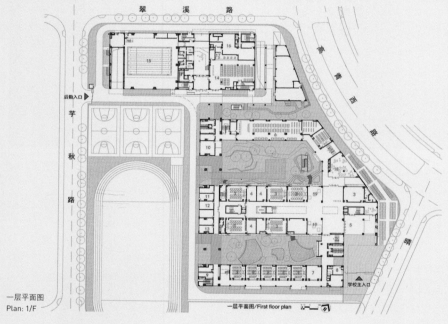

一层平面图
Plan: 1/F

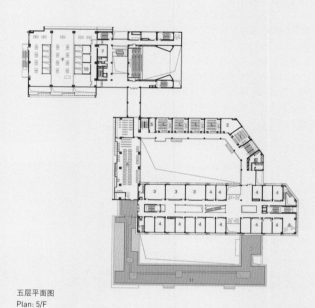

五层平面图
Plan: 5/F

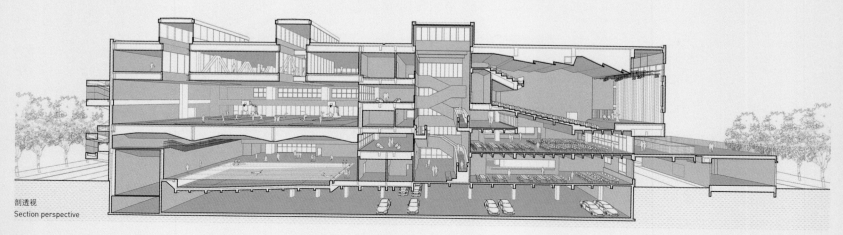

剖透视
Section perspective

西南侧鸟瞰
Bird's eye view from the southwest

整体鸟瞰
Bird's eye view

前滩媒体城西区
New Bund Yuan

项目地点：浦东新区前滩国际商务区
设计单位：同济大学建筑设计研究院（集团）有限公司
合作设计单位：KPF建筑设计事务所（设计顾问）
建设单位：上海耀体实业发展有限公司
设计时间：2020.6-2020.12
建成时间：在建
建筑面积：266292m²
用地面积：40954m²

Location: The New Bund International Business Zone, Pudong New Area
Designer: Tongji Architectural Design (Group) Co., Ltd.
Collaborator: KPF (Design Consultant)
Constructor: Shanghai Yaoti Industrial Development Co., Ltd.
Design Period: June-December, 2020
Date of Completion: Under Construction
Gross Floor Area: 266,292m²
Site Area: 40,954m²

项目位于浦东新区三林镇，用地性质为商业服务业、文化用地、商务办公用地。项目建成后将成为"新外滩"的门户，为前滩区域乃至全上海打造一个全新的夜生活和文化目的地。建筑天际线以水为灵感，提升浦江沿岸天际线形象，力图成为上海的CBD新标杆。

The property is locatedin Sanlin Town, Pudong, on a piece of land reserved for commercial services, cultural, and office uses. When it is completed, it will be the gateway to the New Bund, a new benchmark of CBD and a destination of dynamic nightlife and cultural events in the New Bund area and in Shanghai. Inspired by water, the skyline seeks to be a beautiful one along the Huangpu.

露台
Terrace

沿江夜景
Panoramic view

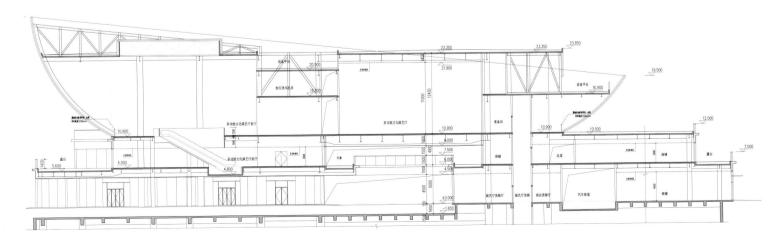
剖面图
Section

晶耀前滩
Crystal Plaza, New Bund

项目地点：浦东新区耀体路308号
设计单位：华建集团上海建筑设计研究院有限公司
合作设计单位：KPF建筑设计事务所（方案）、王欧阳（香港）有限公司（执行建筑师）
建设单位：上海仁陆置业有限公司、上海仁耀置业有限公司
设计时间：2013-2016
建成时间：2019.5
总建筑面积：456695m²
用地面积：55378m²

Location: No. 308, Yaoti Road, Pudong New Area
Designer: Shanghai Architectural Design & Research Institute Co., Ltd., Arcplus Group PLC
Partner: KPF (Scheme design), WO Architects (Executive architect)
Constructor: Shanghai Renlu Real Estate Co., Ltd., Shanghai Renyao Real Estate Co., Ltd.
Design Period: 2013-2016
Date of Completion: May 2019
Gross Floor Area: 456,695m²
Site Area: 55,378m²

晶耀前滩是一个综合超甲级办公楼、高端购物中心及高品质住宅的多业态综合开发项目。项目由四个地块组成，包含六栋80～120米高的办公塔楼、商业裙房，以及南侧的四栋高层住宅楼。商业裙房环绕着场地中央开放的城市广场展开，在商业的二、三层设有环形廊桥，连接四个地块上的办公和商业功能，也提供了与周边地块的空中连接。办公楼从商业平台上空升起，犹如矗立在花园中的水晶。塔楼总体布局在统一中追求变化，六栋沿城市道路布置的办公塔楼，以江耀路为中轴线近乎对称布置，为简化租赁类型统一形态。通过塔楼的旋转变化将各办公楼望向场地的视角打开，在紧凑的地块内创造了丰富的空间关系，同时又塑造出基地的边界和活跃的城市临街气氛。

塔楼的设计灵感源自水晶，打破常规的矩形形态，借鉴水晶丰富多变的形态和晶莹剔透的质感。立面设计通过竖向金属杆件的角度变化，产生微妙的光影关系，在视觉上强化建筑的动感。顶部延续晶体的形态，屋面高低错落，构成丰富的第五立面。这些设计手法也同样运用在裙房及大堂等室内空间中，创造出独特的空间体验。沿基地南侧布置的四栋高层

The Crystal Plaza is multi-functional, consisting of international Grade-A offices, exclusive residences, and quality retail amenities. Located in the heart of the New Bund – an emerging business center in Pudong, it has been designated by the municipal government as one of six priority development areas in Shanghai's 12th Five-Year Plan. Together with the Expo area and Xuhui section of Huangpu waterfront, the region will be one of the most important commercial clusters in Shanghai upon completion. The property consists of six office towers 80 to 120 meters tall and four residential towers situated on the southern-most portion. Positioned like crystals in a garden, the office towers rise atop a podium interlaced by a network of pedestrian axes, encircling an open, multifaceted urban plaza that houses retail and amenity activity at multiple levels. The heart of the complex, this central courtyard is connected by pedestrian bridges to the surrounding city and features a grand set of stairs that invites entrance. This feature exudes a unique spatial quality throughout the site and provides a distinguished, elevated urban experience. The tower mass, ascension of height, and rotation create a sense of enclosure while enhancing views across the site.

The office towers provide Class-A work spaces, with 4.5-meter floor-to-floor heights. Each building has an independent drop-off area, deliberately separated from ground-level pedestrian circulation. Landscape elements such as plants and trellises are

东北侧鸟瞰
Bird's eye view from the northeast

南侧街景
View from the south

住宅与办公塔楼和商业裙房相邻,同时又相对独立,保持私密性。住宅立面在满足居住需求的前提下,通过折面屋顶等处理手法与办公塔楼呼应。在交通组织上,为了提升商业内街的安全性与舒适性,主要的车辆流线沿基地外侧展开,每栋办公楼均设有独立的落客区,并与人行流线分隔。项目的四个地块地下室互相连通,共享汽车坡道和停车空间,增强了地块间车辆出入和泊车的灵活性,留给地面充足的绿化和景观空间。依托绝佳的地理位置,晶耀前滩项目凭借一流的设计品质及地标效应,为前滩地区提供高品质的办公、居住、购物及休闲生活立体空间。

strategically positioned on and around the podium to create a comfortable and attractive outdoor and semi-outdoor environment. Breaking from rectangular geometries to create a more crystalline character, the tower facade design utilizes vertical and diagonal metal fins that reflect light in different directions at different view angles to reinforce the dynamic forms of the buildings. The same logic is carried over to the podium facade and into the interiors, such as the lobby, where surfaces are sculpted and carved to create unique spaces. Crystal Plaza is ideally situated, directly connected to the interchange of three subway lines and a short driving distance to existing central business districts, the airports and the Expo waterfront. Its development strategy will create the highest and best use of the site given prevailing market conditions. By embracing environmentally progressive building principles and taking full advantage of the convenient access to a 10.4 hectare waterfront park along the Huangpu River and parks directly adjacent to or in its vicinity, the property will be built and marketed as a "green" development.

东侧人视
Panoramic view from the east

中庭
Atrium

东北侧鸟瞰
Bird's eye view from the northeast

室内
Interior

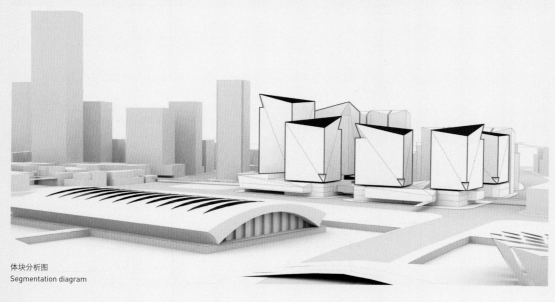
体块分析图
Segmentation diagram

整体鸟瞰
Bird's eye view

上海长滩
SHANGHAI LONG BEACH

整体鸟瞰
Panoramic view

项目地点：宝山区上港十四区
设计单位：华建集团上海建筑设计研究院有限公司
合作设计单位：贝诺、DLR、PE
建设单位：上港集团瑞泰发展有限责任公司
设计时间：2011.5-2018.12
建成时间：在建
总建筑面积：1398094m²
用地面积：77.62hm²

Location: Lot 14, Shanghai Pot, Baoshan District
Designer: Shanghai Architectural Design & Research Institute Co., Ltd., Arcplus Group PLC
Partners: Benoy, DLR, PE
Constructor: Shanghai Port Group Ruitai Development Co., Ltd.
Design Period: May 2011 – December 2018
Date of Completion: Under Construction
Gross Floor Area: 1,398,094m²
Site Area: 77.62 hm²

上海长滩城地公园装置景观
Installations in Shanghai Long Beach Park

西侧鸟瞰
Bird's eye view from the west

上港滨江城总体开发建设项目，位于上海市宝山区东北角——原上港十四区，利用其滨江港口区位优势、自然资源和工程条件优势，逐渐从交通基础性设施向城市复合型生活功能转型，形成新的城市活力空间。

设计着重区域城市结构的整体开发，着眼于区域城市机能的复活，地块从以往的港口改造成为居住社区及社区配套、城市公园、区域公共活动中心、文化集中区、及游憩商业区，力求打造集邮轮集散、游艇产业、旅游、居住、消费为一体的多功能复合型城市滨水带，塑造具有城市特色的城市生活空间和形象，提高淞宝地区的城市空间品质。

地块的空间结构可用八个字概括——两极一轴，一带六区。

两极：即两组地标建筑，对整个滨水空间起到提纲挈领的作用，主要功能为综合展示板块内的观光塔及精品酒店办公区、主题活动板块内的巨构中心；一轴：即东西向的中心视觉长廊，贯穿国际社区板块和精品住宅板块，使得各个地块既独立，又紧密联系；一带：即滨水休闲绿化带，是整个区域的公共活力和记忆延续的纽带。通过保留部分港口装置成为景观元素来延续老工业码头的历史记忆；六区：即六个功能板块。由西向东分别为主题活动区、国际社区、滨江精品居住区、滨江休闲绿地区、综合展示区、游艇活动区。各个功能板块相互配合、协同，共建国际港口活力新城区。

The project site is located in the northeast corner of Baoshan District, Shanghai - the former Lot 14 of Shanghai Port, which is taking advantage of its riverside port location, natural resources and engineering conditions to gradually transform from a transportation infrastructure to a complex urban living function, forming a new urban vitality space - Shanghai Long Beach.

The project design focuses on the overall development of the regional urban structure and the resurrection of regional urban functions. The site is transformed from a former port into a residential community and community support, a city park, a regional public activity center, a cultural concentration area, and a recreational and commercial area, aiming to create a multi-functional composite urban waterfront zone integrating cruise ship distribution, yacht industry, tourism, residence and consumption, to shape an urban living space and image with urban characteristics, and to improve the quality of urban space in the Songbao area.

The spatial structure of the site can be summarized in eight words - two poles and one axis, one belt and six districts. Two poles: two groups of landmark buildings, which play a leading role in the whole waterfront space, The main function includes one sightseeing tower and boutique hotel office area in the comprehensive exhibition block, and a giant structure center in the theme activity block. One axis: the east-west central visual promenade running through the international community block and the boutique residential area block, making each plot both independent and closely connected. One belt: the waterfront leisure green belt, which is the public vitality and memory continuity link of the whole area. By retaining part of the port installation as a landscape element to continue the historical memory of the old industrial wharf. Six zones: six functional zones. From west to east, they are thematic activity area, international community, riverfront residential area, riverfront leisure green area, comprehensive exhibition area and yachting activity area. Each functional block cooperates and synergizes with each other to build a new vibrant

滨江精品居住区
Riverfront residential area

项目设计还注重滨水区域的立体开发，旨在将地块打造成为开放的富有活力的公共活动场所。其通过各个地块不同标高的垂直与水平联系，在滨水区域设置休闲、娱乐、观光、餐饮、购物等多种设施，产生具有吸引力的城市氛围；同时以滨水区域连通公园为活动舞台，创造出多样化的开发空间，举办公共性的节庆活动；还强调利用水资源环境的优势，建造不同性格的"围合村落"，结合滨水区域形成多种公共空间和半公共空间等。这些立体混合式滨水开发设计策略使上海长滩的滨水区域产生持久的活力。

international port city.
The project design also focuses on the three-dimensional development of the waterfront area, aiming to turn the site into an open and dynamic public activity place. Through the vertical and horizontal connection of different elevations of each parcel, it sets up various facilities for leisure, entertainment, sightseeing, dining and shopping in the waterfront area to generate an attractive urban atmosphere; at the same time, it uses the waterfront area connected to the park as a stage for activities to create diverse development spaces and hold public festivals; the advantage of using the water resource environment is also emphasized to create "enclosed villages" with different characters, and to create a variety of public and semi-public spaces in combination with the waterfront area. These three-dimensional hybrid waterfront development and design strategies enable the waterfront area of Shanghai Long Beach to generate lasting vitality.

上海长滩瀑布公园实景
Waterfall-theme Park in Shanghai Long Beach

总平面图
Master plan

东侧夜景
Bird's view from the east

设计·师说
MASTERS' WORDS

传承 延续
让·努维尔

浦东美术馆栖身于黄浦江边，面向黄浦江对岸拥有百年历史的外滩建筑群，描画着陆家嘴金融城这一著名的当代城市天际线。正是这样一个前所未有的机遇，使设计师能够参与深化和丰富场地的内涵，继续讲述这片土地的故事——一个属于21世纪的上海的故事。

我们深感自己的职责，因为美术馆将成为这座城市历史构成的一部分，这令人激动又紧张。我们融入梦想、想象力、诗意，与理性、勇气一起塑造浦东美术馆的建筑形象。然而，真正能使浦东美术馆成为一个传奇的是它体现出的场所精神，即场地与城市的共生、共鸣。正是这种远瞻性的视角和对历史的充分认知才能创造出真正的吸引力。人的情感会因为厚重与轻盈的共存而被激发，就像感受上海的昨天与明日，在设计中得以体现延续与升华。

我们将浦东美术馆视为一片领地，而领地本身即是浦东美术馆。它占据了40000平方米的土地，并呈现为一个由暖白色花岗岩构成的媒介物，其不规则的组合方式使人联想到"冰裂纹"。领地周围环绕着树林——其中包括从原有场地移栽过来的树木，形成起伏的景观和巨大的风障，将领地与周围的高楼分隔开来，将美术馆衬托得更加明亮、通透。在这个领地中，不论是在葱郁的林地景观中，还是在石质的"冰裂纹"体块、前广场、露台亦或是滨水步道上，均遍布着各种各样的雕塑和当代艺术品。前广场和露台采用具有相同材料和纹理的基座和树池，以呼应领地的存在感。

在靠近黄浦江、与水面相接的位置，石块的颜色变得更白。在与水面垂直的白色岸壁上，创造出一个兼具餐厅功能的凉廊，尽可能地亲近外滩上来往的船只和人群。两块垂直的屏幕分别位于美术馆的上方和通风塔的顶部，它们环抱着场地，像两个括弧（符号）一样突显着美术馆的位置。两块屏幕在夜间发出的镭射光将美术馆上空照亮。

朝向外滩的西立面是美术馆的主展示面。这是一个带有冰裂纹的双层玻璃立面；它是一面镜子，变幻莫测——白天是艺术装置或表演的场所，夜晚则成为播放映像的空间。它是一台看外滩的双重全景机，根据光线强度不同，反射出双重全景，也将美术馆中陈列的艺术品的反射片段叠印到外滩的映像上。它展现出美术馆内部的活动和观众，还可以增强这些活动和观众的层次。

从沿江的堤岸步道看美术馆的西立面，它像一处伏笔，层出不穷的新影像与上海和外滩对话，令人着迷，成为21世纪艺术的见证。

美术馆表现出可供人们同时欣赏展品和外滩美景的包容尺度，保护人们在任何天气下都可悠闲地观展，以及一种随时间而变幻色彩，随机反射高楼天际线图像的神秘感。浦东美术馆将一切能够激发人们情感的元素汇聚于此，以此来向上海这座城市致敬，向这片拥有伟大历史的场地致敬，向为我们创造了这些建设伟业的前辈们致敬。同时，也让我们继续前辈们的工作：为未来制造属于我们这个时代的记忆！

Perpetuation and Enhancement
Jean Nouvel

The Museum of Art Pudong is ideally situated along the Huangpu River, a mighty waterway with the century-old Bund on one side and the contemporary skyline of Lujiazui Financial Hub on the other. It is an unprecedented opportunity for us designers to celebrate and enhance the significance of the site by perpetuating the story of the land - a story of the 21st century Shanghai.

We are acutely aware of our responsibility, as this museum will become a part of the city's history. It is a project that is both exciting and nerve-wracking. We seek to imbue our dreams, imagination, poetic rein, reason and courage into its architecture. However, what really makes the Museum a legend is the spirit this venue embodies, or in other words, its symbiosis and resonance with the city. It is such visionary perspective and full cognition of history that have made it truly appealing. One's emotions are aroused by the justaposition of heaviness and lightness, just in the way one feels the past and the future of Shanghai: This is what we try to perpetuate and enhance in our design.

We see the Museum of Art Pudong as a domain, and so it is. It occupies 40,000 m² of land and is presented as a medium formed by granite in warm white color. It is irregularly assembled in a manner reminiscent of "ice cracks." The domain is surrounded by woods - including trees transplanted from the original site - creating an undulating landscape and a large windbreak that separates the domain from the surrounding skyscrapers and make the museum look all the more bright and airy. A variety of sculptures and contemporary artworks are scattered throughout the site - in the lush woodland landscape, in the stone "ice crack" blocks, on the front plazas, terraces, and waterfront walkways. The front plaza and terrace feature pedestals and tree ponds of the same material and texture to echo the presence of the domain.

At the space close to the Huangpu River where it meets the water, the stones become even whiter than those elsewhere. On the white wall perpendicular to the water, a loggia that serves as a restaurant is created to get close to the boats and people that come and go on the Bund to the utmost degree. Two vertical screens, one above the museum and one on top of the ventilation tower, encircle the site, highlighting the museum's location like a pair of brackets. The two screens emit laser light at night to illuminate the sky above the museum.

The west façade facing the Bund is the main surface of the museum for display. It is a double-glazed façade with ice cracks, an unpredictable mirror - a place for art installations or performances during the day, and a space for reflections at night. It is a double panorama machine facing the Bund, reflecting a double panorama depending on how strong the light is, and also overlapping the reflected fragments of the artworks at the museum onto the reflection of the Bund. It shows the events and viewers inside the museum, and also delivers their multiple layers.

When viewed from the promonade, the museum's west facade looks like a foreshadowing touch. It suggests a fascinating dialogue of Shanghai's new images with the historic Bund and makes for a fascinating testament to 21st century art.

The museum exhibits an inclusiveness that affords the enjoyment of the exhibits and the view of the Bund at once; it also guarantees a leisurely visit regardless of weather, and provides mysteries arising from both changing colors over time and random reflections of images of the skyline formed by skyscrapers. The Museum of Art Pudong brings together all the elements that can inspire emotion to pay tribute to the city of Shanghai, to this site of glorious history, and to the people who have built these marvelous structures for us. At the same time, we are allowed to continue the work of our predecessors: to create memories of our time for the future!

HUANGPU RIVER
SUZHOU CREEK

THE WATERFRONT SPACE & ARCHITECTURE IN SHANGHAI

活力水岸连接
DYNAMIC WATERFRONT CONNECTION

日晖港步行桥 Footbridge on Rihui River

游客集散中心 Tourist Distribution Center

龙华港桥 Longhua Creek Bridge

洋泾港步行桥 Yangjing Creek Pedestrian bridge

民生轮渡站 Minsheng Ferry Station

东昌栈桥 Dongchang Elevated Passage

白莲泾 M2 游船码头 Bailianjing M2 Tourist Terminal

倪家浜桥 Nijiabang Bridge

为公共性赋形：景观基础设施与浦江两岸贯通

张 斌

在中国进入以存量更新为特征的、追求高质量发展的城市化新阶段，面对土地资源紧张、城市高密度与可持续发展，我们急需改变单一功能的传统基础设施对于城市公共空间和社区空间的阻隔，用协同共生的方式引入以统筹城市开放空间系统发展为目标的景观基础设施这一新的发展范式，促使基础设施各要素之间形成功能与效益的最大化，以实现区域内生态网络、交通网络和户外慢行网络的一体化运作。景观基础设施作为一种多功能的媒介和载体，通过提供各种生态过程和市政工程之间流动交换的共生界面，催生和协同两者之间的融合与优化。它对于开放空间和公共生活的支持，体现在我们把城市不只看作为一个高性能的功能机器，更看作为一个生命混合体。

近年以来，致正建筑工作室 Atelier Z+ 在浦江两岸贯通中进行了一系列以整合城市公共空间资源、活化社区与公共生活为目标的景观基础设施建设与实践，涉及生态修复、防洪防汛设施景观化、绿道、滨水公共空间、慢行系统、工业遗产活化等方方面面，而且往往是多种景观与基础设施系统的复合和共生。

东岸望江驿：作为触媒的日常生活基础设施

正如伯纳德·屈米所言，建筑不仅是容纳活动的中立空间，而且是塑造新的生活方式并激发社会变革的工具。在一系列城市更新的实践中，基础设施可以是一种触媒，它从功能必需品变成可以影响城市物流形式和聚集方式的工具。触媒概念在将建成环境视为系统化的基础设施的假设下，可以重塑建筑学的方法。基础设施作为空间干预的溢出效应，证明建筑学的社会效应增强。

在黄浦江两岸打造世界级水岸空间的进程中，我们将目光投在与公众日常生活联系最为紧密的小型公共设施上。浦东滨江公共空间中的东岸望江驿，让我们有机会思考如何在一个大尺度的滨江公共空间中介入小尺度的公共基础设施，使其成为一处可以提供日常性使用的休息驿站；如何将这些日常性的公共基础设施与滨江公共空间原有的景观、地形、肌理融合，在成为滨江景观观看平台的同时，自身也成为景观要素，成为滨江公共空间景观的锚固点与放大器；如何在公共空间与城市地域性格与特质之间进行联结，将诸多提供日常服务的驿站转化为新的城市公共生活的空间触发器；如何在极短的设计和施工期内，构建一种自发生成的建造体系，能够极好地在施工工期、空间品质以及空间体验三者之间达成平衡。同时，望江驿建成后也成为观察市民公共生活的生动场所，通过对 22 个望江驿空间的自由享用，人们确认了闲暇和娱乐不必须和消费关联，感到自己和这个城市的密切相连。这样非消费性、独立自主的公共空间是颇具"上海性"的城市空间类型，我们在望江驿中看到上海市民的主体意识、公共意识和参与意识，看到多元并置、包容开放的现代性。

东昌栈桥：协调多方利益的地形化能动形式

随着景观基础设施的兴起，研究地表起伏的地形学进入建筑学语境的讨论，它描述了一系列创造人工地形重新组织空间，引导人流并形成新的设计语言的城市设计实践。地形成为一种包括景观、建筑与各种构筑物的，整合的空间形式，关注消解构筑物与场地的二元对立关系，并重塑现代主义所忽视的建成环境的连续性。基础设施代表的一系列技术、协议、规则和各种社会关系，在当代可以成为一种"能动形式"激发特定地区的连锁反应，改变关于空间的协议，进而产生衍生的空间行为。建筑学在这样的实践中可以突破形式游戏的单一功能，也突破信息与符号的界限，产生社会效益。

在浦江两岸公共空间贯通开放工程中，东岸有 12 座跨越河道、轮渡站等主要空间断点的慢行桥。东昌栈桥是其中之一，兼容跑步和骑行的高架慢行景观步道跨越多种用地，涉足周边若干土地所有方，横穿过江隧道并纵贯源水管道等复杂的地下条件，实现通过复合型的景观基础设施，协调多方利益、促进城市公共空间品质提升及公共性最大化等复合诉求，并构建滨江公共空间中承载多样空间行为的新地形。

杨树浦六厂公共空间更新：后工业语境下的滨水空间修复与多维整合

景观基础设施涉及自然与人工、绿色和灰色的组合，代表一种不再区分自然与人工演变，把自然理解为人类社会政治经济的产物——新的自然与生态观念。对于自然的理解可以分为四个层次，即作为原始自然的第一自然；作为人类生产生活改造后的第二自然；作为对于第一和第二自然进行美学再现的第三自然；以及作为被损害及生态恢复的第四自然。景观基础设施实践更多地涉及第二自然与第四自然。

第二自然是人与自然相互适应的结果，也是人居环境的重要组成部分。而第四自然基本是一个后工业的话题，面对被人类活动过度干预后废弃的自然，用自然进程的自我修复或人工干预的生态修复方式为被损害的自然重新带来生机。近年来随着郊野公园、湿地保育与修复、河道治理与滨水绿道、工业废墟的活化与再利用等新型景观基础设施在中国的不断涌现，我们需要拓展对自然的认识，重新思考自然地含义，才能在实践避免简单化、符号化地理解自然，去探讨生态与自然的多样性。

杨浦滨江六厂的公共空间更新与改造，关注后工业语境下的滨水空间修复与多维整合，在对于自然、野趣和原真性的保育诉求之下，试图通过设计介入更多地保存未列入保护名录的工业遗产，并活化利用为开放的文化体验空间和市民服务空间；景观设计中以考古式的方式保存场地地形特征与格局，将留存的厂区码头、栈桥、界墙及隐匿的建筑墙基等作为景观设计的平等要素与新的设计相融合，保留并延续杨浦滨江工业遗产的肌理与场地体验。

我们试图希望通过这样的新型基础设施，改变人们对于传统基础设施隐形、冷峻、功能性的刻板印象，多重呈现更具活力、公共、综合的形式，并以可感知的方式成为城市空间中新的行为载体与审美对象。景观基础设施在拓展建筑学边界的同时也是城市新的公共性媒介，成为新的、中国建筑师参与社会进程的实践方式。

而公共空间的本质其实是关乎人的权利，关乎人的连接，它应该兼具包容与开放，承载包容丰富的城市生活和公共性特征，适用于宏大的庆典模式，更重要的是兼容日常生活以及对于不同人群的开放与尊重。公共性的议题必须在生态（自然的或社会的）可持续的语境中，以一种多元协同的机制形成。一方面是针对凋敝的、割裂的、破碎的城市空间进行整饬，另一方面则是对于城市生活、文化氛围和地域特征的培育，只有这样才能达成社会形态、城市文化和物理空间的关联性，孕育出多样的市民生活和城市文化。

Built for the Public: The Landscape Infrastructure and the Huangpu River Waterfront Revitalization
Zhang Bin

Now China has entered a new stage of urbanization characterized by renewal of existing buildings and the pursuit of high quality development. In the face of limited land resources, high urban density and sustainable development, we urgently need to change the single-function traditional infrastructure that hampers the development of urban public spaces. Instead, we should introduce a new development mode of landscape infrastructure that aims to integrate the development of urban open space systems in a synergistic and symbiotic manner. That can maximize the functions and benefits of the various elements of infrastructure, and to realize the integrated operation of the ecological, transport and outdoor slow traffic systems in this area. Landscape infrastructure act as a multifunctional medium and vehicle. They catalyze and synergize the integration and optimization of various ecological processes and municipal works by providing a symbiotic interface. Its support for open space and public life is reflected in the fact that we see the city as a vital hybrid rather than as a high-performance machine.

In recent years, Atelier Z+ has been engaged in a series of landscape infrastructure projects. These efforts seek to integrate citywide public space resources, and activate community and public life. They involve ecological restoration, beautifying flood control facilities, construction of greenways, waterfront public spaces and non-motorized traffic networks, and redevelopment of industrial heritage. Landscaping and the infrastructure often come together in such development practices.

Pudong East Bund Riverview Service Station: A catalyst to facilitate daily life

As Bernard Tschumi observes, a building is not only a neutral space to host events, but also a tool to shape new ways of life and to inspire social change. In a range of urban regeneration practices, infrastructure facilities can be a catalyst. They can turn from a functional necessity to tools that can influence the way urban logistics works. The concept of the catalyst can reshape architectural approaches under the assumption that built environment is seen as systematized infrastructure. The spillover effect of infrastructure as a spatial intervention is evidence of the enhanced social effects of architecture.

While developing a world-class waterfront on both sides of the Huangpu River, we focused on the small public facilities that are most closely connected to the public's daily life. The Pudong East Bund Riverview Service Station gives us the opportunity to consider how to intervene in a large swath of waterfront with small facilities, making them resting venues for daily use.

When these daily public facilities are adapted to the original landscape, topography and fabric of the waterfront, they can be sightseeing platforms and landscaping elements that beautify the waterfront landscape. We also have to explore how to match the underlying spirit of the public space with that of the city, how to transform the Stations into activators of civic life and how to build a spontaneous construction system within a short design and construction period. Such a system can help strike a balance between construction period, spatial quality and spatial experience. At the same time, the Pudong East Bund Riverview Service Stations have also provided venues to observe civic life. Through their unrestrained use of the 22 Service Stations, people can realize that leisure and entertainment do not necessarily mean consumption, and they can feel they are closely associated with the city. Such non-consumptionist, independent public spaces are a type of urban spaces that is quintessentially "Shanghainese". The Service Stations are perfect illustrations of the civic- and public-mindedness and social engagement of the locals, offering a glimpse of the pluralism, tolerance and openness that characterize modern civil society.

Dongchang Elevated Passage: Topographical active form for multiple stakeholders

With the rise of landscape infrastructure, the study of terrain topography has become a much-discussed topic of architecture. It describes a series of urban design practices that create artificial terrain to reorganize space, guide people and form a new design language. Terrain becomes an integrated spatial form that includes landscape, architecture and various structures. It focuses on the dissolution of the binary opposition between structures and sites, and reshapes the continuity of the built environment neglected by modernism. Infrastructure represents a series of technologies, agreements, rules and various social relations; it can become an "active form" in contemporary times to stimulate the chain reaction at a specific region, change the agreement about space, and then generate the derivative spatial behavior. With such practices, architecture can be more than a game of forms; it is no longer an incarnation of information and symbols but yield benefits for the society.

In the Huangpu River Waterfront revitalization campaign, there are 12 bridges of non-motorized traffic on the east bank. They span major spatial breakpoints, such as the creeks and ferry stations and the Dongchang Elevated Passage is one of them. With both running and cycling trails, the elevated passage passes a variety of sites and involves multiple stakeholders; it crosses several creeks, tunnels and complex underground facilities such as the source water pipelines. The project has met the complex demands and balanced the interests of multiple stakeholders, promoting the quality of urban public space and maximizing their public nature. It has also built a new space that can host s a variety of behaviors.

Redevelopment of the Six-Plant Joint Site: Waterfront space restoration and multidimensional integration in post-industrial context

Landscape infrastructure involves the combination of natural and artificial, green and gray, representing a new concept of nature and ecology that no longer distinguishes natural and artificial evolution and understands nature as a product of human society, politics and economy. The understanding of nature can be divided into four levels, that is, the first nature being the original nature; the second nature being the one after the transformation of human production and life; the third being the aesthetic representation of the first and second nature; and the fourth being the one that has undergone ecological restoration. Landscape infrastructure practices mostly deal with the second and the fourth nature.

The second nature is the result of the mutual adaptation between human and nature, and also an important part of human living environment. The fourth nature is basically a post-industrial topic. In the face of an abandoned natural environment after excessive human intervention, the self-repair of natural process or the ecological repair of artificial intervention can bring vitality to the damaged nature. In recent years, thanks to the advent of country parks, wetland conservation and restoration, river management and the redevelopment of green roads, and former industrial sites, new landscape infrastructure has popped up in China en masse. Hence we need to develop new understanding of nature and reconsider the meaning of nature. Only by doing so can we avoid understanding nature in our practice in a superficial and symbolic way. Only on this basis can we explore the ecological diversity of nature.

The redevelopment of the Six-Plant Joint Site pays attention to the repair and multi-dimensional integration of waterfront in the post-industrial context. In response to the pursuit for unexploited state of nature, we tried to save more industrial facilities not on the list of heritage protection, and repurpose them as open, cultural spaces and the civil service facilities In our landscape design. We have explored the site's topographic features and preserved them in an archeological way, and have treated docks, trestle bridges, boundary walls and hidden architectural wall foundations as equal elements and integrated them in our landscape design. Consequently, we have preserved the fabric and site experience of industrial heritage.

Through this repurposing the infrastructure, we seek to change their stereotyped impression as inaccessible and impersonal, and present a dynamic, public and integrated form. We made the facilities to be a perceptible new behavioral bearer and aesthetic object in urban space. In addition to pushing the boundary of architecture, landscape infrastructure can also be a new public media for the city, becoming a new way of Chinese architects to be engaged in social process.

Essentially, public space is about the rights of people, about human connection, and it should be both inclusive and open, carrying the richness of urban life and public character, suitable for grand modes of celebration, and more importantly, compatible with everyday life and openness and respect for different people. The issue of publicness must be shaped by a pluralistic and synergistic mechanism in an ecologically (natural or social) sustainable context. On the one hand, it is about reorganisation of the fragmented and fragmented urban space, and on the other hand, it is about the cultivation of urban life, cultural atmosphere and regional identity, in order to achieve a correlation between social form, urban culture and physical space, and to nurture a diverse civic life and urban culture.

日晖港步行桥
FOOTBRIDGE ON RIHUI RIVER

远眺步行桥
View from distance

项目地点：徐汇区日晖港与黄浦江交界处
设计单位：大舍建筑设计事务所
合作设计单位：大野博史+张准（结构）、上海市城市建设设计研究总院（深化）
建设单位：上海徐汇滨江开发投资建设有限公司
设计时间：2012-2015
建成时间：2016

Location: Intersection of Rihui River and Huangpu River, Xuhui District
Designer: Atelier Deshaus
Partner: Ohno Hirofumi (Structure), Shanghai Urban Construction Design & Research Institute (M&E)
Constructor: Shanghai Xuhui Waterfront Development Investment and Construction Co., Ltd.
Design Period: 2012 - 2015
Date of Completion: 2016

步行桥夜景
Footbridge in panoramic view

步行桥结构细部
Footbridge (details)

日晖港步行桥跨越约70米宽的黄浦江支流日晖港，连接黄浦区和徐汇区，两岸的场所特征也风格迥异。徐汇区一侧延续西岸公共开放空间的宽阔平坦，是广场般的公共尺度；而对岸黄浦区南园则是以园林般的小尺度呈现的公园，地形起伏复杂，道路蜿蜒。桥梁的置入需转换两种空间尺度并恰当地连接不同的标高和流线，而且，桥梁本身不能阻碍河流上的船只通行，也不能影响岸边的人行通道的高度。
步行桥采用带有推力平衡索的空间三叉钢箱拱梁结构，使得桥梁形态整体且富有力度，也强化了桥梁作为结构体对于其下方城市空间的覆盖和限定。通过桥梁分叉的宽度变化，转换两种空间尺度，并巧妙地连接不同的标高和流线。它的形式是功能考虑的结果，同时也形成合理的结构。抽象的结构设想呼应场地以及使用特点，最终成为人们漫步滨江的活动和行为的发生器。日晖港步行桥积极地塑造着场所，也展现着桥梁作为一种沟通城市空间的结构体的魅力。

The Rihui Creek Footbridge crosses the 70-meter-wide namesake tributary of the Huangpu River, spanning the waterfront spaces in Xuhui and Huangpu districts.
The designer of the bridge had to meet a series of challenges: The ground plots on the two ends of the bridge are at different heights, and the landscape designs are different, too. Meanwhile, the bridge should be high enough to accommodate the passing boats and pedestrian paths underneath. As a solution, the footbridge has adopted a trigeminal steel box arch girder structure pulled by thrust balance cables. Such a design makes the bridge both stable and visually interesting, while properly addressing its influence on the underlying space. Its changing width due to the branching acts as a transition between the two spatial scales, and tactically connects different heights and circulation. Its physical form is the result of functional constraints, but also reacts to the structure in a reasonable way.

步行桥
Footbridge

游客集散中心
Tourist Distribution Center

全景 Panoramic view

项目地点：徐汇区龙吴路1900号
设计单位：上海创盟国际建筑设计有限公司
合作设计单位：同济大学建筑设计研究院（集团）有限公司
建设单位：上海徐汇滨江开发投资建设有限公司
设计时间：2018.5-2019.10
建成时间：2020.12
总建筑面积：12110m² （地上4310m²）
用地面积：4934m²

Location: No. 1900, Longwu Road, Xuhui District
Designer: Archi-Union Architects
Partner: Tongji University Architectural Design Research Institute (Group) Co., Ltd.
Constructor: Shanghai Xuhui Waterfront Development Investment and Construction Co., Ltd.
Design Period: May 2018 - October 2019
Date of Completion: December 2020
Gross Floor Area: 12,110m² (above ground 4,310m²)
Site Area: 4,934m²

顶视图
Top view

游客集散中心是实施黄浦江45公里贯通中重要的公共服务设施，在三港线"港口渡口"原址新建一处服务设施，保留了原有的轮渡功能，同时整合为游客服务的配套设施以及为浦江游览预留的游船码头，是沿江区域面对多主体协同、多功能复合、多条件约束的创造性实践。

建筑设计秉持公共利益最大化的原则，在"三道"贯通的基础上，最大限度地提供了共享空间和观江视线，通过穿插的流线组织，在满足轮渡、游客服务、游船三部分功能独立管理的基础上，又能保持空间相互联系。

西北角鸟瞰
Bird's eye view from the northwest

The Tourist Distribution Center is located in the ecological leisure area of the 1.5 km section of the run-through project. The former ferry station has a gross floor area of 908m² and a site area of 4,934m². As an important function node of the run-through project, the ferry station is an important carrier to promote water transportation, sightseeing and enhance the waterside vitality. On the basis of the main navigation function of the original port ferry station, some new composite functions such as cruise ship sightseeing and tourist service, etc. are added. Combined with the project site conditions, the riverside circulation line is run through from south to north, and the water transportation berths are built through the overall consideration of the upstream and downstream shorelines.

According to the basic requirements of the three-line running through of the run-through project, combined with the building scheme design, the pedestrian traffic line is guided up to the second floor platform, and reaches the opposite side of the building after crossing the ferry passage. Thus, the pedestrian traffic line is separated from the non-motor vehicle traffic line. The continuous curved surface is used to solve the crossing problem of the non-motor vehicle traffic line and the pedestrian traffic line. In the architectural design, the traditional courtyard space is combined with the flowing ferry passage, forming a spatial form that combines landscape, publicity and functionality.

东侧鸟瞰
Bird's eye view from the east

内部楼梯
Interior stairs

建筑因外观被称为"带跑道的飞船码头",直接反映了不同功能在这一节点的高度融合。在兼顾成本控制和满足空间跨度的前提下,采用钢龙骨外包玻璃纤维增强混凝土的建造体系,通过钢结构的悬挑优势在二层沿江公共平台处形成一整排无柱空间。一体化的屋顶造型创造了轻薄的感受,进一步强调出水平的延展态势。模块化的预制建造大大缩短了现场的工作周期。

The prefabricated GRC material-glass fiber reinforced concrete is used as the surface material, which has the advantages of good durability, high tensile strength, better plasticity and convenient construction. The architectural appearance reflects the concise and fluent aesthetic language of modern buildings, with cutting-edge parametric design means creatively used to reflect the simplicity and fluency of buildings. The project is the first prefabricated public traffic building in Shanghai.

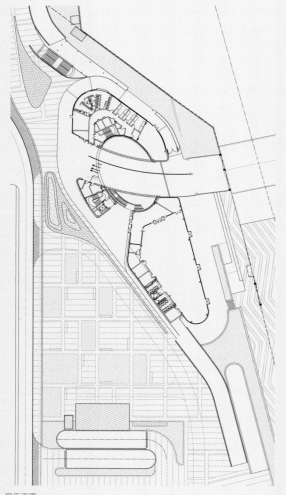

一层平面图
Plan: 1/F

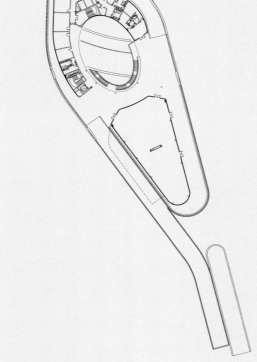

二层平面图
Plan: 2/F

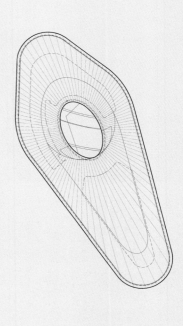

屋顶平面图
Roof plan

立体交通组织
Stereoscopic road network

龙华港桥
LONGHUA CREEK BRIDGE

西侧鸟瞰
Bird's eye view from the west

项目地点：徐汇区龙华港与黄浦江交汇处
设计单位：马达斯班建筑设计事务所
合作设计单位：上海市政工程设计研究总院（集团）有限公司
建设单位：上海徐汇滨江开发投资建设有限公司
设计时间：2009
建成时间：2010

Location: Confluence of Longhua Creek and Huangpu River
Designer: Shanghai Mulnicipal Engineering Design Insititute (Group) Co., Ltd., Madas.p.a.m
Partner: Shanghai Mulnicipal Engineering Design Insititute (Group) Co., Ltd.
Constructor: Shanghai Xuhui Waterfront Development Investment and Construction Co., Ltd.
Design Period: 2009
Date of Completion: 2010

下层桥面的景观视野
Bridge (details)

钢箱梁与桁架结构结合的全钢双层桥
All-steel double deck bridge composed of steel box girders and truss structures

作为龙华港汇入黄浦江的重要节点，龙华港桥不仅承载着市政交通功能，也是连接龙华港南北两岸城市商务区和滨江公共开放空间的重要纽带。桥梁采用钢箱梁与桁架结构组合的全钢双层桥造型，总长达345米。基于生物仿生学原理，主桥模拟生物脊柱（又名"龙之脊"），为两跨变截面连续钢结构，并采用全焊接板桁结合形式。结构立面的变化结合景观线形的要求设计，自中间向两侧按曲线过渡，使桥墩与桥梁形成有机的整体。

上层桥面为双向四车道及预留有轨电车道，近期兼做观景平台，下层桥面为非机动车及步行通道。人车分流的组织方式，有效提升通行效率，并降低桥梁的纵向起伏。桥身侧向规则几何形开孔，营造出桥体通透、轻盈的效果，更为桥上通行的人群，提供最佳的观景窗口。轻盈的桁架同时形成最佳的灯光载体，为夜间的桥梁披上五彩霞衣，与南岸耸立的海事塔形成纵横对比，完美呼应，构成黄浦江畔最亮丽的风景线。

Situated at the confluence of the Longhua Creek and the Huangpu River, the Longhua Creek Bridge not only sustains traffic but also serves as a landmark of the Xuhui Riverfront Waterfront. With a total length of 345 meters, the fully steel double-deck bridge uses both steel box girders and truss structures.

The bridge structure emulates the spine of creatures according to the principle of biomimicry. With a two-span continuous steel structure with variable cross-sections, the main body of the bridge adopts a fully welded plate-truss combination. The two-way four-lane upper deck has tram lanes reserved, and serves as a viewing platform recently. The lower deck serves as a passage for non-motor vehicles and pedestrians. The diversion between people and vehicles effectively improve traffic efficiency while reducing the bridge's longitudinal fluctuation of the bridge.

The changes in design of structural facade follows the requirements of landscape linearity, as the transition from the middle to duo curved sides holistically combines the pier with the beam into an organic whole. Regular geometric openings on the side make the bridge body look transparent and light, offering the best view for passerbys on the bridge. The light trusses act as the optimal light carrier, covering the bridges at night with colorful lights as perfect resonance to the Maritime Tower on the southern bank, forming an eye-catching scenery along the Huangpu River.

非机动车及步行通道
Non-motor vehicle and footway

Lighting effect

海事塔
Maritime Tower

项目地点：徐汇区龙腾大道，近龙华港
设计单位：马达斯班建筑设计事务所
建设单位：上海徐汇滨江开发投资建设有限公司
设计时间：2008
建成时间：2009

Location: Near Longhua Creek, Longteng Avenue, Xuhui District
Designer: Mada s.p.a.m
Constructor: Shanghai Xuhui Waterfront Development Investment and Construction Co., Ltd.
Design Period: 2008
Date of Completion: 2009

海事塔位于龙华港桥南侧，建于20世纪80年代，塔高约41.2米，为海事局瞭望、监控黄浦江船只安全航行的功能塔，是徐汇滨江公共开放空间的制高点。

改造后保留海事塔瞭望功能，以上海市市花"白玉兰"、中国传统细口瓷器和稻穗为概念，力求形态动感有机，并体现出中国传统、上海特征以及自然状态。灯塔用一张充满弹性的金属网"外壳"包裹塔身，在不同的高度有不同的半径，也拥有不同的张力。在不同的张力作用下，表皮表达出不同的肌理特征，远看仿若一朵含苞待放的白玉兰。如今，它不仅是江上航标，也是徐汇滨江的景观标志。

The Maritime Tower, located at the south of the Longhua Creek Bridge, was built in the 1980s. The Maritime Safety Administration then used it to observe river traffic of the Huangpu. Rising to 41.2 meters, it is the highest structure at the Xuhui Riverfront.

The renovation has preserved the artifact's observation function. Fusing the shapes of "magnolia denudata"—Shanghai's city flower—traditional Chinese narrow neck porcelain bottle and rice ears, the inspired design features quintessential Chinese and Shanghainese elements and embrace the charm of mother nature. The tower is wrapped with an elastic metal mesh. The mesh is of different radius at different heights, each suggesting a specific measure of resistance to reflect a nuanced texture. When looked from a distance, the whole tower looks like a magnolia denudate bud that is about to bloom.

When they are turned on, more than 3,000 lights installed on the facade puts on an engaging light show, making it a chic landmark of the riverfront space along with the Longhua Creek Bridge.

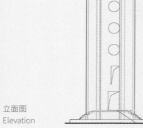
Before renovation

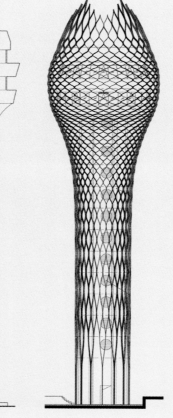
立面图
Elevation

全景鸟瞰
Panoramic view

洋泾港步行桥
YANGJING CREEK PEDESTRIAN BRIDGE

整体鸟瞰
Panoramic view

项目地点：浦东新区洋泾港与黄浦江交汇处
设计单位：刘宇扬建筑设计顾问（上海）有限公司
合作设计单位：上海市政工程设计研究总院（集团）有限公司（水工及结构）
建设单位：上海东岸投资（集团）有限公司
设计时间：2016-2017
建成时间：2018

Location: Confluence of Yangjing Creek and Huangpu River, Pudong New Area
Designer: Atelier Liu Yuyang Architects
Partner: Shanghai Municipal Engineering Design Institute (Group) Co., Ltd. (Hydraulic Engineering & Structure)
Constructor: Shanghai East Bund Investment (Group) Co., Ltd.
Design Period: 2016-2017
Date of Completion: 2018

洋泾港步行桥跨洋泾港、指陆家嘴，清水一道为泓，结构如弓，隐喻浦江东岸第一桥蓄势待发之张力，是为"慧泓桥"。设计将结构、功能及造型相结合，利用高差分流骑行、跑步及漫步动线，且各自拥有良好的观景视野。主桥部分采用异型桁架结构，民生码头段引桥均采用钢箱梁，骑行道引桥总长85米，宽4.5米；慢行道引桥长37米，宽6.5米。桥体以优雅的曲线回应周边景观，将视线引导至其东北侧的杨浦大桥，以及西南侧的陆家嘴中心建筑群。为了减少对南侧民生码头遗留厂房建筑的影响，桥体选址在更靠近黄浦江的一侧，通过桥型的调整避让沿江侧的防汛墙。

利用高差隔开骑行/漫步道
The cycling path and footway separated by height difference

Located at the eastern tip of the East Bund Mingshen Wharf, the Yangjing Pedestrian Bridge crosses over the Yangjing Canal where it flows into the Huangpu River, connecting the Mingshen Wharf with the Yangjing Park which is visually connected to the majestic long-span Yangpu Bridge. Also known as Huihong Bridge, its futuristic and graceful form represents a comet, or "Hui", streaking across the starry sky, bridging a clear and deep water, or "Hong", and pointing towards Lujiazui CBD. The bow-like structure metaphorically foretells the launching of the first slow-traffic bridge in the East Bund. Integrated with structure, function and form, the design utilizes the height difference of the structure to divert the various usage flows: cycling, jogging and walking, ensuring safe passages and good view.
The main span crossing the river is an abnormity steel-truss structure, while the approach bridge from the Minsheng Wharf utilizes the steel box girder, with the cycling path of 85 meters long and 4.5 meters wide. The spiral jogging and walking approach is 37 meters long and 6.5 meters wide. With elegant curves, the bridge responds to the surrounding landscape, guiding the sight to the Yangpu Bridge on the northeast and the Lujiazui CBD complex on the southwest. To minimize the impact on the industrial buildings on the Minsheng Wharf site, the bridge is intentionally placed closer to the Huangpu River instead. Through the adjustment of the bridge form, the flood control wall along the river stays intact.
The truss structure spans 55 meters and the height is 4 meters. In order to adapt to different grade requirements of transportation, the convex and concave curves have been well-adjusted. Cycling path is placed at the gentle concave part with less-than-4% slope to provide a comfortable and safe riding

骑行和漫步道各自拥有良好的观景视野
The cycling path and footway of noninterference

起伏变化的栏杆
Distinctive handrail

桁架结构跨度55米，高4米。骑行道布置在平缓的桁架下弦，坡度控制在4%以内；慢行道布置在上弦，跑步及步行能适应10%的坡度，局部坡度较大处设置缓步台阶，桥面局部放宽形成驻足休憩的空间。桥面铺装采用统一的环氧树脂材料，深灰色为骑行道，红色为跑步道。桥面放宽处，使用深灰色划分出漫步区域。

栏杆部分采用氟碳喷涂圆钢管，拼接出起伏变化的三角形变截面连续体量。如鳞鳞波光般的不锈钢绳网，通过绕绳法的固定方式与杆件合二为一。断面设计在栏杆一侧为过桥管线预留空间，保证桥身桁架空间的完整性。夜间的灯光照明设计，将点光源和线性光源结合，照亮栏杆网面及结构桁架。整体桥身喷涂铂金色，造型轻盈，如彗星划过天际，勾勒出浦江东岸的崭新气象。

experience, flanked by a protective interface made of vertical bars towards the river side. The slow traffic path is arranged in the convex part with less-than-10% slope which is suitable for walking and jogging. Steps are placed in area where a bit steeply sloped, the width of the bridge is widened accordingly to form a belvedere. Epoxy paving goes through the whole bridge with the dark grey for cycling and red for the jogging track. Dark grey is again used at the belvedere to represent walking path.

The railing part adopts fluorocarbon-coated round steel tube, joining to shape an evolving triangular-cross-section continuous volume, while glittering stainless steel rope nets are reeved. In design, space is reserved for the bridge pipeline on the railing side to ensure the integrity of the bridge truss space. In the night, the combination of point light source and linear light source light up the railings and structural trusses. The whole bridge is painted with platinum silver, as a comet crossing the sky, bringing out the new profile of East Bund of Huangpu River.

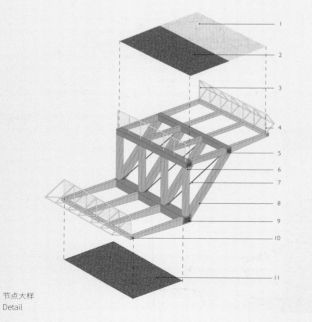

栏杆细部
Handrail (details)

1 镀锌钢管扶手，喷铂金色 / Galvanized steel pipe handrail in platinum
2 金属丝网 / Meshed wire
3 LED 灯带 / LED strip
4 10 厚钢板边 / Steel plate edge (10mm)
5 200mm×100mm 边梁 / Spandrel beam (200mm×100mm)
6 30 厚塑胶地面 / Plastic floor (30mm)
7 20 厚水泥砂浆 / Cement mortar (20mm)
8 100 厚桥面结构 / Bridge structure (100mm)
9 变截面次梁 / Variable cross section beam

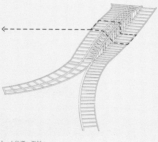

节点大样
Detail

1 人行道：石材
2 跑步道：黑色塑胶地面
3 丝网护栏
4 护栏立杆：钢杆形成三角桁架 @800mm
5 箱形主梁 300mm×250mm
6 箱形主梁 700mm×300mm
7 拉钢结构 φ70~100mm
8 桁架斜杆 200mm×400mm
9 结构箱形主梁 400mm×700mm
10 次梁 100mm×200mm
11 骑行道：骑行专用黑色塑胶地面

1 Pavement: stone
2 Jogging trail: black plastic floor
3 Meshed wire fence
4 Guardrail rod: triangle truss formed by steel pole @800mm
5 Box beam 300mm×250mm
6 Box beam 700mm×300mm
7 Stayed structure φ70 mm~100mm
8 Inclined chord truss 200 mm×400mm
9 Box girder 400 mm×700mm
10 Beam 100 mm×200mm
11 Cycling track: black plastic floor

桥体与场地景观相融
Bridge and landscape in good harmony

民生轮渡站
MINSHENG FERRY STATION

鸟瞰图
Panoramic view

轮渡站与贯通桥相连
Ferry station and the bridge

项目地点：浦东新区民生路1号
设计单位：刘宇扬建筑设计顾问（上海）有限公司
合作设计单位：上海中交水运设计研究有限公司（水工结构）、上海东方建筑设计研究院有限公司（施工图）
建设单位：上海东岸投资（集团）有限公司
设计时间：2016-2018
建成时间：2019

Location: No. 1 Minsheng Road, Pudong New Area
Designer: Atelier Liu Yuyang Architects
Partner: Shanghai China Communications Water Transportation Design& Research Co., Ltd. (LDI & Structure), Shanghai Oriental architectural design and research institute Co., Ltd. (Construction)
Constructor: Shanghai East Bund Investment (Group) Co., Ltd
Design Period: 2016-2018
Date of Completion: 2019

轮渡站登船通道
Boarding channel of the ferry station

民生轮渡站位于民生路尽端，北临黄浦江，西侧连接新华绿地，东侧通过慧民桥接入民生艺术码头。作为联系浦江两岸的水上交通基础设施，以及东西两侧滨江贯通的重要景观节点，整体设计将建筑融入周边景观，并通过上下层功能分离的设计策略，将出入轮渡站的人流同滨江三条慢行道合理分流。建筑共分两层，高度9米，建筑面积为645平方米。一层主要为轮渡站候船厅及站务用房，主要流线为南北向，二层为配套设施，可服务周边贯通道，提供便民设施。建筑高度9米，上部为覆盖金属网的景观构筑物，结合夜景灯光变化成为独特地标建筑。

The Minsheng Ferry Station is located at the end of Minsheng Road and to the east of the Xinhua Green Space. It is linked to the Minsheng Wharf waterfront by the Huimin Bridge. As the meeting point of two sections of the Pudong section of Huangpu waterfront, it combines architecture with landscape while separating ferry passengers from the slow traffic flow on the watefront. Nine meters high and offering a gross site area of 645 square meters, the building has two floors: The first serves as the ferry terminal, while the upper floor connects the open spaces on its flanks. The building is covered by stainless mesh net on the top which establishes itself as a spectacular sightwith its iridescent light at night.

圆形天窗下的一层空间
Skylight

一层售票厅
Ticket hall on 1/F

设有云亭的二层平台
Platform under cloud-shaped pavilion: 2/F

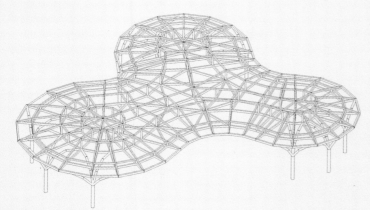

钢构架轴测图
Axonometric drawing of steel frame

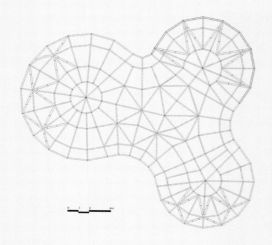

钢构架平面图
Steel frame plan

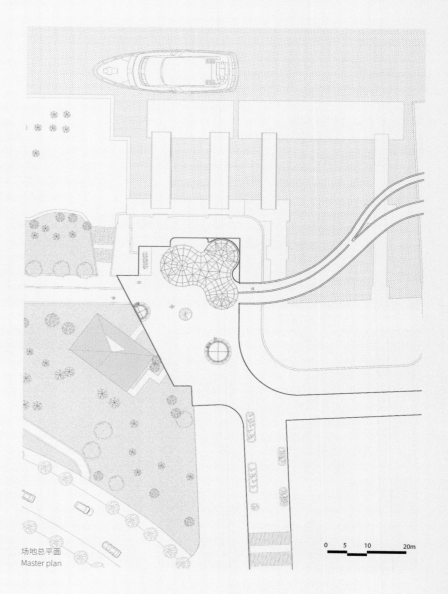

场地总平面
Master plan

平台江景
View to the Huangpu River from the platform

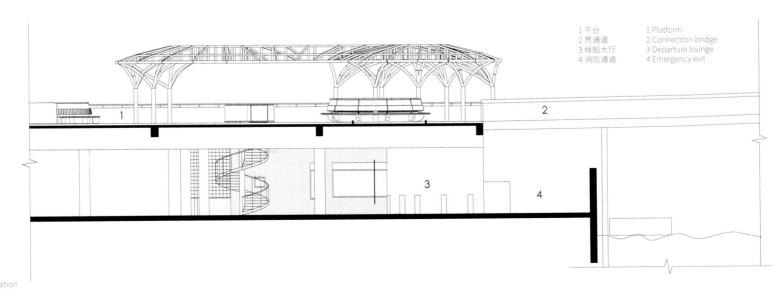

1 平台	1 Platform
2 贯通道	2 Connection bridge
3 候船大厅	3 Departure lounge
4 消防通道	4 Emergency exit

北立面图
North elevation

东昌栈桥
DONGCHANG ELEVATED PASSAGE

全景鸟瞰
Panoramic view

螺旋环道
Spiral loop

Space below the bridge

剖面图
Section

项目地点：浦东新区东昌轮渡站前广场及周边地区
设计单位：致正建筑工作室
合作设计单位：上海市政工程设计研究总院（集团）有限公司
建设单位：上海东岸投资（集团）有限公司
设计时间：2017.3-2018.6
建成时间：2018.6-2019.4

Location: Front Square and Surrounding Area of Dongchang Ferry Terminal, Pudong New Area
Designer: Atelier Z+
Partner: Shanghai Municipal Engineering Design Institute (Group) Co., Ltd.
Constructor: Shanghai East Bund Investment (Group) Co., Ltd.
Design Period: March 2017 - June 2018
Date of Completion: June 2018 - April 2019

东昌栈桥面临下列城市空间矛盾：首先，东昌路轮渡站大量的通勤人流、旅游进出站客流与贯通后的人流在站前广场的冲突；其次，轮渡站西北侧游艇俱乐部的封闭场地使滨江公共空间无法贯通；再者，轮渡站、游艇俱乐部及其两端水上餐厅连续密集的建筑体量造成近400米滨江岸线上无江景可看。

这一兼容跑步和骑行的高架慢行景观步道，跨越交通设施、商业空间和公共绿地，涉及周边多家土地所有方，并面临横穿轮渡广场下方的人民路过江隧道和纵贯场地北段的源水管道等复杂的地下条件。栈桥设计用前期漫长的多方案比较与优化协调多方利益与诉求，最后形成的桥形设计与周边现有建筑、场地条件相互作用，蜿蜒逶迤，微妙介入的状态。

整个栈桥由南往北分为五段。南引桥段，跑步道和骑行道分叉，以减小桥的体量，婉转穿行在树木密集的滨江绿带中；在接入轮渡站前广场之前，在建筑夹缝中设置面江、凸出、放大的观景平台，结合花坛、凉棚和座椅供人驻留。轮渡站前广场段，整体向东昌路、面陆家嘴中心区方向凸出，跑步道和骑行道间设行人休憩设施，成为栈桥中最宽的一段，兼为下部站前广场遮阴避雨，还在双道间为广场留出采光天井。游艇俱乐部段，跨越俱乐部入口及停车前场，整体相对平直，在中段渐变加宽，布置双道间的行人休憩空间，并设置大台阶与轮渡站前广场衔接。水上餐厅绿地平台段，利用进深较大的场地优势。回应滨江贯通规划方案中立体化复合诉求，在大型架空绿化平台下设置公共停车场；保证双道在平台背江路侧通过，并设计一条向面江侧大幅度迂回的跑步道及漫步道，形成出挑9米的C形悬浮环道。虽然有水上趸船餐厅临时阻隔，但强化了面江的态势；平台中部的景观设施顺应两侧直道与弯道的特征，以渐变的曲率塑造流线形特征。北引桥段有最开阔的城市视野，顺应场地面宽较小、进深足够的特征，设置360°大螺旋环道为市民提供盘旋其上步移景异的城市体验。环道外侧疏竹掩映间贴建一条引自平台段漫步弯道，直接连接北侧临江步道的台阶步道，与环道一起成为观赏外滩浦江落日的绝佳视点。

为避让过江隧道及地下源水管，栈桥设计选择占地较小的Y形墩柱作为支撑结构，以跨度较大的钢箱连续梁桥体作为主体结构。经过缜密计算，在隧道复线夹缝中唯一可行处精确落桩，利用三组紧密贴合、不同方向、高低错落的Y形墩柱形成树状分叉支撑结构，在轮渡站前广场上实现桥面跨越。

桥栏板采用间隔密排的铝方通，既保证安全感又保持通透性。方通正面采用与桥体相同的银灰涂装，侧面为浅绿色，在不同天候条件下保持明亮温暖的光感。栈桥上容纳了不同尺度的绿化景观空间及座椅、凉棚等街道家具设施，为骑行、跑步、漫步的市民提供休憩的可能。建成后的东昌栈桥形式新颖，融于城市，与现有建筑与景观和谐共生，创造丰富多变的城市景观和自由包容的开放空间。

The urban space contradictions solved by Dongchang Elevated Passage consist has eased traffic conflict between significant numbers of commuters, heavy streams of tourists in Dongchang Ferry Terminal and the crowds after the Public Space Connection is opened to the public. The closed area of a yacht club on the northwest side of the Ferry Terminal has cut off the riverside public space; the rolling buildings of Ferry Terminal, Yacht Club and floating restaurants on wharf boats have become a solid wall, citizens can enjoy the river view along this near 400-meter stretch.

The whole passage can be divided into five sections from south to north. The south approach bridge section bifurcates running and cycling paths, reducing the size of passage and winds through the dense greenbelt along the riverside. Before accessing to the front square section of the Ferry Terminal, an enlarged platform is set up protruding toward the river, which offers a good river view for the passing people. The section over the square of the Ferry Terminal protrudes towards Dongchang Road and Lujiazui Central District. It also provides a large sun and rain shelter for under-bridge square in front of the Terminal, and sets aside a light well between running and cycling paths. The section over the Yacht Club spans the entrance and the parking porch of the club. In the middle, it is gradually enlarged in width to provide passersby with a resting area between two trails. Moreover, a big step is set to connect the front square of the Ferry Terminal. Further north is the Greenbelt Platform section facing a floating restaurant. By taking advantage of the greater depth of the site, a large overhead greenbelt platform with a public parking lot underneath is built. At the same time, a running and walking path is designed to detour to the riverside, and then a 9-meter overhanging C-shaped hollow suspension loop is formed on the platform. The landscape facilities in the middle of the platform conforms to the dual features of straight and curved paths on both sides, and uses the gradual curvature as design language to create streamlined features. The north approach bridge section boasts the widest city view. A large 360 degree spiral ring path is set up to invite citizens to a fantastic urban experience in which the scene is always changing with walking at the outskirts of the ring path, there is also a step footpath directly connected with the river-facing walkway to the north side, which is drawn from the walking path of the platform section. Together with the ring path, it becomes an excellent view point to watch the Huangpu River sunset facing the historic Bund.

In order to avoid the river-crossing tunnel and underground water source pipes, this elevated passage is designed with Y-shaped composite piers with less land occupation as supporting structure, and the steel-box continuous girder bridge with a large span as its main structure. At the same time, the bridge deck crossing is achieved at the Ferry Terminal front square.

The trestle slabs are arranged at tight intervals with Aluminum Rectangular Tubes. This design not only ensures a sense of security but also maintains its permeability. The front of Aluminum Rectangular Tubes adopts the same silver-gray coating as the passage body and its side is painted light green. Such coating gives it a bright warm light sensation under different weather conditions. The elevated passage accommodates green landscape of different scales and street seats, awning; allowing cyclists, runners and walkers to rest. The Dongchang Elevated Passage achieves a perfect integration into the cityscape with its novel form. It harmoniously coexists with existing buildings and landscape, creating a rich, varied urban landscape as well as a relaxing and inclusive open space.

白莲泾 M2 游船码头
BAILIANJING M2 TOURIST TERMINAL

顶视图
Top view

连续的筒拱
Twin arch bridges

项目地点：浦东新区世博大道970号
设计单位：同济大学建筑设计研究院（集团）有限公司原作设计工作室
合作设计单位：华建集团华东建筑设计研究院有限公司
建设单位：上海东岸投资（集团）有限公司
设计时间：2016.8-2017.8
建成时间：2018.12
建筑面积：7320m²

Location: No.970 Shibo Avenue, Pudong New Area
Designer: Original Design Studio (A subsidiary of Tongji Architectural Design (Group) Co., Ltd.)
Partner: East China Architectural Design & Research Institute, Co., Ltd., Arcplus Group PLC
Constructor: Shanghai East Bund Investment (Group) Co., Ltd.
Design Period: August 2016 - August 2017
Date of Completion: December 2018
Gross Floor Area: 7,320m²

码头，是陆地与水面之间的界面，人与货物在这里或是开启，或是终结一段旅程，这使得码头既是空间上的又是时间上的转换口。上海是个飞速发展的城市，白莲泾M2游船码头的出现成为一个锚固点，架起水面与陆地之间的桥梁，嵌入城市的历史与未来。设计始于为了织补滨水公共空间的一个断点。

M2游船码头位于曾经的上海世博会核心地带，基地北临黄浦江，东靠白莲泾公园，西侧为亩中山水园，南部被城市主干道世博大道所切割。设计有两个关键性出发点：其一是在东西方向上衔接两侧的城市公园，与城市滨江景观体系编织在一起，承担滨水公共空间的休憩功能；其二是在南北方向上打开面江的景观视线，从城市腹地快速抵达江边码头，共享亲水氛围。由此，设计问题转化为如何在尽可能压低层高的前提下，既满足候船大厅作为公共建筑的净高要求，又能满足上方景观廊道植物的覆土需求。

建筑近景
Details

Terminal, a place where people or cargo start or end a journey, is an interface between water and land. Hence force, it is both a spatial and chronological transmission. M2 Tourist Terminal at Bailianjing, Shanghai, emerges as an anchoring point for the fast-developing metropolis, which connects water and land, history and future of the city. Its design is meant to weave a breaking point in the waterfront public space.

M2 Tourist Terminal is located near the former site of Expo 2010, Shanghai. It faces the Huangpu River to the north, Mu Zhongshanshui Garden to the west, and Shi bo Avenue to the south. The project belongs to the Program for Improvement of Huangpu River Public Space. The design starts from two key assumptions: first, it means to weave itself into the landscape system of Huangpu riverfront by connecting the existing parks to the west and to the east; second, it intends to open a landscape corridor from the city directly to the waterfront.

The building is an earth-sheltered structure with a continuing barrel vaulting system. The vault form provides enough height for the waiting room below while the gap between the vaults saves space for flower beds above. In this way, it minimizes the entire height of the structure to open a view to the river. Meanwhile, three courtyards are inserted into the homogeneous structural grids so as to introduce the natural light and to create interaction between above and below.

设计采用连拱的方式。对于下方的候船空间而言，拱券从起拱点到拱顶的高度变化最大限度地为室内空间争取了净高，缓解高度限制下大空间可能造成的压抑感，形成更为丰富的空间效果；而对于上方的景观体系来说，能够利用拱与拱之间下凹的部分进行覆土，获得更多的覆土高度。

M2游船码头候船大厅以混凝土壳屋盖＋钢索框架-屈曲约束为结构体系，结构设计的重点从单个拱转向拱的体系，从降低拱板厚度的角度，使用与之相匹配的拉杆、梭柱和支撑，形成轻盈的系统。

拱的使用不仅在结构层面上实现了城市策略，还表达了空间特性，通过钢索框架-屈曲约束的使用，将承重墙抽象为点式支撑，几乎去掉了所有分隔墙造成的阻碍，创造了空间的连续性，与候船大厅的大运量交通功能相吻合。当人们进入候船大厅头顶上方连续的筒拱，以及混凝土薄壳上清晰的木模印痕标定出纵深的方向，引导人们前往码头。而长达300多米的筒拱体量却是横向的，使得人们往来穿梭于候船大厅的不同部分。

同样，拱上减少侧推力的拉杆是纵向的，保证抗震和整体结构稳定性的支撑是横向的。它们都在反复诉说着码头在东西方向上对滨江景观带的接续，以及在南北方向上由城市到江面的转换。只有细长的梭柱是没有方向的，它们和拉杆、支撑一道为白漆裹覆，在混凝土粗粝木模的对照下，举重若轻地将拱的构造形式、建筑体量与结构方式标定出一个三维的矩阵，在这个矩阵里是柯·罗笔下一片片"透明的"浅空间，暗示了经由身体运动去对建筑进行体验。

The vaulting system is composed of cast-in-place concrete vaults, steel-cable framework and buckling-restrained brace (BRB), which is the contemporary transformation of sheer shell structure. The vault is 200mm-thick shell. Its upturned beams formulate a grid system on the roof top, which works as flower beds, flooring and water drainage. The steel-cable is welded to the capital and balances the side thrust of the vaults, which allows the 180mm-diameter spear pillar bear merely the vertical axial force. BRB is enforced to the capital as well so as to fulfill the anti-seismic requirements. It thus builds up an all-sided structural system.

The employment of arch not only achieves urban strategy in a structural way, but also articulates its spatial character. The structural system eliminates all kinds of barriers caused by the supporting walls in the traditional vaulting. It makes continuous space possible and complies with the mass transit function of the terminal. Once entering the waiting room, one is exposed to a world composed of arch, steel-cable framework and buckling-restrained brace. The scars in the surface of the arch imply the longitudinal movement while the 300m-long volume hints the transversal flow.

Meanwhile, the steel-cable framework enforces the transversal flow and the buckling-restrained brace articulates a longitudinal movement. Only the spear pillars wrapped in white stucco has no explicit directions, which form an abstract grid. Together, they constitute a layered shallow space identified by Colin Rowe, which implies the body movement along the building.

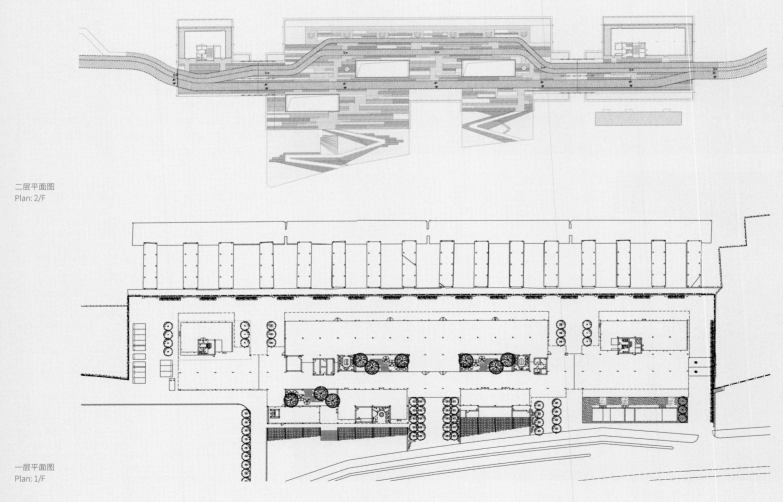

二层平面图
Plan: 2/F

一层平面图
Plan: 1/F

整体鸟瞰
Panoramic view

倪家浜桥
Nijiabang Bridge

错落的双桥
Staggered twin bridges

项目地点：浦东新区世博后滩公园南侧
设计单位：上海高目建筑设计咨询有限公司
合作设计单位：上海市政工程设计研究总院（集团）有限公司、华建集团华东建筑设计研究院有限公司
建设单位：上海东岸投资（集团）有限公司
建成时间：2017.12

Location: South side of Expo Beach Park, Pudong New Area
Designer: Atelier GOM
Partners: Shanghai Municipal Engineering Design Institute (Group) Co., Ltd., East China Architectural Design Institute Co., Ltd., Arcplus Gtroup PLC
Constructor: Shanghai East Bund Investment (Group) Co., Ltd.
Date of Completion: December 2017

两桥之间眼睛般的空隙
An eye-like gap in between

人行拱桥
Pedestrians arch bridge

Suspension bridge for cycling
自行车吊桥

倪家浜是一个宽约25米、废弃的"盲肠"河道，南北两侧没有任何建设现状，也未纳入世博会场地的规划，因此设计采用"原型设计"的切入方法。即跟场地无甚关系，只与功能构成参数有关，并且可以用在很多地方的设计。

功能构成参数主要由三股交通构成，人行、自行车、跑步，对坡度和宽度的要求各不相同，在其他桥梁里有三线合一也有两线合一的做法。通过分析坡度和速度参数，以及三种行为的停驻愿望，将本该是一座的桥梁拆分为两座——将自行车和跑步两股合一，采用平缓的吊桥形式通过；而单独设置漫步道，用坡度略大的拱桥形式通过和停驻，一拱一吊，平添趣味。除了在竖向设计中将拱桥和吊桥分开以外，在平面上也将两座桥体在中部分开，并分别采用直线倒角和弧线的不同方式。这一处理减少了桥梁中部的荷载，从平面上看，像一只望向天空的眼睛。

The Nijiabang creek is a defunct, dead-ended waterway about 25 meters wide. There are no buildings under construction or to be constructed on either side of it, leaving only the residual sites from the World Expo. The approach of "prototype design" is adopted, which has nothing to do with the sitebut relates to the functional parameters.
The functional parameters are mainly three traffic streams, namely, pedestrian, bicycle and running, which have different requirements on slope and width. In the design of other bridges, the methods of three lines in one and two lines in one are adopted. After analyzing the parameters of slope and speed, as well as the desires of stop of the above three behaviors, we divide the bridge which has been supposed to be one bridge into two, with one of them combining bicycle and running functions into one, and the form of gentle suspension bridge adopted; However, the footpath is set up separately, with an arch bridge having a slightly larger slope adopted for passing through and stop. Thus, the arrangement consisting of a suspension bridge and an arch bridge adds to the fun. In addition to separating the arch bridge and suspension bridge in the vertical design, the two bridges are also separated in the middle on the plane, with different ways of straight chamfering and arc adopted respectively, which reduces the load in the middle.

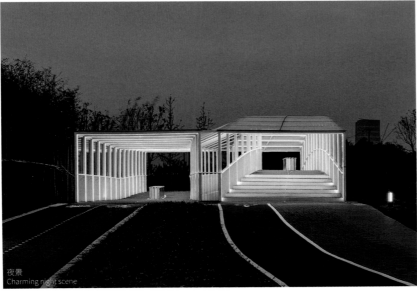
夜景
Charming night scene

设计·师说
MASTERS' WORDS

舒适人居，趣味城市
吴 蔚

世界上很多有活力的大城市都是滨水城市，河流湖泊在交通、市政、景观、生态等方面服务着城市，也形成众多城市滨水空间。随着历史的发展，水和城市的关系也在变化，人们对水的感受和认知也在改变，而城市功能的变迁为城市更新创造了条件，也提出了挑战。

特别是在集装箱运输盛行后，许多欧洲城市，例如哥本哈根、阿姆斯特丹、伦敦、汉堡等都经历了船运业、造船业的变迁，在城区"突然"出现"空闲"的空间，而且多为滨水地区。给了城市和市民一个重塑滨水空间、重构产业结构的机会。这同时也是一个难题，因为投资的逐利本性在进一步升级，如何塑造一个人性化的、宜居的城市，是摆在各地政府决策部门和规划师、建筑师面前的难题，其实也是一个价值取向的问题。如何打造一个高品质的空间，是在考验管理部门、投资者、设计师的眼光和远见。

欧洲的滨水空间系统性改造始于20世纪70年代，相比那些城市，上海对滨水空间的思考、开发和再利用稍晚，但也为上海提供了很多有价值的范例和经验。现代的都市人群还愿意在密不透风的钢筋混凝土森林中行走吗？八车道的大街还是城市中心需要的吗？白领职员还愿意在低矮的没有自然采光、通风的格子间里工作吗？市民还满足于坐在弄堂口休闲吗？这些问题的回答其实就是我们的设计引领。

上海的城市底蕴、历史文化、市民素质决定了她关注的是高质量的发展、市民的获得感，所以从政府决策者到规划管理部门、规划师、建筑师、业主单位很容易接纳好的事物、先进的经验，并结合上海的实际情况。在过去的十几年中，我们沿着黄浦江设计实施了多个项目，在不断了解这座城市的过程中，也在试图将自己以往在滨水城市中的经验与上海的现实结合起来。

织补、重构城市肌理

上海沿黄浦江两岸，特别是浦西有着自己既有的城市肌理，这些肌理是城市空间的骨架和底图，也是构建城市记忆的重要载体，而将这个肌理在城市更新中织补和重构对我们而言是一个义不容辞的责任。

在外滩SOHO项目上，分析了周边的城市肌理，形成六座建筑组合的方案，使得建筑的体量和周边建筑包括外滩风貌保护区的建筑形成一个恰当的关系，而建筑之间进一步延续了从老城厢往外滩的空间。在南外滩滨江岸线公共空间方案设计时，我们对垂直于外马路，并由于黄浦江弯曲而形成的放射状巷子进行发掘和梳理，在城市空间设计上提出重构原有巷子空间的概念，恢复历史城市肌理。

重塑城市空间、升级城市功能

黄浦江两岸的贯通工程对上海的城市发展、城市功能、市民生活产生了深远的影响。沿外马路从董家渡至南浦大桥段的南外滩滨江岸线将封闭的、主要服务城市市政的沿江空间释放出来，对大众开放，这也为后面腹地的建筑、城市空间创造了新的条件。这个区域不应再是封闭式的空间，而是一个集办公、休闲、文化、商业于一体的新的城市空间。而沿江岸线休闲空间的标高为旁边的建筑以及公共空间既提供了一个有利条件，同时也出了一道题。新建建筑的首层和二层、架高平台，以及与滨江岸线公共空间的连接都会为城市功能的多样性和复合性提供新的契机，并可以为公众提供更多与水的视线联系，这也是我们在这个区域的城市设计研究或竞赛中一直想表达的。

当代城市形象的"绽放"

正如吴志强院士说的，上海现在的城市形象是过去各个历史时期"绽放"的集成，城市中可以看到各个历史时期的建筑。如何在滨江城市更新过程中延续我们目前所处时代的"绽放"也是一个需要建筑师仔细思考的问题。

外滩SOHO之所以从一开始的"议论纷纷"到建成后的"欣然接受"就是恰当地处理了历史和现代之间的关系。从建筑立面的比例划分、横竖线条的应用、石材和玻璃的组织、细部处理都是在仔细拿捏，细心处理当代建筑与整体外滩风貌的关系。位于民生码头旁边的陆家嘴滨江中心采用简洁的建筑语汇，通过贯穿所有建筑的、有序的外露框架隐喻这个地块在历史上的码头功能。

在沿外马路的南外滩地区有不少优秀的历史建筑，包括一些工业、仓储建筑。我们建议将具有价值的历史建筑保护下来，赋予其新的功能，同时进一步通过仔细研究其他现有建筑，发掘有价值的建筑。还有，就是在新建筑设计时充分考虑所处时代的特征。这也是我提出"保护真古董，认定新古董，杜绝假古董"的用意所在。

规划学者约翰·丰德史密斯在1988年提出了一个口号：把城市做得有趣，这句话在汉堡港口新城规划中经常被人提及。这个有趣指的并不是建一座迪士尼乐园，而是将城市的功能进一步混合，增加公共空间，并结合文化艺术休闲活动。上海的滨水空间已初具形象，特别是浦江两岸的贯通工程为整个城市注入新的活力，市民的获得感进一步提高。城市让生活更美好，而城市也会越来越有趣、有活力，人们的感受也越来越舒适，上海的实践不仅实现了趣味城市，同时还达到人居，这两句如果翻译成中文便是：舒适人居，趣味城市。

Making cities fun, giving people comfort
Wu Wei

Many of the world's most dynamic cities are waterfront cities, where rivers and lakes provide favorable conditions for transport, municipal engineering, landscape and ecology, and form many urban waterfronts. The relationship between the water and the city has changed over time, and so have people's perceptions of waterways. The changing functions of the city have created new conditions and brought challenges to urban regeneration.

After the extensive use of containers in the transport industry, European cities such as Copenhagen, Amsterdam, London and Hamburg underwent a change in their shipping and shipbuilding industry. That led to an abrupt emergence of redundant spaces in urban areas, mostly on the waterfront. This has given the cities and their citizens an opportunity to repurpose the waterfront and restructure their industries. This is also a challenge, as the profit-seeking nature of investment is further escalating, how to create a humane and livable city is a great challenge for government decision makers, planners and architects everywhere. It is in fact also a matter of meeting values: how to create a high-quality space is testing the vision and foresight of administrators, investors and designers.

The systematic transformation of waterfronts in Europe began in the 1970s. Compared to those cities, Shanghai is a late comer in considering, developing and reusing waterfronts, but it also provided Shanghai many valuable examples and experiences. Does the modern urban people still want to walk in a dense jungle of steel and concrete? Is an eight-lane street still what the city center needs? Do white-collar workers still want to work in low-ceilinged cubicles without natural light and ventilation? Are people still content to sit and relax at the entrances of alleyways? The answers to these questions are actually our design guidance.

Shanghai's historical and cultural legacy and the quality of its citizens mean that the city needs quality development and in turn, it should give its citizens benefits. For this reason, it is easy for government decision-makers, planning administrators, planners, architects and owners to accept experience and best practices and also adapt them to the actual conditions of Shanghai. Over the past decade, we have designed and implemented several projects along the Huangpu River. As we continue to better understand the city, we are also trying to apply our previous experience in waterfront redevelopment to the realities of Shanghai.

Reconstruction of the Urban Texture

Shanghai along the banks of the Huangpu River, especially on the Puxi side, has its own existing urban textures. This textures are the skeleton and foundation of the city's urban space, serving as an important depository of urban memory. For this reason, it is imperative that we reconstruct this texture fabric through urban regeneration.

In the Bund SOHO project, the urban fabric of the surrounding area was analyzed and a scheme of six buildings was developed to forge a proper relationship between the volume of the building and the surrounding buildings, including those in the Bund Landscape Conservation Area, while the buildings further continue the space from the old Shanghai town to the Bund. In the design of the public space along the South Bund, we have explored and sorted out the radial lanes perpendicular to the Bund and formed by the bend of the Huangpu River, and proposed the concept of reconstructing the original lane space in the urban space design, in a bid to restore the historical urban fabric.

Reshaping Urban Spaces and Upgrading Urban Functions

The project to build a continuous waterfront on both sides of the Huangpu River has a profound impact on Shanghai's urban development, city functions and the lives of its citizens. The South Bund riverfront along the section of Waima Road from Dongjiadu to the Nanpu Bridge has released the previously inaccessible riverfront and opened it up to the public. This has also created conditions for the buildings and urban spaces in the neighboring areas. This area should no longer be a closed space, but a new urban space that features offices, leisure spaces, and cultural and commercial facilities. The elevation of the leisure spaces along the riverbank provides both a favorable condition and posed a question for the adjacent buildings and public spaces. The first and second floors of the new buildings, the elevated terraces, and the connection to the public space along the waterfront provide new opportunities for diversity and complexity of urban functions and can provide the public with more visual connections to the water. This is also what we are always trying to express in urban design studies and competitions.

Showcasing the Contemporary Glory of the City

As Prof. Wu Zhiqiang, Academician of the Chinese Academy of Engineering, said, Shanghai's current urban image is the highlight of the glory of all previous periods, and buildings from all periods can be seen in the city. How can we perpetuate the glory in the regeneration of the waterfront? This is a question every architect has to consider seriously.

SOHO on the Bund, once a subject of criticism, became warmly received. The reason is that it has properly handled the relationship between the old and the new. The proportions of the facade, the use of horizontal and vertical lines, the use of stone and glass, and the detailing are all carefully considered, and the relationship between contemporary architecture and the overall Bund landscape is carefully treated. The Lujiazui Riverfront Centre, located next to the Minsheng Wharf, adopts simple architectural languages, metaphorizing the site's historic function through the orderly exposed frames that run through all the buildings.

There are a number of historic buildings in the South Bund area along Waima Road, including some industrial facilities and warehouses. We propose to preserve the valuable historical buildings and have them take on new functions, while further discovering other valuable buildings through careful stud. It is also important to take into consideration the characteristics of the period in which the new building is designed. This is also my intention in proposing to "protect the authentic cultural relics, identify the new, and say no to the fake".

"Making cities fun," the slogan that planning master John Fondersmith proposed in 1988, is often referred to in the case of Hamburg's new Hafencity. This "fun" does not mean building a Disneyland, but further mixing the functions of the city, expanding public spaces and qualifying them for cultural, artistic and leisure activities. Shanghai's waterfront has already taken shape, especially with the completion of the Huangpu River Waterfront project, which has injected new vitality into the city and further increased citizens' sense of well-being. "Better city, better life," and the city will become more interesting and vibrant, and people will feel more comfortable. Shanghai's practice is not only "making cities fun," but also "giving people comfort."

苏州河篇
SUZHOU CREEK

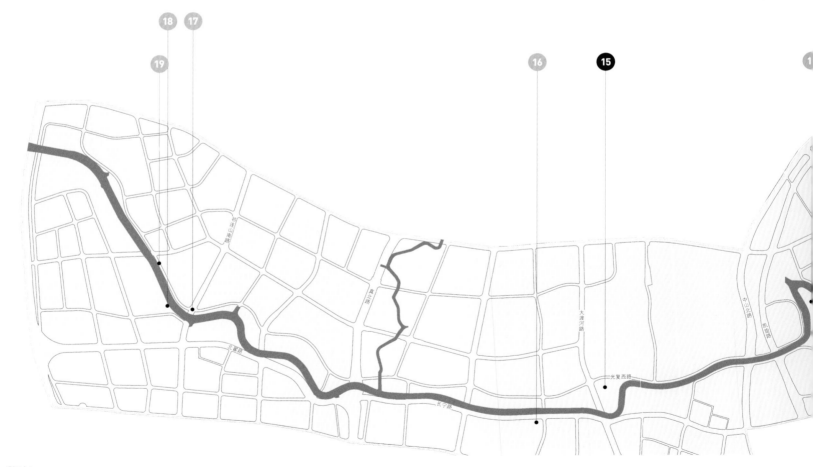

普陀区 Putuo District

长宁区 Changning District

虹口区 Hongkou District		
01	海鸥饭店 Seagull Hotel	342
02	中美信托金融大厦 Sino-American T. & F. Tower	344
03	虹口区北苏州路滨河空间 North Suzhou Road Waterfront	296

静安区 Jing'an District		
04	华侨城苏河湾 OCT Suhe Creek	332
05	四行仓库 Joint Trust Warehouse	336
12	蝴蝶湾花园 Butterfly Beach Park	314
13	昌平路桥 Changping Road Bridge	358

黄浦区 Huangpu District		
06	洛克·外滩源 Rockbund	330
07	介亭 Pavillion Inbetween	300
08	外滩源 33# Bund 33	328

09	中石化一号加油站 Sinopec No.1 Gas Station	304
10	飞鸟亭 Flying Bird Pavilion	306
11	九子公园 Jiuzi Park	310

长宁区 Changning District		
14	华东政法大学长宁校区 Changning Campus of East China University of Political Science and Law	316
16	天原河滨公园 Tianyuan Riverside Park	320
17	苏州河桥下空间 Spaces Under Suzhou Creek Bridges	360
18	江森自控亚太总部大楼 Johnson Controls Asia Pacific Headquarters Building	352
19	临空滑板公园 Airport Skatepark	322

普陀区 Putuo District		
15	上海市少年儿童图书馆新馆 New Library of Shanghai Children's Library	348

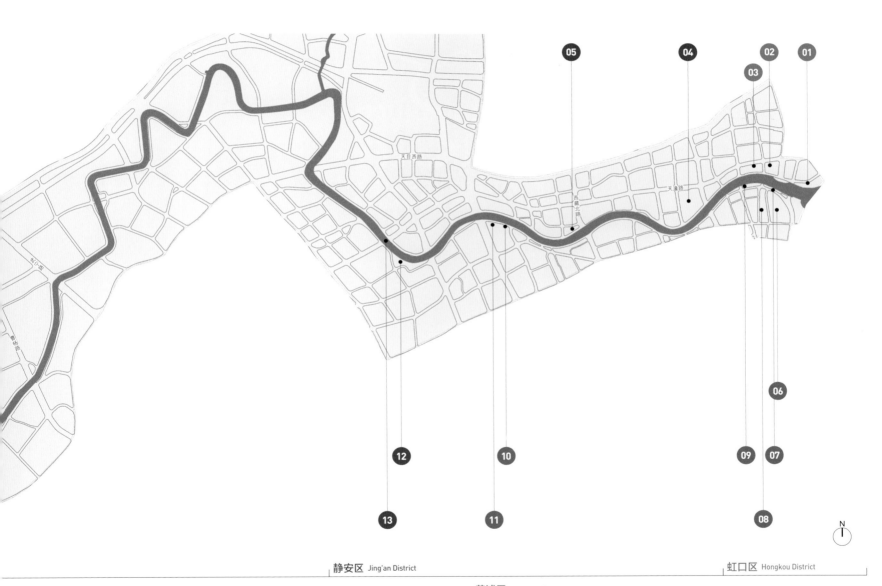

苏州河篇 SUZHOU CREEK

HUANGPU RIVER
SUZHOU CREEK

THE WATERFRONT SPACE & ARCHITECTURE IN SHANGHAI

滨水空间重塑
REVITALIZATION OF WATERFRONT

虹口区北苏州路滨河空间
North Suzhou Road and North Bund Waterfront

介亭 Pavillion Inbetween

中石化一号加油站 Sinopec No.1 Gas Station

飞鸟亭 Flying Bird Pavilion

九子公园 Jiuzi Park

蝴蝶湾花园 Butterfly Beach Park

华东政法大学长宁校区
Changning Campus of East China University of Political Science and Law

天原河滨公园 Tianyuan Riverside Park

临空滑板公园 Airport Skatepark

虹口区北苏州路滨河空间
NORTH SUZHOU ROAD WATERFRONT

苏州河虹口段全段贯通
Hongkou section of Suzhou Creek

项目地点：虹口区北苏州路（外白渡桥-河南北路段），江西北路（北苏州路-天潼路段）
设计单位：上海安墨吉建筑规划设计有限公司（城市设计与景观概念方案）、上海市建工设计研究总院有限公司（施工图）
合作设计单位：上海北斗星景观设计设计院有限公司（绿化景观）
建设单位：上海市虹口区建设和管理委员会
设计时间：2018.11-2020.4
建成时间：2020.12
用地面积：20714m²

Location: North Suzhou Road (from Garden Bridge to North Henan Road), North Jiangxi Road (from North Suzhou Road to Tiantong Road), Hongkou District
Designers: Shanghai AMJ Architecture, Planning and Design Co., Ltd. (Urban design & landscape concept) Shanghai Construction Design & Reserch Institute Co., Ltd. (Construction)
Partner: Shanghai Triones Landscape Design Institute Co., Ltd. (Plant & landscape)
Constructor: Hongkou Construction and Management Commission
Design Period: November 2018 - April 2020
Date of Completion: December 2020
Site Area: 20,714m²

苏州河虹口段位于一江一河交汇处，东起外白渡桥，西至河南北路，拥有得天独厚的区位优势、丰富的文化资源和深厚的历史积淀。

Starting from the Garden Bridge in the east and ending at North Henan Road in the west, the Hongkou section of the Suzhou Creekis last stretch of the waterway before it meets the Huangpu River. The waterfront has unique geographical advantages, rich cultural and historical resources.

苏州河篇 · 滨水空间重塑　SUZHOU CREEK · REVITALIZATION OF WATERFRONT

Cultural garden section of Broadway Mansions Hotel Shanghai

北苏州路滨河空间贯通提升工程因地制宜打造"一岸四段"的和谐美景：外白渡桥至乍浦路的上海大厦文化花园段，主要依托上海大厦和外白渡桥打造多层次全视角观景平台。乍浦路至四川北路的宝丽嘉酒店休憩观景段，主要以滨河廊道+共享街道为特色，突出休闲服务功能。四川北路至江西北路的邮政大楼风貌展示段和江西北路至河南北路的河滨大楼特色风情段，则以历史建筑为背景，通过改造步道、提升绿化、优化建筑底层功能等手段，形成宜人的滨水游憩景观步道，营造高品质人文生活氛围，以花为脉，文化为轴，打造一个融入当地文化的城市公共形象空间，提供人们心中向往的最美会客厅花园。

The redevelopment project of the waterfront space of North Suzhou Road started in April 2020. By taking measures such as prohibiting motor vehicles driving on the North Suzhou Road and running-through of major breakpoints, etc., a harmonious beauty of "one bank and four sections" was created according to local conditions: "Cultural Garden Section of Broadway Mansions Hotel Shanghai" from Waibaidu Bridge to Zhapu Road, which mainly relies on Broadway Mansions Hotel Shanghai and Waibaidu Bridge to create a multi-level full-angle viewing platform; "Rest and Viewing Section of Bellagio Shanghai" from Zhapu Road to Sichuan North Road, which is mainly characterized by riverside gallery+shared street, highlighting the leisure service function; "Post Office Building Style Exhibition Section" from North Sichuan Road to North Jiangxi Road and "Riverside Building Characteristic Style Section" from North Jiangxi Road to North Henan Road, where a pleasant waterfront recreation landscape walkway is formed and a high-quality humanistic living atmosphere is created through walkway renovation, landscape improvement and optimizing the functions of the ground floors of buildings based on historical buildings.

Rest and viewing section of Bellagio Shanghai

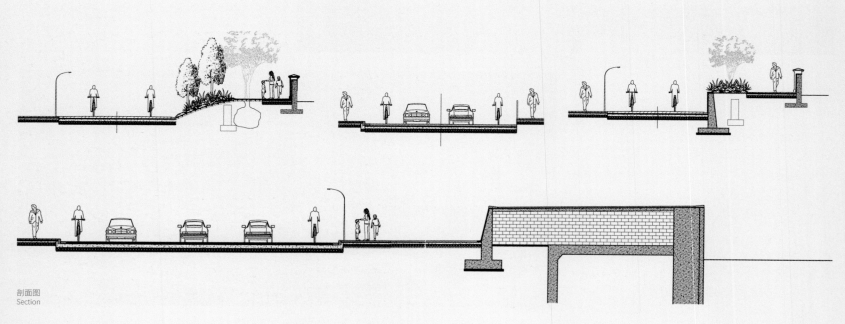

剖面图
Section

外白渡桥与苏州河虹口段滨河空间
Waterfront between Waibaidu Bridge and Hongkou Section of Suzhou Creek

介亭
PAVILLION INBETWEEN

沿街立面
Façade

项目地点：黄浦区南苏州路76号
设计单位：同济大学建筑设计研究院（集团）有限公司原作设计工作室
建设单位：上海市黄浦区市政工程管理所
设计时间：2019.12
建成时间：2020.10

Location: No. 76, South Suzhou Road, Huangpu District
Designer: Original Design Studio (A subsidiary of Tongji Architectural Design (Group) Co., Ltd.)
Constructor: Shanghai Huangpu Municipal Engineering Administrative Office
Design Period: December 2019
Date of Completion: October 2020

2009年11月吴淞路闸桥拆除，现场遗留下吴淞路闸桥西侧钢箱桥墩。这座矗立场地中斑驳却具有一定几何形状的混凝土方墩承载着非同寻常的城市历史。

一座钢质长亭建于遗留的桥墩结构上方，从桥墩西侧起始贯穿至东侧。桥墩西侧现存一株枝繁叶茂的构树，以桥墩为基础，将钢栈道环绕构树悬挑绕行后拾级而上到达桥墩顶部。近4米的悬挑栈道彰显了介亭轻盈的结构美感，此刻无论是桥墩还是构树都成为介亭与场地环境互动的重要组成元素。介亭东侧栈道通过钢柱支撑"悬浮"于桥墩之上，纤细精巧的

"悬浮"于桥墩之上的栈道
Walkway above the piers

After the Wusong Road Sluice Bridge was demolishedin November 2009, the steel-box pier on the west bank remained. The mottled concrete square pier with a certain geometric shape standing on this site actually preserves the memory of the city.

A steel pavilion was proposed to be built above the pier structure to connect the west and east sides of the former bridge pier. Given the Broussonetia papyrifera on the west side of the pier, a steel plank walkway is built with the bridge pier as the foundation, and it is cantilevered around the Broussonetia papyrifera and then climb

沿街透视
Section perspective

苏州河篇·滨水空间重塑　SUZHOU CREEK · REVITALIZATION OF WATERFRONT

环绕构树拾级而上
Spiral loop

钢结构与敦实厚重的基础再次形成了鲜明的对比。介亭总长约50米，栏杆由50毫米宽扁钢框架嵌入钢丝网构成，地面由花纹钢板与钢格栅组成，在确保安全与隐私的同时，提供了通透且灵巧的视觉形象。介亭如同一艘沿着苏州河缓缓行驶船只，载着上海文化历史的血脉并肩负城市未来发展目标与希望，朝着太阳升起的地方乘风破浪。

upward step by step to the top of the pier. The nearly four-meter wide cantilever plank walkway highlights the light structural beauty of Pavilion Inbetween. Both the bridge pier and the Broussonetia papyrifera have become important elements of the pavilion and the site environment. The plank walkway on the east side of Pavilion Inbetween is "suspended" above the pier by steel column supports, and the slender and delicate steel structure is in sharp contrast with the solid and heavy foundation again. The total length of Pavilion Inbetween is about 50 meters, and the railing consists of a 50-mm-wide flat steel frame embedded with steel meshes. The floor is made of checkered steel plates and steel grating, which not only ensures safety and privacy, but also provides transparent and dexterous visual image experience. Visually, the Pavilion Inbetween is like a boat moving slowly in theSuzhou Creek, carrying the hope of Shanghai's culture and history.

剖面图
Section

立面图
Elevation

介亭与苏州河
Pavilion in-between and Suzhou Creek

中石化一号加油站
SINOPEC NO.1 GAS STATION

中石化一号加油站沿街立面正视
Front view of façade: Sinopec No.1 Gas Station

中石化一号加油站沿街立面透视
Perspective view of façade: Sinopec No. 1 Gas Station

项目地点：黄浦区南苏州路198号
设计单位：同济大学建筑设计研究院（集团）有限公司原作设计工作室
建设单位：中国石化销售股份有限公司上海石油分公司
设计时间：2019.4
建成时间：2020.10
建筑面积：222m²
用地面积：520m²

Location: No. 198, South Suzhou Road, Huangpu District
Designer: Original Design Studio (A subsidiary of Tongji Architectural Design (Group) Co., Ltd.)
Constructor: China Petrochemical Marketing Co., Ltd. Shanghai Petroleum Branch
Design Period: April 2019
Date of Completion: October 2020
Gross Floor Area: 222m²
Site Area: 520m²

中石化一号加油站站房立面
Façade: Sinopec No. 1 Gas Station

站房二楼咖啡厅
Café on 2/F

站房内部楼梯空间
Interior

中石化一号加油站，原址为1948年建成的中国第一个自营的加油站。加油站为经典的超市办公加上钢结构罩棚的模式，加油的机动车流线同公众流线之间缺乏适当的分流。设计的重点在于对原有加油站模式的突破，形成一个公共通透、动线合宜、功能复合、当代语境的基础设施建筑。加油站的设计从滨河景观带的梳理开始，将原有的南侧混合动线拆解为南北两条动线，行人从靠近河边的北侧绕过加油站，加油车辆从南侧出入。拆分流线后，加油站确定了南北贯通、通透的建筑形象，最大限度地消解加油站建筑对于滨河景观的阻隔。

加油站分为一虚一实两个体量，虚的为加油棚架，实的为站房。不同于以往两个异质的、完全功能化的处理方式，在设计中，两个体量更加统一，共同形成建筑的形态。建筑被构想成为一高一低两组从地面翻折而起的折板，高的一组折板容纳了两层站房的功能，一楼为超市，二楼为咖啡厅；略矮的一组折板覆盖加油的区域。结构理性和视觉特征在方案中得到了一体化的显现。折板一端顺势延长落地，另一端以一排细柱支撑。考虑到钢结构直接落地会有生锈风险，折板材料与地面接触的部分，由钢结构改为清水混凝土。两组折板体量在不同的高度发生材料改变，在保证形态连续性的同时，显露的铰接点增强了结构的可读性。

折板的形态更加纯粹，显示出轻盈和简洁的状态，落地的混凝土折墙则在对比之下，更加呈现出精致有力的空间感受。折板的概念，附和了对老建筑折墙的呼应、苏州河浪花的想象，以及折扇般的精致感。结构的纯粹性和意义的多义性赋予建筑当代性，成为这个富有历史记忆的场所中"文化加油站"最合理的注解。

The Shanghai Sinopec No.1 Gas Station is in the waterfront near the mouth of the Suzhou Creek. The plot was formerly occupied by the first state-run gas station in post-1949 China back dating to 1948. Before the renovation, the original pattern is the classic supermarket + office + steel structure canopy that lacked publicity and permeability. With respect to the overall streamline organization, there is no proper division between the refueling both vehicle and pedestrian streamline. As of functionality, the excellent landscape resources can not be used by a single supermarket and a refueling function resulting in wastage. In regard to cultural resources, it is difficult for cultural resources to be embodied differently in the original pattern gas station. Therefore, the design team has focused on how to make a breakthrough in the original pattern of the gas station to form an infrastructure with public transparency, appropriate circulating streamlines, complex functions, and contemporary context. The design of the gas station starts from combing the riverside landscape. The mixed streamline on the south is split into two streamlines, namely, the pedestrian streamline by passes the gas station from the north side near the river, and the vehicle refueling streamline enters and exits from the south side. After the splitting of the streamlines, the public traffic direction of the gas station is also changed from the original south side to the north and south sides, in sync with the transparent direction of the city riverside landscape. Thus, the building is determined to be in north-south penetration, eliminating the barrier of the gas station building to the riverside landscape to the best extent.

The gas station is divided into two volumes, one virtual and other real. The virtual one is the refueling scaffold and the real one is the station building, which is different from the previous treatment method (two similar fully functional blocks). In the design, the team hopes that the two volumes are more unified to form the shape of the building together. Due to the functional requirements of refueling, the scaffold of the gas station must have a certain span. Considering the folded board shape has the function of increasing span, and has the characteristics of structural rationality and integration of visual features, the building is conceived as two sets of folded board folded from the ground, with the higher set of folded board accommodating the two-floor station building (the first floor houses a supermarket, and the second floor a coffee shop), and the slightly shorter set of folded board covering the refueling area. One end of each set of folded board is inserted into the ground, and the other end is supported by a row of thin columns, which further highlights the evolving nature of the folded boards. At the early stage of the design, the folded board was conceived to be of steel structure. However, considering the risk of rusting caused by direct landing of the steel structure, the material of the folded board part in contact with the ground was changed to light coloured concrete. The folded board became a structural body combining steel folded board and concrete folded boarded, and the materials of the two sets of folded boards change at different heights. While ensuring the continuity of shape, the steel folded board was anchored on the concrete folded board by hinges.

飞鸟亭
FLYING BIRD PAVILION

飞鸟亭正立面远观
Front view of facade, Flying Bird Pavilion

铝结构单元
Aluminum frame

项目地点:黄浦区南苏州路1305号
设计单位:同济大学建筑设计研究院(集团)有限公司原作设计工作室
建设单位:上海市黄浦区市政工程管理所
设计时间:2019.12
建成时间:2020.10

Location: No. 1305, South Suzhou Road, Huangpu District
Designer: Original Design Studio (A subsidiary of Tongji Architectural Design (Group) Co., Ltd.)
Constructor: Shanghai Huangpu Municipal Engineering Administrative Office
Design Period: December 2019
Date of Completion: October 2020

飞鸟亭位于苏州河南岸、一处半地下倒班房的屋顶平台之上,平台下是开阔的小广场和亲水驳岸,独特的位置使其成为滨河空间的制高点。初到这里,站在平台之上,苏州河从脚下静静流过,举目四望,视野极佳。
改造后的休息平台与两条平行于河岸的漫步道相通,并通过两端的大台阶分别与东侧的小广场及南侧的跑步道连接。原场地上凉亭的四根斜柱裁切到人们落座的高度作为新结构支撑,依其定位沿弧线阵列排出铝结构单元。弧线外侧为漫步道,弧线内侧形成一个半包围的休憩空间。铝结构单元继续向内向上生长,放大聚合成雨棚,人们落座于结构单元"缝隙"之间,面水而观,谈笑风生。

The Flying Bird Pavilion is to the south of the Suzhou Creek in the stretch between the Wuzhen Road and Xinzha Bridges. It is built on the rooftop platform of a partial-underground shift workers' duty room. In front of the platform are a small balcony and a revetment. The unique location of the pavilion perches it at a "commanding height" of the riverside space. Visitors standing on the platform can enjoy the quiet flow of the Suzhou Creek and the breathtaking landscape spreading before their eyes.
Therefore, retaining the original structure of the shift workers' duty-room and maintaining the original site height became essential conditions for the design.

飞鸟亭沿街立面透视
Perspective view of façade: Flying Bird Pavilion

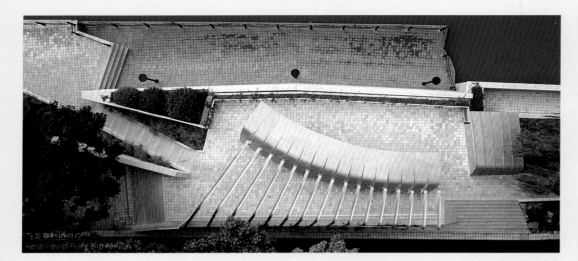

飞鸟亭航拍俯视
Aerial view of Flying Bird Pavilion

铝结构单元向外向下蔓延，围绕出漫步空间，台下的爬藤顺着延展的铝条向上攀爬，包裹出一条绿荫小道，人们游走其中，捕捉藤叶间的斑驳。而连接内外的铝结构单元，自身也变成充满韵律感的"隧道"，成为小朋友嬉戏玩耍的小天地。这样一个长达14米的全铝结构整体，重量完全承托在原先的四根斜柱上，雨棚向上往河边悬挑，铝条向下通过栓接轻巧地搭在平台的边沿，整体维持着巧妙的平衡。铝结构自身的金属光泽在粼粼波光的映衬下熠熠生辉，如同羽毛般的结构韵律，如悬浮般的体量支撑。新的凉亭像是银色丝带织成的小鸟，在苏州河岸边振翅待飞，因此而名"飞鸟亭"。

There was a pavilion of tensile membrane structure on the platform, with four inclined steel columns crossing the structure top slab connecting with the underground foundation. As a landscape sketch of the waterfront space, the original membrane structure pavilion has poor retention, its volume is insufficient to become the space background of the small piazza and is also becomes difficult to meet the requirements of "commanding height" for recognition.

亲水	观河	休憩	玩耍	漫步
Waterfront	Platform for distant view	Lounge	Activity area	Walkway

倒班房
Duty room

剖面图
Section

飞鸟亭沿河远观
Panoramic view of Flying Bird Pavilion

轴测爆炸图
Exploded axonometric diagram

长达14米的全铝结构
14-meter all-aluminum structure

飞鸟亭黄昏灯光照
Flying Bird Pavilion in the dusk

九子公园
JIUZI PARK

九子公园-纸鸢屋园区入口
Entrance of Paper Kite House of Jiuzi Park

Paper Kite House of Jiuzi Park

项目地点：黄浦区成都北路1018号
设计单位：同济大学建筑设计研究院（集团）有限公司原作设计工作室
建设单位：上海市黄浦区绿化和市容管理局
设计时间：2019
建成时间：2020
用地面积：6939m²

Location: No.1018, North Chengdu Road, Huangpu District
Designer: Original Design Studio (A subsidiary of Tongji Architectural Design (Group) Co., Ltd.)
Constructor: Shanghai Municipal Huangpu District Landscaping and Appearance Bureau
Design Period: 2019
Date of Completion: 2020
Site Area: 6,939m²

WC of Jiuzi Park

九子公园得名于九种弄堂里玩的游戏，包括扯铃子、跳筋子、滚轮子、打弹子、掼结子、跳房子、圈子、抽坨子、顶核子。改造设计以"城市公园开放化"为总体策略，打开公园围墙，通过绿化与城市之间形成弱空间限定，提升公园的可达性，让公园成为市民日常生活的一部分。

通过设置二级防汛墙，形成亲水平台，公园空间进一步扩展至河边，将公园与苏州河更紧密地链接在一起。公园的铺装与滨河公共空间的铺砖相同，采用多彩的琉璃水磨石预制块，不同颜色暗示不同功能——橙色代表主要动线，绿色代表绿化空间，蓝色代表亲水空间，灰色为各种颜色之间的过渡色彩——共同形成如同"马赛克"的抽象拼贴画，呼应河岸的精致性。代表公园主要动线的橙色跨越城市道路，与滨河的橙色铺装相接，强化公园开放性。

公园驿站又名"纸鸢屋"，具有公园驿站与管理办公的多重功能。建筑通过倾斜的折叠清水混凝土折板获得结构刚度，巨大的挑檐在城市空间中具有标志性，强化公园的节点形象。公厕改造在保留原有功能面积的基础上，以正交的折叠方式，形成集融灰空间、院落空间、通廊空间于一体的亭子般的空间效果，器械租借管理间与厕所空间则成为用阳极氧化铝包裹的"小盒子"，穿插于折板间。

沿城市道路设置了完整的雨棚休息空间，游客在经过时可临时就座休憩，不必进入公园也可享受公园氛围，是公园空间向城市空间的另一种外化。折墙浮台与后方的竹林掩映着泵房基础设施，被想象成一个集矮墙、座椅、桌子于一体的"超级清水混凝土家具"，人们可攀爬，可穿越，可坐卧，可聚会，多义的空间启发着人们使用方式的想象力。

The Jiuzi Park literally gets its name from nine (jiu) traditional toys (zi) played in Shanghai's traditional lane neighborhoods. Guided by the principle of "a wall-less park in a city", the walls have been removed and the park integrated into the waterfront space, forming a minimal space limit with the green space. This allows the accessibility of the park to be improved, so that visiting the park can be a regular feature of the citizen's lives.

The riverside space on the north of Jiuzi Park is made into a hydrophilic platform section by setting the secondary flood control wall. When the flood control gate of the secondary flood control wall is opened, the park space can be further extended along the river. This does not only enrich the garden experience, but also closely links the park with Suzhou Creek. The pavement of the park is the same as that of the riverside public space, using colorful glazed terrazzo precast blocks. Different colors imply different functions - orange represents main traffic lines, green represents green spaces, blue represents hydrophilic spaces, and gray is the transitional color among various colors. All these colors form an abstract mosaic collage, which echoes the exquisiteness of the riverbank. The main traffic lines of the park in orange cross the urban roads and connect with the orange pavement of the riverside, which also strengthens the openness of the park. The Jiuzi Park is rich in narrative color. Buildings and structures including park stop station, pavilion and toilet, folded walls and floating platform, etc. form a unified movement for fair-face concrete.

The Park's service station, also known as the Paper Kite House, gets structural rigidity by obliquely folding the lightly coloured concrete folded boards. The functions of the park stop station and the management office are integrated. The huge overhanging eaves are symbolic in the urban space, which strengthens the node image of the park. In the renovation of public toilets, with the intention of retaining the original functional area, a pavilion-like space integrated with the transition space, the courtyard space and the corridor space is shaped in an orthogonal manner. The administration room of equipment for rent and toilets are built into "small boxes" coated with anodized aluminum and are arranged between the folded boards.

A canopied area for rest and relaxation is arranged along the main street, where passers-by can sit down for a rest and enjoy the environment without entering the park. It is kind of extending the park space. The folding wall floating platform and the bamboo grove behind it can conceal the pump house, and all these resemble a piece of super furniture made of fair-face concrete. The low wall, benches and tables enable people to climb, cross, sit down, lie down, and get together. The multi-purpose space encourages imaginative ideas for usage.

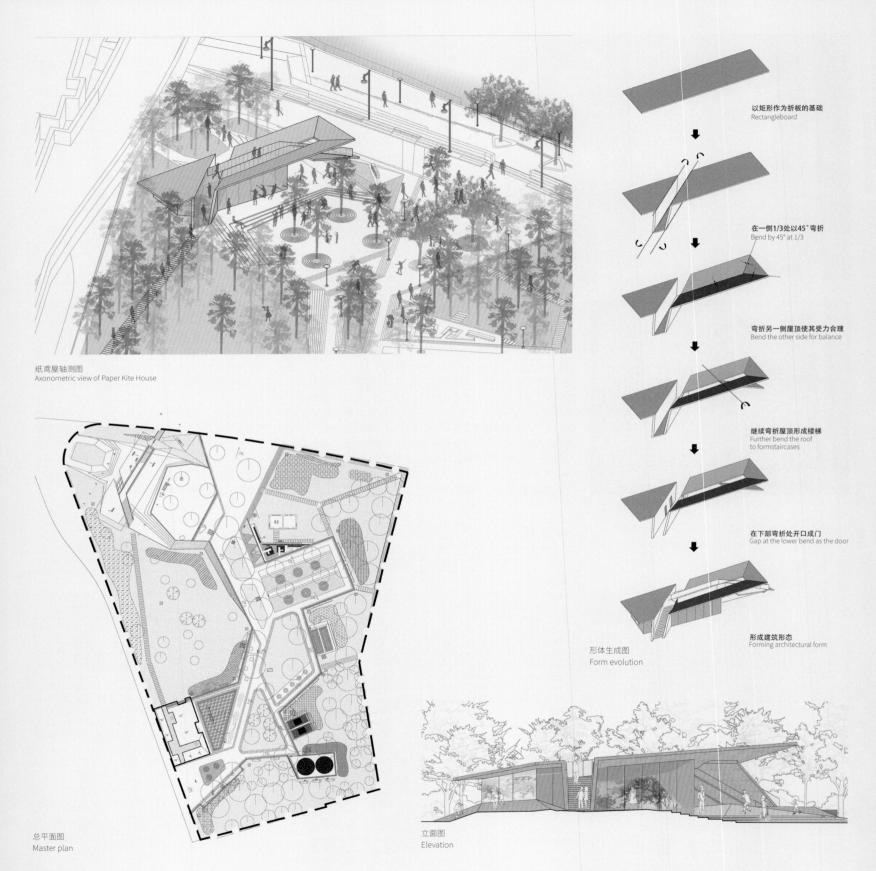

九子公园鸟瞰俯视
Aerial View of Jiuzi Park

蝴蝶湾花园
BUTTERFLY BEACH PARK

蝴蝶湾花园
Butterfly Beach Park

总平面图
Master plan

项目地点：静安区苏州河沿岸（康定东路—泰兴路）
设计单位：上海光华建筑规划设计有限公司、上海青同建筑设计工程有限公司、上海市园林设计研究总院有限公司
建设单位：（原）静安区绿化管理局、（原）静安区市政和配套管理局
建成时间：2008.5
用地面积：16670m²

Location: Waterfront along Kangding Road East and Taixing Road, Jing'an District
Designer: Shanghai Guanghua Building, Shanghai Qingtong Architecture Design Co., Ltd., Shanghai Landscape Design & Research Institute Co., Ltd.
Constructors: (Former) Jing'an District Greening Adminisration, Jing'an Disctrict Urban Management and Supporting Facility Administration
Date of Completion: May 2008
Site Area: 16,670 m²

沿苏州河一侧采用高低两级防汛墙设计
Secondary flood control wall along Suzhou Creek

蝴蝶湾花园位于静安区苏州河畔，地下为大型市政排水泵站，因公园地形如蝴蝶，故名"蝴蝶湾花园"。公园广场正面为石阵跌落水池，西侧为阵列式水池，通过水体的流动循环产生负离子净化周边空气。水池往北延伸两段弧形艺术景墙，将市政设施巧妙地隐于其中。花园内的桁车钢架外饰解构模块，具强烈雕塑感，功能结合艺术简约而不夸张。公园沿苏州河一侧采用高低两级防汛墙设计，一级防汛墙标高4.2米，为下沉式亲水平台，周围种植水生鸢尾、水杉、金丝柳等水生植物或喜水性的植物。二级防汛墙标高5.7米，为永久性防汛墙，后退3米，形成亲水景观平台。二级平台和廊架结合，竹林、高大的白玉兰、华盛顿棕榈、榉树、银杏等植物点缀其间。高低错落的设计，增加了临水观赏的层次，丰富了公园的竖向景观。蝴蝶湾花园在注重景观效果的同时，也充分重视公园所承载的功能，针对周边居民对绿地的休闲使用需求，在公园东块范围内兴建了两个半场的篮球场，并铺设红色塑胶的健康步道，提供舒适的慢跑空间。

春日樱花
Spring blossoms

The Hudiewan Park (literally, "Butterfly Beach Park") is located along the stretch of the Suzhou Creek in the Jing'anDistrict, at the intersection of Taixing Road and East Kangding Road. It covers an area of 16,670m². There is a large municipal drainage pump station underground. Hudiewan Park has its name due to the terrain being like a butterfly, which is in W-shape. On the park square, there is a stone plunge pool in the front and an array pool on the west, which generate negative ions to purify the surrounding air through the flow and circulation of water bodies. The pool extends northward, with two arc-shaped artistic landscape walls set to subtly conceal the municipal facilities. The external decoration module for the crane travel frame in the park has a strong sense of sculpture, and looks simple but not exaggerated when the use function is combined with an art. The flood control wall along the Suzhou Creek side in the park is designed into a high and a low flood control walls. The primary flood control wall has an elevation of 4.2m, and has a sinking platform, with aquatic plants such as water iris, metasequoia, golden willows, etc. planted around. It is normally open and closed when flooded in flood season. The secondary flood control wall has an elevation of 5.7m, and is a permanent flood control wall, with a setback of 3m. It is also used as a waterfront landscape platform. The secondary platform is combined with the gallery frame, with bamboo forest, tall magnolia, Washington palm, beech tree, ginkgo and other plants scattered. This high-and-low design increases the waterfront viewing levels and enriches the vertical landscape of the park. The design of the Hudiewan Park not only focuses on the landscape effect, but also the functions carried by the park. Aiming at the needs of surrounding residents for green leisure space, two half-court basketball courts are built on the east block of the park, with red plastic health trails laid to provide comfortable jogging space, thus both green space ecology and functions are achieved.

华东政法大学长宁校区
CHANGNING CAMPUS OF EAST CHINA UNIVERSITY OF POLITICAL SCIENCE AND LAW

华东政法大学长宁校区
Changning Campus of East China University of Political Science and Law

项目地点：长宁区万航渡路1575号
设计单位：上海安墨吉建筑规划设计有限公司（校园规划和景观方案）、上海林同炎李国豪土建工程咨询有限公司（施工图）
合作设计单位：上海市交通大学设计学院城市更新保护创新研究中心（规划顾问）
建设单位：上海市长宁区建设和管理委员会、华东政法大学
设计时间：2021.3-2021.5
建成日期：2021.9
用地面积：93600m²

Location: No. 1575, Wanhangdu Road, Changning District
Designer: Shanghai AMJ architectural, Planning and Design Co., Ltd. (Planning & landscape scheme design), Lin Tung-Yen&Li Guo-Hao Consultants Shanghai Co., Ltd.(Contruction)
Partner: International Research Center for Urban Regeneration, Preservation and Innovation, School of Design,Shanghai Jiao Tong University (Consultant)
Constructor: Shanghai Changning Construction and Management Committee, East China University of Political Science and Law
Design Period: March 2021 - May 2021
Date of Completion: September 2021
SiteArea: 93,600m²

桃李园实施后滨河步道
Waterfront Footway of Taoli Garden after renovation

设计旨在将华东政法大学这所"百年校园，苏河明珠"塑造成为苏河沿岸最开放的公共空间，最高雅的历史风貌，最美丽的校园景观。通过保留保护空间格局、历史建筑、景观环境的方式，打造别具一格的"园中院、院中园"的景观形式，将华政校园整体风貌作为苏州河沿线景观的一部分，将景色开放给人民。

滨河的设计充分挖掘现有资源，将原本相互独立的华政校园与滨河步道结合贯通，依托华政特有的中西合璧式的优秀历史建筑，以"凸显国宝建筑风貌，优化滨河景观品质；挖掘校园人文元素，激活滨河人文空间；保障校界界面安全，建立安保管理体系"三大设计理念为核心，对原本狭窄逼仄、景观单一、缺少公共服务设施的滨河步道进行全面梳理与重新布置，将原有单一的"人行步道+植物绿化"的景观形式改造成为"校园景观+共享空间+滨河步廊"的全新模式。

思孟园实施后景观长廊
Gallery of Simeng Garden (after renovation)

The design concept is to shape East China University of political science and law, a "century old campus and the Pearl of the Su River", into the most open public space along the Su River, the most elegant historical style and the most beautiful campus landscape. The landscape goal is to create a unique landscape form of "courtyard in the garden, garden in the courtyard". By retaining and protecting the spatial pattern, historical buildings and landscape environment, the overall style of campus is taken as a part of the landscape along the Suzhou River to open the scenery to the public.

The riverside design fully excavates the existing resources, combines the originally independent campus with the riverside footpath, and relies on the unique Chinese and western excellent historical buildings of the campus to "highlight the style of national treasure buildings, optimize the riverside landscape quality; excavate the campus humanistic elements, activate the riverside humanistic space; ensure the safety of campus interface and establish a security management system" With the three design concepts as the core, the riverside footpaths that were originally narrow, narrow, single landscape and lack of public service facilities were comprehensively combed and rearranged, and the original single landscape form of "pedestrian footpath + plant greening" was transformed into a new model of "campus landscape + shared space + riverside promenade".

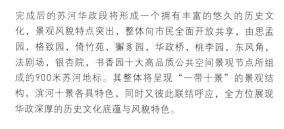

完成后的苏河华政段将形成一个拥有丰富的悠久的历史文化，景观风貌特点突出，整体向市民全面开放共享，由思孟园，格致园，倚竹苑，獬豸园，华政桥，桃李园，东风角，法剧场，银杏院，书香院十大高品质公共空间景观节点所组成的900米苏河地标。其整体将呈现"一带十景"的景观结构，滨河十景各具特色，同时又彼此联结呼应，全方位展现华政深厚的历史文化底蕴与风貌特色。

After completion, this section will form a rich and long history and culture, with outstanding landscape features and overall opening and sharing to the public. It is a 900 meters landmark of the ten meters from the Simeng garden, the Gezhi garden, the Yizhu garden, the Xiezhi garden, the Hua Zheng bridge, the Taoli garden, the Dongfeng Jiao, the French theatre, the Ginkgo biloba and the Shuxiang Garden. The landscape structure of "one belt and ten scenic spots" will be presented as a whole. The ten riverside scenic spots have their own characteristics. At the same time, they are connected and echoed with each other, showing the profound historical and cultural heritage and style characteristics of Chinese politics in an all-round way.

华东政法大学长宁校区总平面图
Master plan

天原河滨公园
TIANYUAN RIVERSIDE PARK

天原河滨公园
Tianyuan Riverside Park

总长400米的天桥
400-meter bridge

项目地点：长宁区水城路735号附近
设计单位：BAU建筑城市设计
建设单位：上海市长宁区建设和管理委员会
设计时间：2008
建成时间：2010
用地面积：约33200m²

Location: No. 735, Shuicheng Road, Changning District
Designer: Brearley Architects + Urbanists
Constructor: Shanghai Changning District Constrction and Administration Committee
Design Period: 2008
Date of Completion: 2010
Site Area: About 33,200m²

在高密度住宅和苏州河之间，通过改造原来三块狭长间隔的200米绿地，形成一条总长延续600米的线性公园，为附近居民提供了一条景观步道和活跃的运动休闲空间。景观规划将运动带、休闲带和自然带并列布置，各功能带的宽度变化使线性公园在连续的同时又富于变化。运动带包括塑胶跑道、各类固定体育运动器械、篮球场、羽毛球场以及其他运动场所。休闲带连接着茶室、临时展厅、小型开放空间以及社交聚集区。绿色自然带通过景观设计实现步移景异效果。三大功能带交叉交织创造出有序多变的城市滨水空间。为实现更好的观景效果，设计方案在跨水城路处架起一座总长400米的天桥，既保持了景观带的延续性，又创造了良好的视觉景观通廊。天桥设计灵感源于苏州河内硅藻的形态，意喻健康良好的生态系统。

天桥设计灵感源于苏州河内硅藻的形态，意喻健康良好的生态系统
Inspired by the form of diatoms in Suzhou Creek andhope for sound ecosystem

The park is formed by three narrow, 200-meter-long stretches tucked between densely packed residential buildings and the Suzhou Creek, through the renovation of the original three 200 m long and narrow green spaces, a linear park with a total length of 600 m is formed, which provides a landscape footpath, active sports and leisure space for nearby residents. In landscape planning, sports belt, leisure belt and natural belt are arranged side by side. The width change of each functional belt makes the linear park with a total length of 600 m be continuous and full of changes. There are plastic trail, various fixed sports equipment, basketball court, badminton court and other sports venues on the sports belt. The leisure belt is connected with the teahouse, temporary exhibition hall, small open space and social gathering area. The effect of varying sceneries with changing view-points is achieved for the green natural belt through the landscape design. The three major functional belts are interwoven to create an orderly and changeable urban waterfront space. In order to achieve a better viewing effect, a 400-meter long flyover is built at the place crossing Shuicheng Road, which not only maintains the continuity of the landscape belt, but also creates an impressive sight in its own right. The design of the overpass is inspired by the form of diatom in Suzhou Creek, which calls for a healthy ecosystem.

苏州河篇 · 滨水空间重塑　SUZHOU CREEK · REVITALIZATION OF WATERFRONT

临空滑板公园
AIRPORT SKATEPARK

临空滑板公园
Airport Skatepark.

玩滑板的孩子
Skate players

项目地点：长宁区临虹路7号东北方向50米
设计单位：BAU建筑城市设计
建设单位：上海市长宁区绿化和市容管理局
设计时间：2015
建成时间：2015
用地面积：约24300m²

Location: 50 meters northeast of No. 7 Linhong Road, Changning District
Designer: Brearly Architects & Urbanists
Constructor: Shanghai Municipal Changning District Landscaping and Appearance Bureau
Design Period: 2015
Date of Completion: 2015
Site Area: about 24,300 m²

沿河景观
Attractive landscape

迷你滑板场
Mini space

临空滑板公园是临空地区六大主题公园之一，以滑板运动及生态保护为景观设计的两大核心理念，成为苏州河沿岸连续公园带景观设计中重要的组成部分。

设计方案将运动带、自然带、休闲带交织布置，利用三条功能带的宽度在500米长线性公园中的变化，围合出3个"海藻"的微生物形态，不仅整合了两个观景亭及沿线休闲设施，也时刻警示人们保护生态环境和苏州河。公园分为专项滑板区、休闲公园区、景观桥、疏林草坡区。其中，专项滑板区包含一个专业滑板场和四个迷你滑板场，可以让滑板运动爱好者畅游整个区域。休闲区包含开放的草坪、避雨亭、亲水空间及丰富多彩的植物，为游客提供一个轻松漫步的游园空间。

The Airport Skatepark is one of the six major theme parks in the area near the Hongqiao Airport. The landscape design of the park focuses on two essential points, namely, providing skateboarding space and protecting the eco-environment. It is an important part of the landscape design of the continuous park belt along the Suzhou Creek. In the design, the sports belt, natural belt and leisure belt are of interweaving layout, and the change of the width of the three functional belts in the 500 meter long linear park to enclose three "seaweed" microbial forms, which not only integrates the two viewing pavilions and leisure facilities along the line, but also constantly warns people to protect the ecological environment and Suzhou Creek. The park space is divided into four parts, for skateboarding, leisure activities, landscape bridge area, and a lawn with sparse woods. The skateboarding area includes a professional skateboarding field and four mini-ones, which allows skateboarders to reach all parts of the park. The leisure activity area includes a big lawn, a rain shelter pavilion, a lake and various plants, providing visitors with a relaxed strolling experience.

HUANGPU RIVER SUZHOU CREEK

THE WATERFRONT SPACE & ARCHITECTURE IN SHANGHAI

历史建筑激活
RENOVATION OF HISTORICAL BUILDINGS

外滩源 33# Bund 33

洛克·外滩源 ROCKBUND

华侨城苏河湾 OCT Suhe Creek

四行仓库 Joint Trust Warehouse

设计·师说
MASTERS' WORDS

苏州河的新生

唐玉恩

穿越上海市区的苏州河是上海的母亲河之一，它是伴随上海城市历史与发展的宝贵资源，曾是城市重要的经济交通命脉、凝聚着数代人的乡愁。百余年来，苏州河见证了上海近代民族工商业在其两岸的崛起、兴盛。与密布的仓库、堆栈、小工厂、简易民宅店铺相对应的，是船只繁忙装卸的黄金水道，几乎串联的驳岸石阶，以及河两岸高密度低层建筑群的城市肌理。它见证了1937年河北岸燃起的抗日烽火；经历了20世纪80年代起的环境恶化、河水黑臭令人避之不及；90年代以来产业转型、河道治理、土地规划转性、板式高层住宅楼紧贴河边造成河道空间受挤压；及至近年的大力治理、优化环境和建设。

因历史原因，原先苏州河两岸用地较紧凑，缺少文化设施与公共空间。如今，向公众开放经保护修复的沿河历史建筑、建设多种功能的商业居住等新建筑群并控制高度、修复生态环境、贯通绿化步道与休闲驿站等，正在逐渐构成上海的滨河公共空间，提升上海城市空间及市民生活品质。

近年，苏州河岸重要的历史建筑、工业遗产得到精心保护与再利用，并向公众开放。如苏州河东端近黄浦江口的外滩源及原英国领事馆建筑群、河南路桥北岸的原上海市总商会和原中国实业银行货栈大楼建筑群、西藏路桥北岸的四行仓库等。本着维护厚重的历史积淀的理念，这些保护修复、再利用工程均在各相关部门和专家的指导下，执行尊重历史的原真性与整体性、最小干预等原则，全面评价建筑价值、重塑总体环境、精心保护修缮各个重点保护部位，科学植入当代所需的新功能、加固结构、更新设备……使历史建筑融入当代社会生活，获得新生，同时激活苏州河两岸社区活力，可持续地走向未来。

上海人民不会忘记，1937年10月27日在苏州河北岸打响了第二次淞沪会战中的四行仓库保卫战。当时，在西藏路桥之西，河的北岸为华界，南岸为租界，中外记者正是在南岸记录下对岸四行仓库及其身后大片城区烽烟滚滚，此战之惨烈震惊世界。之后该处长期作仓库、批发部用。2014年，设计团队怀着敬意为保护修复四行仓库，纪念抗战遗址、重塑功能而努力。其中，以各种技术设施寻找平射炮造成的西墙洞口、复原主要洞口并确保建筑长期安全，是设计重点和难点。同时复原沿苏州河的南立面原貌、拆除西墙外的后期搭建、开辟"晋元纪念广场"，68米长的西墙犹如卧碑，警示勿忘历史。大楼内西部作抗战纪念馆，东部为创意办公，西面的纪念广场也是苏州河北岸新增的富有内涵的公共文化活动空间。四行仓库附近还有几座建造时间相近的仓库，它们的保留修缮将与四行仓库组成独特的沿河工业遗产建筑群，为滨江注入新功能、释放新活力。当代新建筑创作理念之一是善待所处环境。苏州河宽40～50米，近年沿岸的新建筑注意在高度、体量上与苏州河的尺度相匹配，其造型现代、简约得体、适度退让、部分架空、临河设不同高度的平台广纳河景，创造从城市到苏州河的视觉走廊……如上海少年儿童图书馆及老建筑群、静安国际中心等。同时，沿河功能多样的现代商业、办公、文化建筑，通过与历史建筑组合，共同构成新老共生、包容融合的临水长卷——缔造和谐舒适、有活力的滨河城市环境。

在蜿蜒流淌的苏州河两岸的绿化中，新建的、造型活泼的休闲驿站和桥下球场，亲切有趣，令人耳目一新。前者均设封闭空间和半开敞空间以方便公众停留休息；后者则利用通常闲置的后场空间资源，变废为宝，打造成有新鲜活力的球场活动场地，还在旁设适宜公众步行的环境。

通过充分利用空间资源，创造激发河岸环境的重塑新生。深入研究苏州河水域空间的特色与价值、与城市的关系、时代赋予的需求等，苏州河沿岸将继续增设开放共享的高品质公共空间，成为更美好的滨河区域。

THE REBIRTH OF THE SUZHOU CREEK
Tang Yu'en

The Suzhou Creek, which winds its way through downtown Shanghai, is one of the city's mother rivers. It has accompanied the city's development being once an important economic and transportation lifeline and has been a place of nostalgic memories for generations.

For over a hundred years, the Suzhou Creek has witnessed the rise and fall of national industry and commerce in Shanghai in modern times. Alongside the densely packed warehouses, small factories, residential buildings and shops, there have been this busy waterway where cargo was loaded and unloaded, the continuous stone steps of both riverbanks, and the low-rise urban housing cluster. The river has always been there when the Chinese people fought the heroic fight against Japanese aggression in 1930s and early-to-mid 1940s; when there was severe environmental degradation and water pollution in the 1980s; when its waterway was narrowed as a result of the industrial transformation, river regulation, new land planning and increasing amount of high-rise residential buildings built nearby in the 1990s; and when vigorous efforts have been made in improving river environment in recent years.

Historically, the land on both sides of the Suzhou Creek was in short supply and lacked cultural facilities and public spaces. Nowadays, opening well-preserved riverfront historic buildings, building new commercial and residential complexes with multiple functions with controlled heights, restoring the ecological environment, connecting green paths and leisure facilities, etc. have all become reality. They have gradually formed the riverfront public spaces in Shanghai, improving Shanghais urban spaces and the citizens lives.

In recent years, important historical buildings and industrial heritage on both banks of the Suzhou Creek have been preserved, reused and opened to the public. For example, the Rockbund and the former British Consulate complex at the eastern end of the Suzhou Creek near the mouth of the Huangpu River, the former Shanghai Municipal Chamber of Commerce and the former China Industrial Bank Cargo Building complex on the north bank of the Henan Road Bridge, and the Joint Trust Warehouse to the northwest of the Xizang Road Bridge. With the aim of preserving historical heritage, these conservation, restoration and reuse projects were carried out under the guidance of various relevant departments and experts. The team followed the principles of respecting the originality and integrity of history and minimal intervention, comprehensively evaluated the architectural value, reshaped the general environment, carefully repaired each key area, gave new functions, strengthened the structure and upgraded equipment required by the contemporary needs. The historical buildings are expected to blend into contemporary social life and gain a new life while activating the community along the Suzhou Creek, and embrace the future.

Shanghainese will never forget the Battle to Defend Joint Trust Warehouse fought on the north bank of the Suzhou Creek on 27 October 1937. The battle was fought to the west of the Tibet Road Bridge over the Suzhou Creek, which separated the International Settlement on the south bank from other parts of Shanghai on the north bank. It was on the south bank that Chinese and foreign journalists recorded the horrendous battle with the smoke of gunpowder rising from the Warehouse. After that, the place was used as a warehouse and wholesale department for a long time. In 2014, the design team was honored with an opportunity to work on the restoration of Joint Trust Warehouse. The team members searched for the holes in the west wall caused by the flat-firing cannons with various technical means, and restored them to ensure the long-term safety of the building. At the same time, the south elevation facing the river was restored. The later structures outside the west wall were all removed and a "Xie Jinyuan Memorial Square" was created, with the 68-metre-long west wall acting as a recumbent monument. The western part of the building is used as a war memorial, the eastern part now serves as a creative office building, and the memorial plaza to the west is a new public space on the north bank of the Suzhou Creek. There are several other warehouses constructed at almost the same time as the Joint Trust Warehouse, and if preserved and restored, they will form a unique industrial heritage corridor along the Suzhou Creek.

One of the concepts behind the creation of new architecture is to take into account the environment context. The Suzhou Creek is 40-50 meters wide, and in recent years new buildings along it have been built to matching their heights and volumes with the width of the creek and they are of modern, simple and elegant shapes. With appropriate setbacks, partially open structures and terraces of various heights along the river, a visual corridor to the river has come into being. The new Shanghai Children's Library and the old building complex and the Jing'an International Centre are all fine examples. At the same time, modern commercial, office and cultural buildings merges with historical buildings to form a long scroll along the waterfront, where the old and the new coexist and integrate - creating a harmonious, comfortable and vibrant urban environment.

What is fascinating is that in the greenery on both sides of the meandering Suzhou Creek, there built new varied service stations and basketball courts, offering a welcoming public destination. The stations are either enclosed or semi-open spaces for the public to stay and rest. The courts, transformed from the originally disused backyard spaces are now a vibrant and refreshing place where walking paths are also placed.

By making full use of spatial resources and creativity, the revitalization of the waterfront environment has been made possible. With further study of the characteristics and values of the Suzhou Creek waterfront, its relationship with the city and the contemporary needs, the waterfront is expected to be a better place with more open high-quality public spaces.

外滩源 33#
BUND 33

整体鸟瞰
Panoramic view

项目地点：黄浦区中山东一路33号
设计单位：华建集团华东建筑设计研究院有限公司
建设单位：上海新黄浦置业股份有限公司
设计时间：2007-2010
建成时间：2012
建筑面积：20000m²
用地面积：22000m²

Location: No. 33, East Zhongshan No. 1 Road, Huangpu District
Designer: East China Architectural Design & Research Institute Co., Ltd., Arcplus Group PLC
Constructor: Shanghai New Huang Pu Real Estate Co., Ltd.
Design Period: 2007-2010
Date of Completion: 2012
Floor Area: 20,000 m²
Site Area: 22,000 m²

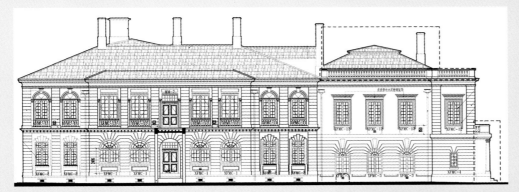
立面图
Elevation

原英国领事馆
Former British Consulate

外滩源33#北临苏州河，项目由外滩源公共绿地及地下空间利用、原英国领事馆主楼和官邸修缮工程、原联合教堂和教会公寓修缮工程、原划船俱乐部历史建筑修缮工程四个部分组成。外滩源33#项目以重现风貌、重塑功能为修缮设计理念。

重现风貌包括：①保护和保留历史建筑，恢复街道空间格局的历史风貌，保留原英领馆主楼及官邸砖木结构；还原官邸连续的敞廊、科林斯柱式和组合柱式的柱廊及精美的装饰；②还原项目区域景观风貌，还原城市绿化空间；③科学规范施工，保护百年古树名木的安全。

重塑功能包括：①完善区域功能，形成文化展示、金融博览、演艺、休闲和艺术体验的场所；②强化区域绿化的公共性和开放性，同苏州河滨水绿地连接；③地下公共空间的建设，确保变电站、架空电线等设施置于地下，并兴建地下车库缓解该区域停车问题。

原划船俱乐部
Former Rowing Club

原联合教堂
Former Union Church

门廊
Porch

Former British Consulate Building and Residence, Bund 33, to the north of the Suzhou Creek mouth, covers an area of about 22,000m². The redevelopment effort consists of four parts: the revamp of the green space along the Bund and its respective underground space; renovation of the former British Consulate building and residence; the Union Church and residence and the Rowing Club house. The property aims to restore its former glory and add new functions to them.

This is done through renovation and restoration effort, ① The historical buildings have reappeared in their former architectural styles while the greenery along the street has reemerged. The original British Consulate main building and the official residence of brick-and-wood structure regain their former grandiosity. The open porch, the Corinthian columns, column colonnade and exquisite decoration of the official residence are all repaired; ② The greenery has been restored; ③ Scientific and standardized construction methods are used to ensure the safety of decade-old trees.

New functions are introduced, including: ① The complex can host cultural, exhibitions, financial forums, performing arts, events and leisure activities; ② The greenery is open to the public and connected to the green space along the Suzhou Creek; ③ Underground spaces are constructed to host substations, overhead cables and other facilities. Underground garages are built to alleviate the problem of parking shortages.

洛克·外滩源
ROCKBUND

Panoramic view of ROCKBUND

兰心大楼
LYCEUM building

项目地点：黄浦区虎丘路146号
设计单位：大卫·奇普菲尔德建筑事务所
建设单位：上海洛克菲勒集团外滩源综合开发有限公司
设计时间：2005-2007
建成时间：2010.5
建筑面积：94000m²
用地面积：7188m²

Location: No. 146, Huqiu Road, Huangpu District
Designer: David Chipperfield Architects
Constructor: Shanghai Bund de Rockefeller Group Master Development Co., Ltd.
Design Period: 2005-2007
Date of Completion: May 2010
Gross Floor Area: 94,000m²
Site Area: 7,188m²

女青年会大楼
YWCA building

新老建筑融合，深厚的历史积淀和卓越的地理位置，使洛克·外滩源项目成为汇聚上海艺术、文化创意、潮流的核心之地。洛克·外滩源历史建筑群历经百年风霜，自2005年起，由洛克外滩接手修缮再造。16年间，相继邀约众多世界顶尖建筑师、设计师参与此项目，知名事务所大卫·奇普菲尔德建筑事务所负责11栋历史建筑的修缮和美丰大楼的整体建筑设计。

In the area, buildings old and new historic buildings are juxtaposed, jointly covering a gross floor area of 94,000 m². The profound historical ambiance and prime location make Rockbund an essential destination of art, cultural events and lifestyle. The buildings are a living memorial of many events in the past century. Rockbund started to undergo a redevelopment in 2005. In the past 16 years, many leading architects were invited to this project. The famous firm, David Chipperfield Architects, was commissioned to renovate 11 heritage buildings and the overall architectural design of the former Mei Feng Bank building.

安培洋行
AMPIRE building

苏州河篇·历史建筑激活 SUZHOU CREEK · RENOVATION OF HISTORICAL BUILDINGS

华侨城苏河湾
OCT SUHE CREEK

原上海总商会
Former Shanghai General Chamber of Commerce

项目地点：静安区河南北路、天潼路、曲阜路、文安路、北苏州路之间
设计单位：华建集团华东建筑设计研究院有限公司、上海联创建筑设计有限公司
合作设计单位：福斯特建筑事务所（规划设计、建筑方案设计）、梁黄顾建筑设计（建筑方案设计）
建设单位：华侨城（上海）置地有限公司
设计时间：2011-2019
建成时间：在建
建筑面积：234091m²（1#）；14995.7m²（41#）；130753.9m²（42#）
用地面积：41984.5m²（1#）；90183.45m²（41#）；20423.4m²（42#）

Location: Block surrounded by North Henan, Tiantong, Qufu, Wen'an and North Suzhou Roads, Jing'an District
Designers: East China Architectural Design & Research Institute Co., Ltd., Arcplus Group PLC, UDG Architectural Design Institute Co., Ltd.
Partners: Foster+Partners (Planning & Architecture scheme design), LWK+Partners (Architecture scheme design)
Constructor: Overseas Chinese Town (Shanghai) Land Co., Ltd.
Design Period: 2011-2019
Date of Completion: Under Construction
Gross Floor Area: 234,091 m² (Block 1), 14,995.7 m² (Block 41), 130,753.9 m² (Block 42)
Site Area: 41,984.5 m² (Block 1), 90,183.45 m² (Block 41), 20,423.4 m² (Block 42)

上海总商会航拍俯视
Aerial view of Former Shanghai General Chamber of Commerce

华侨城苏河湾41街坊原怡和打包厂（俯视）
Suhe Creek project of the OCT, block 41: packing facility of the former Jardine Matheson (aerial view)

原上海总商会大门
Gate of Former Shanghai General Chamber of Commerce

华侨城苏河湾项目涉及苏河湾地区三个街坊，集合商业、酒店、住宅、办公功能，包含原上海市总商会、原中国实业银行仓库等上海市优秀历史建筑，以及原怡和打包厂等三幢保留工业仓储建筑。项目对保留的历史建筑均进行了修缮复原，不仅恢复了建筑外立面的历史风貌，同时对内部建筑空间进行适应性改造、建筑结构加固，并配备现代化的设施。重点保护，局部创新。街坊内新建的商业建筑与历史建筑融合，力求在历史建筑的缝隙中渗透出新的建筑形式，表现出时代变迁下建筑的自然生长与发展，打造繁华城区中的历史文化地标，创造苏州河沿岸优美、舒适、和谐的建筑形态和城市空间。
原中国实业银行货栈大楼由通和洋行设计于1931-1932年，是上海苏河湾一带重要的大型金融货栈滨水建筑之一，2005年10月31日公布为上海市第四批优秀历史建筑。此次提升工程对建筑进行全面修缮，根据历史资料和图纸复原暖灰色水刷石饰面及米黄色黄沙水泥粉刷，并修缮装饰线脚。通过结构加固，重点保护全楼无梁楼盖体系，还原具有历史印记的独特室内空间。恢复历史上西侧主入口，并增设东侧出入口与街坊内其他商业空间及户外庭院相连，重塑滨河景观、恢复街区内部活力。街坊相邻的多层建筑在体量上与优秀历史

华侨城苏河湾41街坊原怡和打包厂
Suhe Creek project of the OCT, block 41: Packing facility of the former Jardine Matheson

The Suhe Creek project of the OCT is composed of Blocks 1, 41 and 42 in the Suhewan area. Embracing commercial, hotel, residential and office functions, it features landmark buildings as the former Shanghai General Chamber of Commerce, the former Industrial Bank of China warehouse, and the packing facility of the former Jardine Matheson. The redevelopment of the property has renovated heritage buildings: It has not only restored the historical facades, but also carried out adaptive transformation of the interiors, strengthened the buildings' structure, and equipped them with modern facilities as per innovation guidelines. At the same time, the new commercial buildings in the neighborhood strive to adopt new architectural forms amidst adjacent historical buildings, showcasing the natural evolution of architecture in a changing era. The old and new buildings will all be cultural landmarks in the prosperous urban area, creating beautiful and harmonious architectural forms and urban space along the Suzhou Creek.

建筑的尺度相衔接，沿河立面根据原址历史建筑图纸复原，延续原有格局及风貌。在局部材质运用上进行创新，体现时代感。改造原有工业立面的窗墙，通过骑楼的建筑形式拓展人行道空间，将历史界面的首层完全向城市打开。新建筑东立面、西立面及北立面采用框架式玻璃幕墙，展现新与旧的对话。

The Industrial Bank of China Warehouse building was designed by the British firm, Palmer and Turner, in 1931-1932. As one of the important large financial warehouses along the Suzhou Creek in Shanghai, it was included in the fourth batch of excellent historical buildings in Shanghai on October 31, 2005. This upgrading project has restored the warm gray brushed stone finish and beige yellow sand cement stucco according to historical documents; repairing decorative moldings. The structural reinforcement focused on preserving the building's flat slab system and restoring the unique interior space with historical markings.

The main entrance on the west side of the building was restored, and an additional entrance on the east side was added to connect with other commercial spaces and outdoor courtyards in the neighborhood. This reshaped the riverfront landscape and restored the internal vitality of the neighborhood. The adjacent multi-storey buildings are in line with the scale of excellent historical buildings in terms of volume, and the façade along the river is restored according to historical drawings, paying tribute to the original pattern and appearance. Innovations are made using local materials to reflect the progress of times. The window wall of the original industrial façade is modified to expand the sidewalk space through the architectural form of arcade, and the first floor of the historical interface is completely opened to the city. The new building adopts framed glass curtain walls on the eastern, western and northern elevations to reflect a dialogue between new and old.

华侨城苏河湾42街坊原中国实业银行货栈改造
Suhe Creek project of the OCT, block 42: former Industrial Bank of China's warehouse

入口
Entrance

内部
Interior

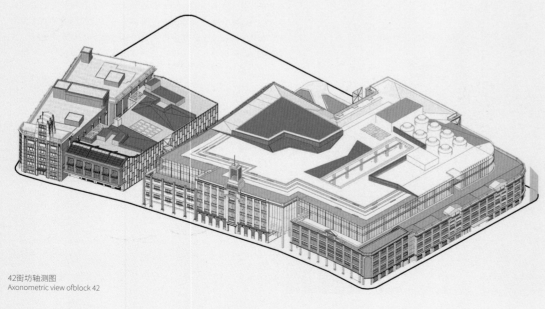

42街坊轴测图
Axonometric view of block 42

东侧人视
The east

四行仓库
JOINT TRUST WAREHOUSE

四行仓库修缮工程
Joint Trust Warehouse

项目地点：静安区光复路1号、21号
设计单位：华建集团上海建筑设计研究院有限公司
建设单位：上海百联集团置业有限公司
设计时间：2014.6
建成时间：2015.12
建筑面积：25550m²
用地面积：4558m²

Location: No. 1 and 21, Guangfu Road, Jing'an District
Designer: Shanghai Architectural Design & Research Institute Co., Ltd., Arcplus Group PLC
Constructor: Shanghai Bailian Real Estate Co., Ltd.
Design Period: June 2014
Date of Completion: December 2015
Gross Floor Area: 25,550m²
Site Area: 4,558m²

四行仓库西墙修缮后弹孔
Bullet Hole Wall (after renovation)

四行仓库 沿苏州河南立面·修缮后
Southern façade along Suzhou Creek (after renovation)

苏州河畔的四行仓库由两座仓库组成，西部的"四行仓库"建于1935年，由通和洋行设计，原高5层，主要为钢筋混凝土无梁楼盖结构体系。这里是1937年淞沪会战中闻名的"四行仓库保卫战"的发生地，如今是市级文物保护单位。此次保护与复原设计以尊重历史、真实性为原则，用多种方法查明西墙在抗战时的炮弹洞口位置，力求准确复原梁柱边的洞口并采取多种创新技术确保建筑安全；拆除后期搭建的七层，六层后退，恢复南北立面历史原貌；恢复原中央通廊特色空间，改作中庭，其西侧设立"抗战纪念馆"，彰显抗战遗址历史意义；其余部分提高舒适度做创意办公等使用。

四行仓库 沿苏州河南立面·修缮前
Southern façade along Suzhou Creek (before renovation)

The Joint Trust Warehouse consists of two parts, and among them the western part dates back to 1935 and is a five-storey structure. Designed by Atkinson & Dallas, Ltd., its main structure embraces reinforced concrete slab-column frame. In the Battle of Shanghai, 1937, the well-known "Lonely Battalion" took a final stand against the Japanese troops before retreat here. Now the warehouse is a Municipality Protected Historic Site in memorial of this battle. History is highly respected and Authenticity is taken as design principle in protection and restoration work. In the battle the west wall of the warehouse was torn by the explosive shells. Several methods have been taken to locate the exact positions of these damaged areas to make sure they are revealed in a way as battle time. At the same time various new techniques have been used to ensure safety of the building. The 7th floor which was an additional construction was removed. The external wall of 6th floor has been moved inward so that the north and south facades could be restored to its original form. The central passage is restored and used as a characteristic atrium. Part of the west side of the atrium is used as a memorial hall to emphasize the significance of this historic site, while the quality of the rest is highly improved to meet the demand for office space.

四行仓库-南立面
Southern entrance (after renovation)

四行仓库-入口中庭-修缮后
Entrance atrium (after renovation)

Southern entrance (before renovation)

四行仓库-入口中庭-修缮前
Entrance atrium (before renovation)

立面图
Elevation

联合仓库远景鸟瞰
Panoramic View of Joint Trust Warehouse

HUANGPU RIVER SUZHOU CREEK

THE WATERFRONT SPACE & ARCHITECTURE IN SHANGHAI

多元融合创新
INNOVATION NEW BUILDINGS

海鸥饭店　Seagull Hotel

中美信托金融大厦
Sino-American T. & F. Tower

上海市少年儿童图书馆新馆
New Library of Shanghai Children's Library

江森自控亚太总部大楼
Johnson Controls Asia Pacific Headquarters Building

海鸥饭店
SEAGULL HOTEL

海鸥饭店改建工程效果图
Sketch of Seagull Hotel Renovation Project

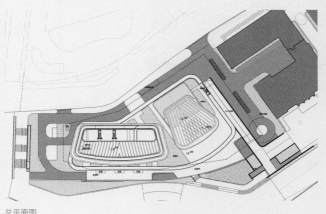

项目地点：虹口区黄浦路60号
设计单位：华建集团华东建筑设计研究院有限公司
建设单位：上海国际海员俱乐部海鸥饭店
设计时间：2018.5-至今
建成时间：在建
建筑面积：29034m²
用地面积：7188m²

Location: No. 60, Huangpu Road
Designer: East China Architectural Design and Research Institute Co., Ltd., Arcplus Group PLC
Constructor: Seagull Hotel of the Shanghai International Seamen's Club
Design Period: May 2018 - Present
Date of Completion: Under Construction
Gross Floor Area: 29,034m²
Site Area: 7,188m²

总平面图
Master plan

原设计始于1981年，1984年建成投入使用，于1999年进行过部分加固改造。此次改造提升力求延续外滩历史建筑，呼应陆家嘴和南、北外滩现代建筑，融合历史、新老共生，彰显历史底蕴，传递时代精神。

项目作为北外滩贯通和综合改造提升工程的组成部分，浦江发展的4.0版，以面向未来的窗口、新老融合的滨水空间、国际视野的文化长廊和带动区域发展的新引擎为目标定位。立足于项目特殊的区位，设计提出既兼顾历史又展望未来的综合解决方案。建筑现代与传统交织，历史与未来辉映，成为北外滩充满当代气息的滨水新地标。

The Seagull Hotel, situated near the mouth of the Suzhou Creek, provides a fantastic view of both sides of the waterway: the fine architecture on the Bund, the Oriental Pearl TV Tower and the breath-taking skyline of Pudong. The hotel, originally designed in 1981, saw the completion of its construction in 1984 and its launch in the same year. In 1999, the hotel was partially reinforced. Its ongoing renovation seeks to blend it into the historical architectural of the Bund and echo the contemporary architecture in Lujiazuiand the North Bund, straddling old style with new taste.

As a part of the grand campaign to revitalize the North Bund in Hongkou, the property seeks to build a future-oriented waterfront where the old meetsthe new. Embracing an international vision, the new waterfront will instill growth momentum for this area. Given the advantageous location of the hotel, a solution was proposed to honor its past and accommodate its future dream.

新老融合的滨水空间
Waterfront space with the new and the old remained

中美信托金融大厦
SINO-AMERICAN T. & F. TOWER

中美信托金融大厦-沿江公寓视角
Fuqiu Apartment of Sino-American T. & F. Tower

沿河立面
Façade facing the creek

项目地点：虹口区天潼路229号
设计单位：华建集团华东建筑设计研究院有限公司
建设单位：上海虹城房地产有限公司
设计时间：1993-2015.4
建成时间：2021.6
建筑面积：122612m²
用地面积：12287m²

Location: No. 229, Tiantong Road, Hongkou District
Designer: East China Architectural Design & Research Institute Co., Ltd., Arcplus Group PLC
Constructor: Shanghai Hongcheng Real Estate Co., Ltd.
Design Period: 1993 - April 2015
Date of Completion: June 2021
Gross Floor Area: 122,612m²
Site Area: 12,287m²

鸟瞰图
Aerial view

项目位于黄浦江苏州河入口北侧，处于上海市国家级和市级文物保护单位建设控制地带。它既是外滩的延伸，又地处苏州河北岸景观走廊。设计以外滩景观的视觉轴线为中心，对项目的体量、高度及立面形态进行综合研究，谋求新建筑与外滩景观的和谐，力求达到保护外滩景观轮廓线的目的。在高度上西高东低，向上海大厦与苏州河沿线风貌保护区致意。

简洁的形体造型使项目谦虚地退隐至后方。从远处看，如同"幕布"一般。搭配现代和色彩对比分明的石材、玻璃，成为区域中较为独特的存在，体现"不矜不伐，虚怀若谷"。

周边优秀历史建筑多为折衷主义风格。项目通过暖灰色系砌体饰面的运用，结合表皮的凹凸变化，形成良好的光影效果，体现建筑的历史感与厚重感。外立面墙面的尺度经过推敲，力求简洁、新颖，富有细部的变化，与周边历史建筑的传统立面尺度取得协调，同上海大厦的

外白渡桥夜景
View on the building and Waibaidu Bridge

Situated to the north of the Suzhou Creek mouth, the property is within the construction control zone that is home to several heritage buildings. Taking the visual axis of the Bund landscape as the center, the volume, height and facade form of the project were comprehensively studied to seek the harmony between the new building and the Bund landscape and to achieve the purpose of protecting the Bund landscape contour. It is high in the west and low in the east, like saluting the Broadway Mansions Hotel Shanghai and the historic conservation areas along Suzhou Creek. The simple shape of the project makes it modestly retreat back. When people look from a distance, it's like a "curtain" on the stage, which doesn't "upstage" nor "compete for attention". At the same time, modern stone and glass with clear color contrast were used to make it a unique but not easy to be ignored building in the area. Thus the goal of "being modest and open-minded" is achieved.

Most of the excellent historical buildings around are in eclectic styles. Through the application of warm gray masonry facing, combined with the concave and convex changes of the surface, a good light and shadow effect is formed, reflecting the sense of history and the massive volume of the building. After careful study, the facade wall was decided to be concise, novel and full of detailed changes, in harmony with the traditional facade scale of the surrounding historical buildings and in contrast with the modesty of Broadway Mansions Hotel Shanghai to highlight its historical features. The open city square space on the second floor makes the whole project "live". It reduces the sense of oppression of the North Suzhou Road, and naturally conforms to the transition from the planar slab-type buildings along the intersection of the Suzhou Creek and the Huangpu River, including the General Post Office Building and the Embankment Building on the west to the modern tower buildings including the Bay Building and the Bund

端庄形成对比，以突现其历史风貌。二层开放的城市广场空间既减少了北苏州路的压迫感，也自然顺应苏州河、黄浦江交汇沿岸建筑，从西侧的邮电大楼、河滨大楼的面状板式至过上海大厦后东侧的海湾大厦、上海滩国际大厦等现代点状形态的过渡，为外滩源景观通廊的端部提供一个新颖的、透气的空间节点。

International Tower on the back east of Broadway Mansions Hotel Shanghai, which provides a novel and breathable space at the Waitanyuan landscape corridor.

沿河远景
Distant view

西侧人视
Two level view from the west

裙房近景
Façade of podium

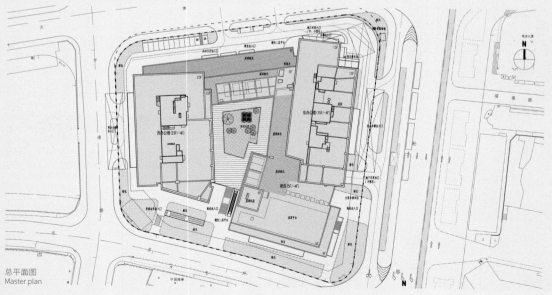

总平面图
Master plan

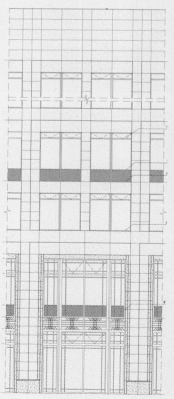

典型墙身大样
Typical wall

上海市少年儿童图书馆新馆
NEW SLTE OF SHANGHAI CHILDREN'S LIBRARY

项目地点：普陀区大渡河路189号正南方向60米长风1号绿地
设计单位：华建集团上海建筑设计研究院有限公司
建设单位：上海少年儿童图书馆
设计时间：2018.8-2021
建成时间：2021
建筑面积：16000m²
用地面积：4558m²

Location: 60 meters to the south of No. 189, Daduhe Road, Changfeng No. 1 Green Space, Putuo District
Designer: Shanghai Architectural Design & Research Institute Co., Ltd., Arcplus Group PLC
Constructor: Shanghai Children's Library
Design Period: August 2018 - 2021
Date of Completion: 2021
Gross Floor Area: 16,000m²
Site Area: 4,558m²

项目与城市公园相衔接，微微起拱的建筑犹如一拱虹桥自然地矗立在安静的湖边，与巨大而五彩缤纷的花园相映成趣，使人联想到莫奈的传世名画《桥》，如孩童般热烈而烂漫。图书馆，作为人类追求智识、成长自我、健全人格的场所，其在抽象意义上，正是一座联系着人类的童年与成熟未来的桥梁，寓意深厚。

The property stretches for several kilometers, adjacent to the Changfeng Park in the north and theSuzhou Creek in the south. With a slightly arched roof, the building contrasts finely with the expansive garden, presenting a lovely view depicted in Monet's *The Japanese Bridge*. The designer envisions the library to be a bridge to a wealth of knowledge and a bridge that spans one's childhood and mature age.

根据不同年龄段的需求，各层公共空间设置各类阶梯式阅读交流空间，同时在屋顶设置充满童趣的儿童乐园。从地面巨大而五彩缤纷的花园，经开放自由的阶梯式阅读学习空间，到达漂浮于花园之上的天空乐园，形成一种螺旋上升的动线系统。整个过程犹如通天巴别塔一般，从大地走向天空，仿佛通往智慧的阶梯，使人充满愉悦与幸福。少年儿童代表人类的未来，犹如初生的树林一般，朝气蓬勃，代表无限的活力与可能性。设计对"树林"这一形态进行抽象提炼，形成简洁有趣的室内空间肌理，结合灿烂的阳光，在室内形成错落有致的效果，营造出"阅读森林"的空间气氛，与室外的绿地花园，达到内外兼修，逻辑一致的效果。

To meet the needs of readers of different ages, various stepped reading and common spaces are set in public spaces on each floor, and the children's park full of children's interests is set on the roof. From the huge and colorful garden on the ground, people can pass through the open and free stepped reading and learning spaces, and reach the sky paradise floating above the garden, which is a spiral moving upward traffic line system. The whole process is like the Babel Tower, moving from the earth to the sky, which makes people full of joy and happiness. It is a ladder leading to wisdom and a palace to create self-accomplishment. Children are the future of mankind, just like a forest full of new trees - full of vigor and vitality, representing infinite vitality and possibilities. In the design scheme, the form of "forest" is extracted to produce a simple and interesting indoor space texture. Combined with brilliant sunshine, the indoor space has graded levels, creating a "forest for readers", a fascinating indoor reading space. Together with the garden outside, the beautiful interior and exterior shows a consistent logic.

轴测剖视图
Axonometric sectional view

模型照片（人视点）
Model (eye level perspective)

模型照片（鸟瞰）
Model (bird's eye view)

总平面图
Master plan

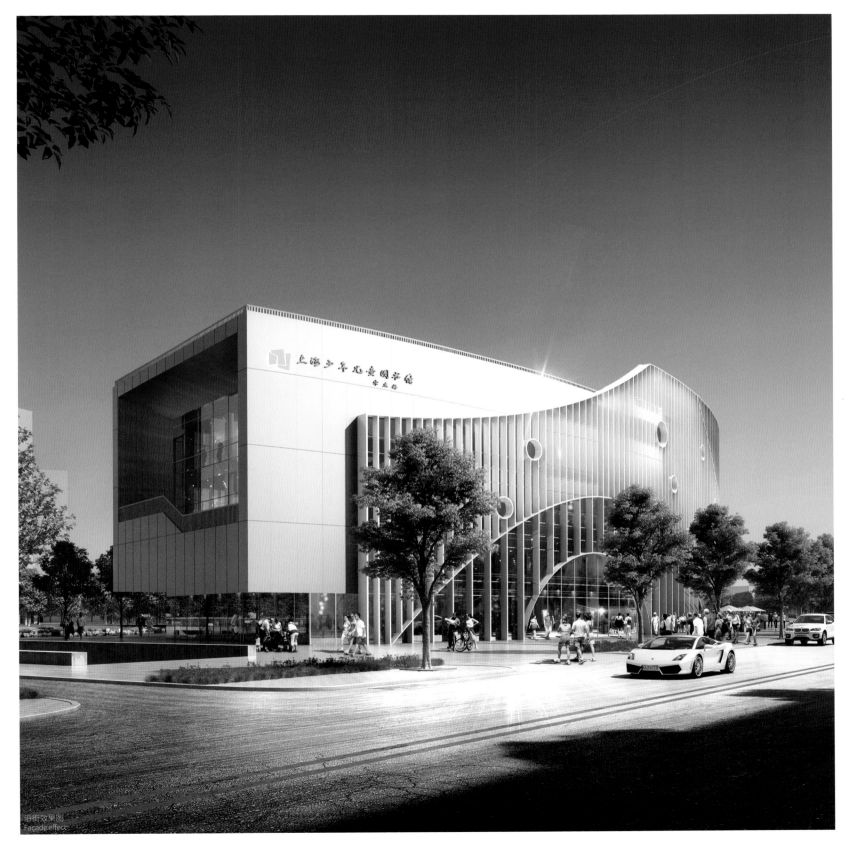

沿街效果图
Façade effect

江森自控亚太总部大楼
JOHNSON CONTROLS ASIA-PACIFIC HEADQUARTERS BUILDING

江森自控亚太总部大楼鸟瞰
Panoramic view of Johnson Controls Asia Pacific Headquarters Building

项目地点：长宁区福泉北路518号
设计单位：Gensler（建筑设计）、华建集团上海建筑设计研究院有限公司（施工图）
合作设计单位：高觅（机电方案）
建设单位：上海新长宁（集团）有限公司
设计时间：2010-2014
建成时间：2017.5
建筑面积：54266m²（9#建筑面积）
用地面积：66661m²（1#-9#总用地面积）

Location: No. 518, North Fuquan Road, Changning District
Designer: Gensler (Architecture design), Shanghai Architectural Design & Research Institute Co., Ltd., Arcplus Group PLC (Construction)
Partners: Glumac (M & E design)
Constructor: Shanghai NCN Co., Ltd.
Design Period: 2010-2014
Date of Completion: May 2017
Gross Floor Area: 54,266m²
Site Area: 66,661m²

建筑细部
Details

项目坐落于虹桥临空园区，毗邻苏州河，四面环水，东侧为配建的公共绿化空间，是临空地区六个主题公园之一。设计专注于建筑健康设计的三个方面：采光与视野、户外空间的可达性和社交空间打造，充分融入绿色、生态、舒适的办公理念，以"水"为内涵，用简洁有力的建筑打造个性、灵动自如的空间。建筑物的整个北翼抬起，规划一条绿道，连接项目地块与河滨公园，联系城市生活与自然，便于人们进入公园，为员工和公众提供延伸的室外环境。

总部建筑的场地非常狭长，由于毗邻机场，建筑高度限制在24米之内。设计师力求在狭长的场地上创建出社区感，同时优化楼面效率。通过建造两个长矩形建筑，狭长的楼板平面有助于引入和最大化自然采光，营造开阔的视野。两个独立的建筑物以相反的方向弯曲，然后连接在一起，形成一个开放动线环绕的大中庭，作为中央社区聚集空间，为跨社区连接、互动和身体活动提供平台，有助于提高生产力和鼓励创新，也契合企业文化。

水面倒影
Reflection

The property sits in the Hongqiao Airport Park, north of Tongxie Road and east of Fuquan Road, adjacent to the Suzhou Creek and surrounded by water. On the east side is a 50m long green space of about 20,000 m²-one of the six theme parks in the airport area.
The building has rich connotation of "water", simple and powerful architectural personality, and flexible exclusive space. Besides, the overall planning is carried out with comprehensive consideration of Hongqiao Fifth Avenue and Impression of the Suzhou Creek, so as to create a culturally-rich, eco-friendly public space. Through the combination of various forms of different points and functions, a continuous public landscape interface is formed from the perspectives of vision, hearing, taste, space, etc.

北翼底层架空
Elevated structure

大楼还应用江森自控先进的智慧建筑产品和技术，诠释未来智能楼宇、城市和社区。项目先后获得三星级绿色建筑设计标识认证、LEEDNC铂金级认证、EDGE设计阶段认证奖优秀高能效设计、2017年度上海绿色建筑贡献奖、2019年度上海市优秀工程设计绿色建筑专业二等奖、RICS年度可持续发展成就冠军、全国绿色建筑创新奖二等奖等多个奖项。

The property has received a series of accreditations and honors: the Three-Star Green Building Design Identification Certification, LEED NC Platinum Certification, EDGE Design Stage Certification Award for Excellent Energy Efficiency Design, the 2017 Shanghai Green Building Contribution Award, Second Prize of the 2019 Shanghai Excellent Engineering Design Green Building Award, the First Prize of the RICS Annual Sustainable Development Achievement Award and the Second Prize of the National Green Building Innovation Award, etc. This property is fully integrated into the green space, ecological and comfortable office.

活动平台鸟瞰
Bird's eye view of platform

悬挑
Overhanging

延伸的绿地
Extended green area

夜景
Night scene

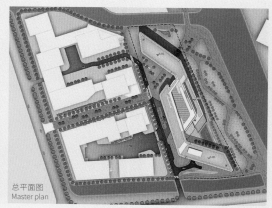
总平面图
Master plan

亲水活动平台
Waterfront platform

HUANGPU RIVER SUZHOU CREEK
THE WATERFRONT SPACE & ARCHITECTURE IN SHANGHAI

活力水岸连接
CONSTRUCTION OF MUNICIPAL FACILITIES

昌平路桥 Changping Road Bridge
苏河桥下空间 Spaces under Suzhou Creek Bridges

昌平路桥
CHANGPING ROAD BRIDGE

昌平路桥全景
Panoramic view of Changping Road Bridge

项目地点：静安区昌平路恒通路跨苏州河处
设计单位：上海市政工程设计研究总院（集团）有限公司
合作设计单位：上海谌稼市政工程咨询有限公司
建设单位：上海公路投资建设发展有限公司
设计时间：2017.12
建成时间：2020.12

Location: Changping Road-Hengtong Road across Suzhouhe, Jing'an District
Designer: Shanghai Municipal Engineering Design Institute (Group) Co., Ltd.
Constructor: Shanghai Chenjia Municipal Engineering Consult Co., Ltd.
Constructor: Shanghai Highway Investment Construction Development Co., Ltd.
Design Period: December 2017
Date of Completion: December 2020

"苏河之眼"
Eyes of Suzhou Creek

工程全长约853米，双向六车道，设计时速40公里/小时，桥梁跨径约60米，采用全钢拱桥一跨过河，上拱的拱轴线，特意高出车行道面的人行道，再配以大跨度的悬臂加劲构件及拱肋间的横向铝板构件，造型宛若明眸凝望着苏州河，因而有着"苏河之眼"的美称。

设计坚持以人为本，在满足桥下通航的同时，最大限度地压低梁高，增加过河的舒适度；通过人行道悬挑让人车完全分离，保障行人安全；人行栏杆均设置二级扶手，即使小孩也能安全、便捷地扶握栏杆驻足远眺；桥头两侧均设置无障碍通道，无障碍通道的栏杆均设置亮化设施。

日景
Day scene

The Changping-Hengtong Road Bridge across the Suzhou Creek is an 853-meter-long, all-steel arch bridge that crosses the river in a 60-meter span. Given the shape of the arch and its reflection in water, the bridge is dubbed "the Eye of the Suzhou Creek" that overlooks the waterway below. With six motor lanes, the bridge can accommodate traffic speed of 40 km per hour.

The bridge enables residents of this area to cross the Suzhou Creek without detour. It is a beautiful sight across the waterway and a structure that affords pedestrians lovely views of both banks while crossing the river.

While satisfying the navigation needs, the design lowers the height of the girders to make crossing a comfortable experience. The sidewalks completely separates the pedestrians from vehicles to ensure safety; the pedestrian handrails are arranged at two levels, one high and one low—tailored for children. On both ends of the bridge are barrier-free accesses where the railings contain lighting facilities.

夜景
Night scene

苏州河桥下空间
SPACE UNDER SUZHOU CREEK BRIDGES

苏州河中环桥下空间
Spaces under Suzhou Creek bridge

桥下篮球场
Basketball court under the bridge

苏河桥下空间（中环苏州河立交）
Space under Middle Ring Road Bridge over the Suzhou Creek

项目地点：长宁区中环苏州河立交桥下
设计单位：上海翡世景观设计咨询有限公司
建设单位：上海市长宁区建设和管理委员会
设计时间：2020
建成时间：2020年（一期）
建筑面积：380m²（驿站）
用地面积：约3700m²（一期）

Location: Under the Suzhou Creek Bridge on the Middle Ring Road, Changning District
Designer: Fish Design
Constructor: Shanghai Changning District Construction and Administration Committee
Design Period: 2020
Date of Completion: 2020 (Phase I)
Gross Floor Area: 380 m² (Post Station)
Site Area: about 37,000 m² (Phase I)

苏州河中环桥下空间的区域交通便利，沿河自然景观优势显著。更新方案通过复合利用桥下空间，提升慢行系统通达性，优化景观和生态三大基本策略，结合苏州河贯通工程，将该区域打造成苏州河特色滨水空间节点。设计方案以象征自由、青春与热爱的"火烈鸟"为主题，采用"火龙果色"作为主色调，打破原本稍显灰暗的高架桥下地带，也与苏州河滨水空间采用的黄蓝相间的步道配色相得益彰。

设计方案综合考虑地区各类公共服务设施和市政设施需求，并结合桥下空间的特点，设置运动场地、公共绿地、苏州河休闲驿站、市政配套设施等功能，实现土地复合利用。休闲驿站结合高架桥柱设置了通往二层平台的旋转楼梯，可在二层平台获得远眺苏州河景观的视觉效果。在景观设计中，设计方案着重考虑将绿地打造成具有公共空间属性的活动场地，并充分运用雨水花园等技术措施减少地表径流，强化绿地的生态功能。

以"火龙果色"为主色调
Eye-catching dragon fruit-color

球场与步道
Courses and trails

The space under the Suzhou Creek Bridge on the Middle Ring Road is known for its convenient road network and impressive waterfront natural scenes. Before the revitalization campaign, it was closed to the public. The renovation scheme follows a three-prong strategy: holistic use of the under-bridge space, improving the accessibility of the slow traffic corridor, and improving the landscape and ecology. As a part of the Suzhou Creek waterfront revitalization project, the area has been crafted to be a dictinctive space. The theme of the design scheme is the flamingo, a bird that symbolizes freedom, youth and love. The flamboyant Pitaya color has enlivened the under-bridge space in gray color, and also matches the yellow and blue color scheme of the waterfront walkway.

The needs of various public service facilities and municipal facilities in the area were considered in the design scheme. Combined with the characteristics of the space under the bridge, sports venues, public green space, leisure spots on Suzhou Creek, municipal facilities and other functions were provided so as to realize composite use of the land. The leisure spots are equipped with revolving staircases leading to the second floor platform in combination with the overpass columns, which can obtain the visual effect of overlooking the Suzhou Creek landscape from the second floor platform. The landscape design focuses on making the green spaces into public activity sites, with rainwater gardens and other technical measures made sufficient use to reduce surface run off and strengthen the ecological function of the green space.

Lovely girl in the mirror

Swing under the bridge

苏河桥下空间（凯旋路、古北路桥）
Space under Suzhou Creek Bridges (Kaixuan Road and Gubei Road Bridges)

项目地点：长宁区凯旋路、古北路苏州河桥下
设计单位：卅吞团队、上海林同炎李国豪土建工程咨询有限公司
建设单位：上海市长宁区建设和管理委员会
设计时间：2018
建成时间：2021

Location: Under Kaixuan Road and Gubei Road Bridges
Designer: SATUN, Lin Tung-Yen & LiGuo-Hao Consultants Shanghai Co., Ltd.
Constructor: Shanghai Changning District Construction and Administration Committee
Design Period: 2018
Date of Completion: 2021

以柠檬、圆形、黄色为设计元素
Lemon, circle, yellow as the design elements

古北路桥和凯旋路桥作为苏州河沿线长宁段的桥下空间更新试点，通过微更新手段对桥下空间进行精细化利用，改善空间割裂的现状，重塑城市、街区关系，并提升城市开放空间的活力和品质。更新方案借用"趟马路"的概念把人们带回旧时一种慢慢走的轻松姿态体验苏州河。"趟"（dang）和"糖"（tang）在上海话中同音，以糖为主题，在凯旋路、古北路等一系列跨河引桥的桥下空间中，以柠檬、圆形、黄色为设计元素，打造艺术亭、桥下剧场、儿童乐园等兼顾实用性的公共艺术品，创造成富有冲击力的"糖果盒子"。古北路桥下空间以西瓜、三角形、红色为设计元素，并融入故事墙、彩色装饰墙和可休憩的绿化带，为市民漫步苏州河带来了崭新的体验方式。

The spaces beneath the Gubei Road Bridge and Kaixuan Road Bridge over the Suzhou Creek are chosen for the pilot projects of the renovation of under-bridge spaces in the Changning section of the Creek. The reuse effort were carried out with micro renovation means, which include using fragmented spaces, redefining the neighborhoods' relationship with the city, and enhancing the quality of open spaces. In the renovation scheme, the concept of "walking leisurely on the road " was borrowed to bring people back to the old time and experience Suzhou Creek with a relaxed attitude of walking slowly". The Chinese pronunciation of "walking leisurely" mentioned above is the same as that of "sweet". Therefore, with the theme of sweet, various kinds of candy boxes are hidden under a number of river-crossing approach bridges on Suzhou Creek, such as Kaixuan Road and Gubei Road, etc., which adds leisure and interesting "sweet" to life. The design elements of lemon, circle and yellow were used for the space under Kaixuan Road Bridge, and the under-bridge space was built into a powerful "candy box" accommodating public artworks with practicability such as art pavilion, under-bridge theater and children's paradise, etc. Watermelon, triangle and red were used as design elements for the space under Gubei Road Bridge, where story wall, colorful decorative wall and recreational green belt are integrated, affording residents a new way to savor the stroll along the Suzhou Creek.

凯旋路桥下空间
Space under the Kaixuan Road bridge

古北路桥下空间
Space under the Gubei Road bridge

古北路
Gubei Road

以西瓜、三角形、红色为设计元素
Watermelon, triangle, red as the design elements

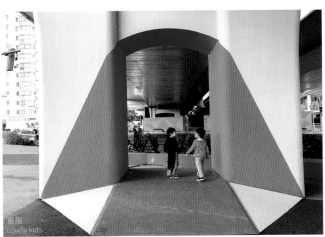
童趣
Lovely kids

设计·师说
MASTERS' WORDS

从标志性到公共性与日常性：
上海城市空间转型的新维度
支文军

千年积淀，百年开埠，上海拥有丰富的人文积淀和长年的富庶繁华。改革开放后，上海再次成为与国际接轨的桥头堡。20世纪90年代以来，房地产市场模式的引入在上海引发了高强度的城市开发。高歌猛进的城市建设过程塑造了上海标志性的天际线。2000年代，全球城市竞争白热化，许多国际大都市先后启动新一轮城市更新计划以吸引资源、提升国际地位，如伦敦市的中央活动区（Central Activities Zone）规划，东京都核心区的连锁再开发项目等。在中国"新常态"和全球城市竞争升级的背景下，上海进入城市建设的转型期，由增量开发逐渐过渡到存量更新。

《上海市城市总体规划（2017—2035年）》提出建设"全球卓越城市"的愿景。存量背景下，城市的卓越不仅在于资源、资本与人力的扩张，也在于用优化现存的资源反哺人民，创造更加美好的生活。我们关注在"全球卓越城市"的愿景下，上海在城市发展思维、空间转型、治理模式、公共空间建设等领域所展开的探索与尝试，探讨上海城市公共价值导向的变迁、城市公共空间如何成为空间转型与城市治理的一种有效途径，及在公共性、日常性等公共空间建设方面的思考与实践。

1 上海城市发展的新阶段与新思维
1.1 快速增量与存量更新

大拆大建阶段，宏大叙事是建设活动的价值导向，繁华商业街和高层建筑群成为上海最具标志性的符号。2000年代，城市开发已趋于土地等自然资源底线，城市更新问题逐渐受到关注。2010年，上海中心城区已基本完成建设，在市域总面积中的占比超过同级国际都市如纽约、东京。2014年，建设用地总规模已超过2020的最高规划目标，促使上海强力遏制增量开发，提出"建设用地零增长，工业用地减量化"的主张。2015年《上海城市更新实施办法》颁布，2016年上海市城市总体规划（2017—2035年）上报，标志着上海城市建设正式进入存量土地利用、城市空间有机更新的新阶段。

从人本角度看，以经济利益为先的城市空间剧烈变动虽是城市建设的必经阶段，但不够重视居民日常需求、忽视地方归属感的做法不利于可持续发展。增量时代，对城市的评价和排名单一地指向经济，如全球化和世界城市研究小组的评价标准。但在存量时代，根据"全球实力城市指数"，评估综合竞争力要看经济影响力、科技创新能力、文化活力、社会治理水平、基础设施与城市服务以及环境与可持续性六个方面。

从增量到存量的转变促使资源和专业智慧逐年流向城市更新项目。如2016年推出的"城市更新四大行动计划"和《上海市15分钟社区生活圈规划导则》，旨在构建居民的慢行生活圈，至今已推广到全市领域。

最具代表性的城市级更新项目是黄浦江45公里全线贯通工程。生产岸线转型为复合功能岸线，滨江空间和沿江社区建立连接，串联建筑遗产、基础设施、人工景观、交通节点、中心城区和龙头产业，所有市民都能在此享受多样的公共资源。习总书记2019年在杨浦滨江考察时提出"人民城市人民建，人民城市为人民"。顺应存量时代，上海城市发展遵循"多目标、多元化和综合化"的原则，更多注重如何改善中观和微观的城市空间品质，近年来成果颇丰。

1.2 城市管理与城市治理

改革开放以来，中国经济改革以对外开放的市场机制连接国际经济，以行政分权方式实现经济自由化和市场化。这种制度让中国近40年来都保持了高速发展。增量时代的上海城市建设以自上而下的管理模式为主：20世纪80年代，政府筹措资金改造旧房、整治居住环境；90年代，政府开始与开发商公私合作，建设城市中心区，尝试典型旧式里弄区的功能产权置换，开发浦东新区；21世纪初的十年，历史遗产得到重视，政府注重保护街区、修复文化建筑，又致力将工业遗产转化为产业基地。

在编制远期规划时，政府与职能部门整合学术机构、咨询机构等专业力量进行城市问题的对策研究与建议报告，针对如土地利用、公共交通、公共服务、城市更新促进政策、改善民生这种战略层面的重要问题。在2010年以前，城市建设由政府部门决策，开发商等大型资本、学术咨询机构等专业力量共同实施影响。

2010年代，上海逐步进入城市更新阶段，城市更新项目实践越来越丰富，精细化城市治理作为一种理念和模式越来越被重视。"治理"的概念始于公共管理学，它是一种以协调而非支配为基础的过程，依赖公私部门持续的相互作用。专业人员、社会组织、个体市民和多样化的资金来源合作完成微更新项目，逐渐形成行之有效的工作机制，即专家负责项目导向、基层政府全程参与、社区组织负责协调、设计团队进行"陪伴式"设计。

项目整合空间、行政、商业和创意资源平衡地实现各利益主体的空间与社会生活愿景，完成以设计为媒介的城市治理。物质空间和协同机制的更新自下而上地带动整体观念的更新，推动城市治理模式向多样性发展。自上而下、统筹规划的城市管理与自下而上、共建共享的城市治理优势互补、双线并行，成为上海城市发展的新模式。

2 公共空间作为上海空间转型与城市治理的一种途径

2.1 公共空间的社会意义及其价值

列斐伏尔（Henri Lefebvre）认为空间既是社会生活的容器，也是社会关系的重要组成。在快速推进城镇化建设的时期，城市空间资源存在分配不均的现象：城市生活丰富性被同质化的建筑空间损伤，街道公共性被封闭的围墙破坏等。同时，城市管理向度稍显单一，如项目各方能动性发挥不足，居民主体感缺失等。存量发展时期要解决这些城市问题，建设具有真正公共性与日常性的公共空间是一种重要的途径。

存量更新时期，公共空间更新与再生产正在优化空间资源。首先，新时期公共空间的实践体现人民至上的理念，将更多的资源与民众共享，城市景观从奇观化、同质化向日常化、个性化转向；其次，城市需要合理的秩序与良好的可达性，各功能片区需要公共空间来连接，小规模、渐进式的微更新项目正织补城市系统，整合闲置资源，成为公共性节点。

公共空间更新与再生产也在丰富城市治理模式，其一，公共空间的微更新促使项目各方探索新合作机制，带动制度灵活化；其二，新公共空间为市民创造精神家园，提高参与感，降低沟通成本。如"一江一河"是上海最丰富的公共空间资源，经过综合整治，黄浦江、苏州河沿岸已是多元复合的城市客厅，市民能充分享受上海发展的成果。

公共空间既承载市民的日常生活需求和文化认同，是城市空间的象征，也是城市治理中权力实践和资源互动的载体。建设公共性与日常性的城市空间在优化空间资源和丰富治理模式两方面都能促进上海向更高的城市能级发展。

2.2 公共空间的公共性价值导向

存量时代，上海的城市公共空间正向真正的公共性转型。玛格丽特·科恩将公共性分为三个核心维度：所有权、连接性、主体间性。马修·卡莫纳将主体间性扩展为"功能"与"观念"，提出"所有权、进入和使用是保证空间公共性的三个方面"。

公共性在空间层面上体现为连接性，城市公共空间能衔接城市系统中本互不兼容的功能，将更多空间资源开放给民众。如上海民生码头地块原本是断裂的工业空间，借2015上海城市空间艺术季的契机重新构筑滨水平台，连接码头与废弃用地。步行可达性的提高吸引了更多的市民进入此地，激发出地区的活力。封闭孤立的历史建筑改造为公共空间，现代产业得以进驻，从而与周边环境产生连接。如杨浦滨江的工业遗产毛麻仓库、船厂等被创意性地改造为当代艺术场馆，成为滨江游憩体系的节点。

在治理层面上，公共性则代表建设过程得到市民充分的参与，赋予市民应有的空间所有权。公共性的形成在于市民离开私人领域，为共同问题开展讨论和行动。政府将行政资源与资金汇聚到引发公共生活的场所，居民前期踊跃提出需求和意见，设计师提炼综合性改造方案，让居民能灵活使用空间。如南京东路街道的贵州西里弄社区改造中，居民的能动性被可见成果逐步激发，他们的监督、维护和利用让社区共享中心等节点切实带动住区整体环境提升。又如苏州河畔的凯旋路桥以"糖"为主题设置公共艺术品和桥下剧场，市民在此展开了丰富的活动。

最终，公共性意味着建设成果能被所有人感知与使用。克莱尔·库珀·马库斯(Clare Cooper Marcus)认为人性场所必须鼓励不同群体使用它。城市公共空间的建设成果通过公共活动的总结真正被广大市民感知进而触发活动。如上海城市空间艺术季融合城市空间更新和国际文化事件，吸引市民亲身体会黄浦江公共空间的发展成果。为方便市民理解，呈现形式从二维的图纸与文本升级到三维的公共空间与艺术作品，如杨浦滨江的起重机被包裹上条纹的《起重机的对角线》，廊柱被涂上黑色方块的《自由方块——方块花园》。展览走到室外，融入滨水空间中。

大到城市级公共空间的塑造、公共事件的参与，小到社区微空间的共同营造，公共性是塑造城市体验感与人民参与感的关键。

2.3 公共空间的日常性价值导向

存量时代，日常生活的微需求正在城市建设的宏大框架下不断生长。日常性公共空间概念见于玛格丽特·克劳福德的《日常性都市主义》一书。她认为日常性的公共空间模糊了公共与私人的边界，其暧昧甚至矛盾的生活方式构成新的公共空间逻辑。而爱德华·索亚用"第三空间"的概念诠释日常性空间的特质，第三空间是物质空间与想象的空间表征的创造性结合。日常性的公共空间既包含物质性的日常生活与功能，又指向人与人基于精神连接的多样交往活动，其上，还有政治、经营等公共事务构成的活动领域，它们在当下都是多元、复杂、有时矛盾却和谐共存的。

在空间层面上，日常性即是融汇不同类型的活动，融合公共与私密，提供多样化的公共空间与服务设施。2016年，上海规划与自然资源局成立城市公共空间设计促进中心，负责"行走上海——城市空间微更新计划"。计划起初着眼老旧小区，创造最贴近生活的公共空间和服务设施，而后拓展类型，如桥下空间、街道空间等。这些空间既有实用性功能，也建立了居民与城市的联系。

日常性的公共空间在得到充分利用时能重建市民的精神连接。高品质的公共空间不仅提供日常游憩的功能，也延续了城市的文化记忆。弗洛姆认为人渴望同他人联合的欲望是人类行为最强的动力之一。苏州河畔的九子公园不仅是社区周边日常运动、休闲、亲水的绿地，还以上海传统的九种健身项目吸引爱好者来此交流切磋，触发爱好与精神的共鸣。

在治理层面上，日常性体现在公共事务处置的灵活与多元化，构建自生自发的微秩序。城市更新项目形成了从设计主体-业主关系展开的利益共同网络，因地制宜地采用不同建设模式。

如如中环桥下空间经过设计，整合了慢行系统、公共服务设施、市政基础设施、城市绿地，还为居民创造了火龙果色的运动场地，成为苏州河滨水空间的节点。设计走出图纸，走向真实的生活。促进中心发起的试点项目也带动了基层自主探索的更新实践，使更大范围的居民能受益。

在"构建15分钟慢行生活圈"的目标下，日常性的公共空间成为生活空间的重要节点，塑造了城市的在地感与市民的归属感。

3 迈向卓越的全球城市愿景下的上海实践

3.1 聚点成网的城市公共空间

城市公共空间是市民可以自由光顾与活动的地方，也是享受城市生活、体认城市风情、彰显城市个性、领略城市魅力之所在，它的更新不仅与居民的生活需求息息相关，也与城市综合形象的树立紧密相连。黄浦江与苏州河是上海的母亲河，为了将一江一河打造成为具有全球影响力的世界级滨水区，政府主导开展了相关编制工作，整体规划沿江45公里的城市滨水空间，突出"布局因地制宜的多样化公共空间，构建整体活动场所空间序列，滨江空间融入城市公共空间网络，提升滨水空间场所感与活跃度、艺术性与舒适性，完善人本关怀服务设施体系"，不仅对沿岸主要活动区域的不同层次公共空间体系作出由点及面的详细指导，也将其纳入城市网格化公共活动场所序列中。

2017年底，"滨江45公里岸线贯通"工程完成，实现了杨浦大桥至徐浦大桥滨江核心全线空间贯通开放。在徐汇西岸滨水空间以"一带一河多节点"的结构布局，注重文化与产业的结合。文化地标如龙美术馆、西岸美术馆、西岸艺术示范区与文化事件不仅有效地串联起物质空间，更在城市公共形象上创造出浓郁的在地文化。杨浦滨江作为重要区域节点，有大量工业遗存。

在更新过程中，统筹旧工厂、仓库、码头等历史遗留场所纳入公共空间与交通系统的考量，通过设置滨江慢行栈道、跑步道、自行车道，新旧建筑与景观巧妙地串联，并与滨江区域外部的交通系统有效连接；景观节点与公共建筑如杨树浦驿站、绿之丘、毛麻仓库遗址等将沿岸空间划分成若干小空间，极具层次感与趣味性，逐渐成为上海城市公共形象的新名片。

目前，"一江一河"沿岸的相关规划仍在有序推进中，在更大范围打造全方位贯通可达、景观优美、层次丰富、设施完善的公共空间系统，将上海的母亲河沿岸打造成共享开放的城市公共客厅。而这种以点带面、聚点成网的城市公共空间系统构建也将从黄浦江、苏州河出发，逐步扩展向其他区域。

3.2 基于生活圈的居住社区公共空间

生活圈的概念最早起源于日本在1965年提出的"广域生活圈"，作为合理安排基础设施和公共服务设施、促进地区均衡发展的规划策略，其任务是从居民行为规律出发，解决不同层面和不同频率的生活需求问题。

上海是国内较早明确提出"15分钟社区生活圈"概念的城市，《上海市15分钟社区生活圈规划导则》中明确提出打造社区生活基本单元，在15分钟步行可达范围内，配备生活所需的基本服务功能与公共活动空间，在居住、就业、出行、服务、休闲五大方面提出具体准则与导引。事实上，社区生活圈不仅是城市居民基本生存与社交的主要空间载体，也是构建城市综合系统的基本单元，其倡导丰富的公共空间类型、关注小型公共空间建设，同时从尺度与边界层面强化公共空间的可达性与使用性，使之回归日常化和生活职能，从而激发社区的自发性与社会性公共活动。

目前，15分钟生活圈的规划和建设试点工作在一些社区与街道的实施中已经取得一些成果。例如，苏州河沿线普陀段利用绿道将公园、绿地、河道与社区串联起来，连点成线贯通可居可游的绿色开放空间，成为附近居民散步休闲的好去处；浦东新区的南码头路街道利用小区荒废的边角地，结合地块条件和周边居民健身习惯，设有运动广场、笼式球场、康体设施等，营造了"花在眼前，绿在身边，健身就在家门口"的开放型参与式环境；陆家嘴街道福山路将沿街一处健身场馆与其邻近的人行道、绿化带、建筑前步行空间整合，不仅在建筑前铺设健身步道，还优化和更新了座椅、路灯、花坛等设施，规范机动车停放，使原本使用效率低下的空间转变为一个面向居民、上班族、过路人的小型聚集场所，在隐秘而温暖地增加社区居民同他人及周边环境互动的同时，提升城市街道的活力。

随着上海市"15分钟社区生活圈"理念的推进与落地，以社区公共空间为基础的更新将会释放更多公共生活的可能性，在提升居民居住体验与幸福感的同时创造更多人的连接，从而营造更富魅力的幸福人文之城。

3.3 作为触媒的城市公共艺术

"公共艺术"作为一个广义概念,很长时间内在城市生活中被看作是新的观念与事物,不仅指艺术作品陈列于城市公共空间,更主要是这些艺术创作紧贴生活,为市民所享用,与城市生活自然交融,其从20世纪80年代开始参与到后工业化城市的复兴计划中。近年来,随着"公共艺术"内涵的不断扩大,其形式也从以观看为主的雕塑转向公共沉浸式的互动行为,并作为"触媒"激发着城市的公共交往与文化潜力。

上海在迈向卓越全球城市的过程中,以打造更富魅力、更有温度的人文城市作为重要目标。上海世博会后,各种艺术活动发展迅猛,从城市空间艺术季的品牌活动到上海当代艺术博物馆、西岸美术馆、浦东美术馆等国际知名顶级艺术展馆IP的落成,不仅体现出政府在规划中对于艺术兴城的重视,也反映出民众对公共艺术活动的参与不断加深。

2015年起,双年展制的城市空间艺术季以展览和公共活动为主要内容,作为上海在迈向卓越全球城市中的重要文化品牌走进市民生活。在积极树立城市形象的同时,赋能公共空间,将建筑空间改造、地区更新、视觉艺术与社会公众参与完整融合在一起,生动演绎了当代上海的城市更新理念,其主题不仅对接世界城市发展的前问题,也将对上海典型空间的讨论置于全球城市网络中,以公共艺术的方式促成"政府—市民—艺术家—专家学者"等多方主体关于城市公共性的对话。

此外,民间与商业机构也是上海城市公共艺术主体的重要组成。这些团队不仅挖掘了城市与建筑所承载的文化内涵,将其以展览与活动的方式与市民的文化公共生活相连接,还致力于用跨界的方式探索城市与乡村公共空间的艺术可能性,并寻求艺术作品与环境之间的联系。

在不同尺度与主导方式下,城市公共艺术以触媒的方式介入城市空间,在与环境不断融合形成新场域的同时,也成为普通人相遇与交往的起点,从而为城市赋能,实现跨领域、多元化的融合,予人愉悦和温暖。

3.4 精细化城市治理下的有机更新

城市如同一个生命体,其有自身持续的新陈代谢与持续更新能力,在发展过程中因各地区独特的自然生态环境、经济生产方式与人文历史发展轨迹形成特有的空间和文化基因。在建设方法上多表现出小规模、渐进式、持恒性特征,这种发展模式为"城市有机更新"。中国城市在经历30年大规模扩张阶段后,逐渐从粗放型增量发展模式转向有机更新发展模式,也对城市精细化治理提出了新的要求。2015年《上海市城市更新实施办法》提出:"对本市建成区城市空间形态和功能进行可持续改善的建设活动",强调"以人为本,激发都市活力""实现协调、可持续的有机更新"。其中,地域环境、文化特色与建筑风格基因是当下上海城市精细化治理中的关注重点,也集中体现在侧重文化历史遗产的建筑项目中。

绿之丘项目作为杨浦滨江贯通工程中的重要节点,在工业建筑改造与城市更新层面均融合了多重思考。作为曾经的烟草仓库,一度被列入拆除名单,设计团队从城市空间角度出发,以"丘陵城市"为概念,在原有结构框架中插入一层层绿色托盘和聚落状的小房子,与既有体量相互交织,不仅连通了城市腹地与滨水空间,也在杨浦滨江的重要节点处打造了集城市公共交通、绿地、服务为一体的城市多功能复合体。此外,艺仓美术馆、1862时尚艺术中心、上海油罐艺术中心、M50文化创意园等案例都通过设计的方式将新功能置入原有空间,巧妙地实现了城市空间形态与功能可持续改善的有机更新,在保留上海特色历史建筑基因的同时实现了现代功能的转译。

在《上海市城市总体规划(2017—2035年)》背景之下,以精细化治理为手段的城市有机更新成为上海未来城市发展的必然趋势。深入了解城市的历史,留存城市的建筑风格基因,并在城市新生命周期里注入新的动能,也将成为规划师与建筑师的长期任务。

3.5 基于人本尺度与体验的设计创新

2015年,上海市人民政府印发《上海市城市更新实施办法》,标志着城市更新逐步从以城市用地再开发为主导的粗放型建设,朝向关注小微空间品质提升与功能复兴的精细化治理转型。城市更新主要面向社会日常生活领域，从人本尺度与体验出发，通过改造、修缮和局部整治等手段，用设计创新的方式对小规模的公共空间或景观进行功能完善、品质提升。

其中，从日常性出发的针灸式介入设计是本轮创新的一大特色。位于黄浦区的贵州西里弄社区面临物质性老化严重，住户生活空间狭小，设施破损等问题。由于居民老龄化严重，经济能力相对较低，社区共识性下降，自我更新能力日益减弱。在此条件下，设计团队从居民日常活动与体验出发，以最小干预的方式，针对现有要素进行重组，植入4个里弄入口空间、3+1处基础环境、3处环卫设施和1800平方米的社区共享客厅与空中书房。不仅为居民提供了必要的生活空间，增强居交流与互动机会，也提升了公共生活的精神品质，加强场所领域的归属感受。

此外，对于公共性的关注与积极融入城市生活是本轮创新的又一大亮点。位于黄浦区苏州河南岸的九子公园是城市密集居住区中一个难得的公共空间，在改造之前是一个典型的有围墙的公园且独立于周边城市空间。设计团队从城市设计尺度入手，基于其与周边功能、交通的连接关系提出"城市公园开放化"的总体策略，打开公园围墙，通过设置亲水平台使之与滨河公共空间融为一体，不仅丰富了游园体验，也将公园与苏州河更紧密地连接。与此同时，设计用绿化地形的方式与城市空间形成弱限定，呈现出积极融入城市的姿态，也使得公园生活成为市民日常性活动的一部分。

从人本尺度与体验出发的设计创新聚焦于以生活需求为导向的公共空间品质提升，在特征与趋势方面呈现出以日常性出发的针灸式介入与以公共性出发的对城市层面的思考与关注，通过细致与人性化的精准介入极大提升居住体验，实现物质与精神双重层面的共建共享。

4 结语

上海进入存量发展阶段后已经有了阶段性成果，如建设用地规模已经得到有效遏制，城市更新计划都在有条不紊地实施，公共空间数量和密度稳步增长，公共活动频次和质量也在上升，各区各社区的探索热情有增无减。《上海市城市更新条例》已自2021年9月1日起正式施行，这标志着上海城市更新进入更深入、系统的阶段，丰富公共性与日常性的内涵。这部法规为城市更新工作提供了有力法治保障。上海将由此更好地践行"人民城市"的理念，提升城市的人居环境、精神文化、公共艺术等方面的软实力。在这个节点回望以往的建设问题与建设成果，同时展望2035，有助于加深对城市空间与治理模式的理解。上海曾有远东第一大都市的荣光，创造过市场经济发展的奇迹，相信2035年的上海也将焕然一新，真正成为创新之城、人文之城、生态之城，市民能充分体验各个尺度、丰富多彩的城市空间。

FROM ICONIC LANDMARKS TO PUBLIC SPACE FOR EVERYDAY USE: A NEW DIMENSION OF SHANGHAI'S URBAN SPACE TRANSFORMATION

Zhi Wenjun

Shanghai has rich cultural and historicalresources. Since China embraced reforms and opening-up in the late 1970s, Shanghai has once again become a pioneer to introduce international practices into this country. Since the 1990s, the introduction of the real estate market operation model has triggered sweeping urban development in Shanghai. In the 2000s, as global competition for cities heated up, many international metropolises launched new rounds of urban regeneration plans to attract resources and enhance their international status, such as the London's Central Activities Zone plan and the chain redevelopment project in the heart of the Tokyo metropolitan area. On the background of China's "New Normal" and the escalation of global urban competition,Shanghai has moved from increment to the regeneration of the inventory.

The Shanghai Urban Master Plan (2017-2035) proposes building a "global city of excellence". Given a large inventory of existing buildings,a city of excellence lies not only in the increase of resources, capital and human resources, but also in optimizing existing resources to serve the people and create a better life. Under the "global city of excellence" vision, this article focuses on Shanghai's exploration and experimentation in urban development strategies, spatial transformation, governance models, and public space construction.It explores the changes in the orientation of public values in Shanghai, how urban public space has been an effective way of spatial transformation and urban governance, and what we can do about constructing public spaces for the service of their public and everyday nature.

1 New stage and strategies of Shanghai urban development

1.1 Rapid increment development and regeneration of inventory

During the phase of wholesale demolition to make room for the new, sweeping change was the guiding principle for construction activities, and bustling commercial streets and high-rise buildings became the iconic symbols of Shanghai. In the 2000s, urban development tended to overtax natural resources, such as land, and the issue of urban renewal gradually garnered attention. 2010 saw the completion of the construction of Shanghai's city center, whose downtown area was more extensive than that of its international counterparts such as New York and Tokyo. In 2014, the total scale of construction land exceeded the highest planning target of 2020, prompting Shanghai to strongly curb increment development and propose "zero growth of construction land and reduction of industrial land". In 2015, the Shanghai Urban Renewal Implementation Measures were promulgated. In 2016, the Shanghai Urban Master Plan (2017-2035) was submitted to the central government, marking a new stage in Shanghai's urban construction with land inventory and organic renewal of urban spaces.

From a people-oriented perspective, although dramatic spatial changes in cities that prioritize economic interests are necessary for urban construction, insufficient attention to the residents' daily needs and neglect of the local belonging sense are not conducive to sustainable development. In the incremental era, cities are evaluated and ranked singularly by their economy, such as the GaWC's criteria. In the inventory era, however, according to the Global Cities of Power Index (GPCI), overall competitiveness is measured by six indicators: economic influence, technological innovation, cultural vitality, social governance, infrastructure and urban services, and environment and sustainability.

The shift from the increment to the regeneration has prompted a flow of resources and professional wisdom to urban renewal projects year by year. For example, the "Four Action Plans for Urban Renewal" and the "Shanghai 15-Minute Community Life Circle Planning Guidelines", launched in 2016, aim to build a slow-traffic living circle for residents and have been extended to the entire city so far.

The most representative citywide regeneration project is the Revitalization of the 45-kilometer Huangpu River Waterfront Campaign. The industry-oriented riverfront has been transformed into a composite one, connecting the waterfront and the communities along the river, linking architectural heritage, infrastructure, artificial landscape, transportation hubs, city center and leading enterprises, where all citizens can benefit from public resources. General Secretary Xi Jinping proposed that "the people build people's city, people's city serve the people" during his visit to the Yangpu section of Huangpu River waterfront in 2019. In response to the new era, Shanghai's urban development follows the principle of "multi-objective, diversification and integration", focusing more on improving the quality of urban space at the meson and micro levels, which has been very fruitful in recent years.

1.2 Urban management and governance

Since China embraced reforms and opening up, her reform programs have been based on an open market mechanism that connects the international economy to the outside world and on administrative decentralization to liberalize and marketize the economy. This system has allowed China to maintain a high rate of development for nearly 40 years. In the 1980s, Chinese government raised funds to renovate old houses and improve the living environment. In the 1990s, the government began to work with developers in a public-private partnership to build the city center, replace the functional property rights of typical old lane neighborhoods, and develop the Pudong New Area. In the first decade of the 21st century, historical heritage was recognized, and the government focused on preserving traditional neighborhoods, restoring heritage buildings, and transforming industrial heritage into bases for creative industries.

While working on long-term planning, the government and functional departments instruct academic and consulting institutions and other professional forces to conduct countermeasure research and recommendation reports on urban issues, targeting essential issues at the strategic level such as land use, public transportation, public services, urban renewal promotion policies, and improvement of people's livelihood. Before 2010, government departments decided on urban construction, and exert influence jointly with academic and consulting institutions and rich stakeholders, such as e.g., developers).

In the 2010s, Shanghai gradually entered the period of urban renewal, and the practice of urban renewal projects are increasingly abundant, and refined urban governance has become more important as a concept and mode. It is a process based on coordination rather than domination, relying on the continuous interaction between the public and private sectors. Professionals, social organizations, individual citizens, and diverse funding sources collaborate to complete micro-renewal projects, gradually forming an effective working mechanism in which experts are responsible for project orientation, district governments are involved throughout the process, community organizations are responsible for coordination and design teams provide corresponding designs.

The projects integrate spatial, administrative, commercial and creative resources to realize the spatial and social life of each stakeholder in a balanced manner and complete urban governance by means of design. The renewal of physical space and collaborative mechanisms leads to the renewal of overall concepts from the bottom up, and leads the development of urban governance modes to a diversified manner. The top-down, integrated planning urban management and bottom-up, shared urban governance complement each other and work in parallel, becoming a new mode of Shanghai's urban development.

2 Public Space as a Way of Spatial Repurposing and Urban Governance in Shanghai

2.1 The social significance and value of public spaces

According to Henri Lefebvre, space is both a container of social life and an important component of social relations. In the period of rapid urbanization, urban space resources are unevenly distributed: the richness of urban life is damaged by homogenized architectural spaces, and closed fences destroy the public nature of the neighborhood. At the same time, the urban management direction is a little bit single, such as the lack of initiative of project parties and the lack of residents' sense of subjectivity. In order to solve these urban problems in the period of inventory development, building public spaces with real publicness and everydayness is a necessary solution.

In the period of repurposing the inventory, public space renewal and reproduction optimize spatial resources. Firstly, repurposing public space in the new era reflects the concept of putting people first and sharing more resources with the people. Urban landscape is turning from exotic and homogeneous styles to familiar and personalized ones. Secondly, the city needs proper order and good accessibility. Areas with different functions need to be connected by public spaces. Progressive micro-renewal projects can reinstitute and complement the urban system, integrating unused resources and serving public interest.

Renewal and regeneration of public spaces are enhancing modes of urban governance. First, the micro-renewal of public space can prompt all parties in the project to explore new cooperation mechanisms, leading to institutional flexibility. Second, new public spaces create a spiritual home for citizens, increasing their willingness to participate and reducing communication costs. For example, the River and Creek is the richest public space resource in Shanghai. After comprehensive improvement, the Huangpu River and the Suzhou Creek have become a multi-faceted civic center where citizens can fully enjoy the fruits of Shanghai's development.

Public spaces accommodate citizens' daily needs and evidence their cultural identification. They are symbols of urban spaces and carriers of urban governance and resource flows. Building public and everyday-use urban spaces can facilitate Shanghai to a higher level in optimizing spatial resources and enriching governance models.

2.2 Value orientation of public space's publicness

In the urban renewal era, Shanghai's urban public spaces gain a genuinely public nature. Margaret Kohn divides the public nature into three core dimensions: ownership, connectivity, and inter-subjectivity. Matthew Camona expands inter-subjectivity into "function" and "perception" and suggests that "ownership, access, and use are the three aspects that ensure the public nature of space."

In the spatial dimension, public nature means connectivity. Urban public space can connect otherwise incompatible functions in the urban system and make more spatial resources open to the public. For example, Minsheng Wharf was a fractured industrial space, but the 2017 Shanghai Urban Space Art Festival redeveloped the waterfront terrace, connecting the wharf to the abandoned plot. The improved pedestrian accessibility has attracted more people to the site and given activities to the area. Closed and isolated old buildings are transformed into public spaces, allowing modern industries to move in and connect with their surroundings. For example, the woolen and linen fabrics warehouse and the shipyards along the Yangpu section of Huangpu River waterfront have been creatively transformed into venues of contemporary art and recreation.

In the governance dimension, publicness means that the building process beckons the full participation of citizens, giving them the ownership sense of the space they deserve. Such a public nature arises when citizens leave their private space to discuss and take actions on common issues. The government brings together administrative resources and funds to the venues where public life occurs, while residents enthusiastically state their needs and opinions, and designers devise comprehensive renovation plans that allow residents to use the space flexibly. For example, in the renovation of the West Guizhou neighborhood on East Nanjing Road, the residents' initiative was gradually unleashed by the visible results, and their supervision, maintenance and utilization allowed the community sharing center to drive the overall environmental improvement of the neighborhood effectively. Another example is the Kaixuan Road Bridge over the Suzhou Creek, where public artworks and a theater under the bridge have been set up to address the theme of sugar. Citizens can hold a variety of activities in these spaces.

Ultimately, publicness means that the construction can be perceived and used by everyone. Clare Cooper Marcus argues that a humanistic place must encourage the use by different groups. The results of the construction of urban public space are truly felt by the general public through public events, which in turn give incentive to activities. For example, Shanghai Urban Space Art Season combines urban space renewal with international cultural events, beckoning the public to appreciate the result of Huangpu River redevelopment. To make it more understandable, the exhibition features not just two-dimensional drawings and texts, but three-dimensional public space and artworks, such as "Diagonals of the Crane" in which the crane on Yangpu Riverfront is wrapped with stripes, and "Free Square - Square Garden", in which the corridors are fixed with black squares. The exhibitions are held outdoors and at waterfronts.

From shaping city-level public spaces and participating in public events to co-creating micro-spaces in communities, the public nature is essential to the sense of urban experience and people's participation.

2.3 Everyday value orientation of public spaces

In the urban renewal era, the micro-needs of everyday life surge in the grand framework of urban construction. The concept of everyday public space can be found in Margaret Crawford's book, Everyday Urbanism (2008). She argues that everyday public space has no clear boundary between public and private spaces, and the ambiguous, even contradictory way of life constitutes a new logic of public space. Edward W. Soja, on the other hand, uses the concept of "third space" to explain the quality of everyday space, which is a creative combination of material space and imaginary spatial representations. Everyday public space contains material life, functions and a variety of people-to-people interactions based on spiritual connections. In addition, there is also a sphere of activities that includes politics and commerce, which are multi-faceted, complex, sometimes contradictory but coexisting at present.

In terms of space, everyday-ness is about integrating different types of activities, blending the public and the private, and providing a variety of public spaces and services. In 2016, the Shanghai Planning and Natural Resources Bureau established the Urban Public Space Design Promotion Center to handle the "Strolling Shanghai - Urban Space Micro-Renewal Program". It initially focuses on old neighborhoods, hoping to create public spaces and services that are relevant to life, and later expands to include under-bridge spaces and neighborhood spaces. These spaces serve both a practical function and provide a connection between residents and the city.

Everyday public spaces, when fully utilized, can rebuild the emotional bonds between them and the citizens. High-quality public spaces not only provide daily recreation, but also preserve the cultural memory of the city. According to Erich Fromm, the desire to associate with others is one of the strongest drivers of human behaviors. The Jiuzi Park along the Suzhou Creek is not only a green space for daily sports and recreation purposes, or simply as a waterfront green space of the community, but also attracts enthusiasts with the nine traditional games of Shanghai, being a gathering place of like-minded people.

In the governance dimension, everydayness means the flexibility and diversity of public affairs disposal and the construction of a self-generated and spontaneous micro-order. Urban renewal projects form a shared network of interests unfolding from the designer-owner relationship. Different construction models are adopted according to local conditions.

For example, the space under the Central Ring Bridge has been designed to integrate a slow-traffic system, public service facilities, municipal infrastructure, urban green space. It also creates a dragon fruit-colored sports field for residents, becoming a Suzhou Creek waterfront node. The design comes out of the drawing board and goes to real life. The pilot projects initiated by the Promotion Center have also led to more independent renewal projects so that a broader range of residents can benefit.

Under the goal of "building a 15-minute slow-traffic life circle", the daily public space becomes an important node of living space, shaping the city's sense of place and citizens' sense of belonging.

3 Shanghai's practice under the vision of a global city of excellence

3.1 Networked urban public spaces

Urban public space is a place where citizens can freely visit and move around, and it is also a place to enjoy city life, recognize city style, reveal city personality, and appreciate city charm. Its renewal is not only closely related to residents' living needs, but also closely linked to the establishment of the city's comprehensive image. The Huangpu River and Suzhou Creek are mother rivers of Shanghai. In order to make the River and the Creek a world-class waterfront area with global influence, the government has carried out the planning. It makes overall planning of 45 kilometers of urban waterfront along the river, highlighting the "layout of diverse public space according to local conditions, the construction of the overall activity space sequence, the riverfront space into the city public space network". The plan not only provides detailed guidance on different levels of the public space system in the main activity areas along the river, but also incorporates the system into the sequence of public activity places in the urban grid.

At the end of 2017, the Huangpu River Waterfront Revitalization Project was completed. The 45-km waterfront from the Yangpu Bridge to the Xupu Bridge was opened to the public. The Xuhui section of the waterfront embraces a layout of "one belt, one river and multiple facilities". It features a range of cultural facilities converted from former industrial ones. Cultural landmarks such as the Long Museum of Art, West Bund Art Museum, West Bund Art Demonstration Zone and the cultural events not only link up different sites on the spaces, but also create a strong local culture on the basis of the city's public image. The Yangpu section waterfront also boasts a wealth of industrial heritage sites.

During the renewal, old factories, warehouses, wharves and other historical sites were considered an overall public space and transportation system. By setting up a waterfront slow traffic trails, the old and new buildings are fully linked with the landscape and effectively connected with the external transportation system of the waterfront; landscape facilities and public buildings such as Yangshupu Post, Green Mound and Mao Ma warehouse site divide the coastal space into several small spaces, which are highly layered and interesting, and gradually become the new business card of Shanghai's urban public image.

At present, the planning of the River and Creek is still in progress. It aims to create a public space system with great accessibility, beautiful landscapes, perfect facilities on a larger scale, and to make Shanghai's waterfront into an open civic spaces of the city. The public space system will extend beyond the waterfronts of the Huangpu River and the Suzhou Creek.

3.2 Living circle-based public space of living community

The concept of the living circle first originated in Japan in 1965 as the "wide sphere of life". As a planning strategy to rationalize infrastructure and public service facilities and promote the balanced development of the area, and its mission is to solve the problem of different levels and frequencies of living needs from the law of residents' behavior.

Shanghai is the first city in China to clearly put forward the concept of the "15-Minute Community Life Circle", and the "Shanghai 15-Minute Community Life Circle Planning Guidelines" clearly proposes to create a basic unit of community living, equipped with basic service functions and public activity space required for living within 15 minutes walking distance, with specific guidelines in five major areas: living, employment, travel, services and leisure. In fact, the community living circle is not only the city's most important element. In fact, the community living circle is not only the main space for the basic survival and interaction of urban residents, but also the basic unit to build an integrated urban system. It advocates rich types of public spaces, focuses on the construction of small public spaces, and strengthens the accessibility and usability of public spaces in terms of scale and boundaries, so that they can return to daily life and living functions, thus stimulating the spontaneous and social public activities of the community.

At present, the pilot planning and construction of the 15-minute life circle have already achieved positive results in some communities. For example, the Putuo section of the Suzhou Creek uses greenways to link up parks, green areas, waterways, and communities, connecting points into a green open space that can be lived in and traveled, becoming a good place for nearby residents to walk and relax. Nanmatou Road Sub-district in Pudong New Area uses the abandoned specks of land it, combining the conditions of the plot and the fitness habits of the surrounding residents, with sports plazas, cage basket courts, and recreation and sports facilities, etc., creating an open and participatory environment of "flowers in front of you, greenery around you, fitness at your doorstep."

Fushan Road of the Lujiazui Sub-district integrated a gym along the street into the fabric made up of the sidewalks, green belt and pedestrian space in front of it, not only paves a running trail in front of the building, but also optimizes and updating the seats, street lights, flower beds and other facilities, regulating the parking spaces of motor vehicles. The originally ineffectively used space has transformed into a small gathering place for residents, office workers and passers-by. The move has facilitated the interaction of community residents and the surrounding environment in an unnoticeable way while enhancing the community's vitality.

With the progress of the implementation of the "15-Minute Community Life Circle" in Shanghai, the renewal based on community public space will release more possibilities of public life and create more connections while enhancing residents' living experience and happiness, thus creating a more charming city of happiness and humanity.

3.3 Urban public art as a catalyst

As a broad concept, "public art" has been seen as a new thing in urban life for a long time, not only in terms of artworks displayed in urban public spaces, but also in terms of relevance of art to life. When art works are enjoyed by citizens, they are naturally integrated into urban life. Since the 1980s, it has been involved in the revitalization of post-industrial cities. In recent years, as the connotations of "public art" have expanded, its forms have shifted from viewing-based sculpture to interactive. Immersion forms, acting as a "catalyst" to stimulate public interaction and unleash the cultural potential of the city.

In the process of becoming an outstanding global city, Shanghai has made it an important goal to create a more charming, culturally rich city. After the Shanghai World Expo, various art events have emerged, from the brand events of the Urban Space Art Season to the completion of internationally renowned art exhibition venues such as the Shanghai Museum of Contemporary Art, the West Bund Museum and the Museum of Art Pudong. They not only reflect the importance the government attaches to art in planning for a prosperous city, but also reflect the public's increasing enthusiasm in the participation of art events.

Since 2015, the Biennale System's Urban Space Art Season, with exhibitions and public events as its main features, has attracted the attention of citizens as an important cultural event while Shanghai marches towards a remarkable global city. While actively establishing the city's image, it gives public space new functions and combines architectural space renovation, regional renewal, visual art with public participation comprehensively, vividly interpreting the concept of urban renewal in contemporary Shanghai. Its themes are relevant to the aforementioned issues of world urban development and bring Shanghai's typical space to a global context, facilitating the dialogue between the government, citizens, artists and experts.

In addition, private and commercial institutions are also significant contributors to urban public art in Shanghai. These institutions not only explore the cultural connotations encoded in cities and buildings and expose them to the public through exhibitions and events, but also devote themselves to exploring the artistic possibilities of urban and rural public spaces in a cross-border manner and seek connections between artworks and the environment.

Under different guiding principles and on different scales, urban public art can influence urban spaces as a catalyst. While continuously merging with the environment to form new fields, it also becomes the starting point for ordinary people to meet and interact with each other, thus empowering the city and achieving cross-disciplinary and diversified integration, giving people much pleasure and warmth.

3.4 Organic Renewal under Refined Urban Governance

Like a living organism, a city has its own continuous metabolism and renewal ability. In its development process, a city forms unique spatial and cultural genes due to its unique natural ecological environment, economic production paradigm, and evolution trajectory. In terms of construction paradigm, "organic urban renewal" is primarily small-scale, gradual and persistent. After 40 years of large-scale expansion, Chinese cities are gradually shifting from a not cost-effective incremental development model to an organic renewal one. This has raised new requirement on refined urban management. The "Implementation Measures of Shanghai Urban Renewal" issued in 2015 suggests that construction should be carried out for sustainable improvement of the urban spatial form and function of Shanghai's built-up area." The document emphasizes that construction activities should be "people-oriented" and "stimulate urban vitality" and "achieve coordinated and sustainable organic renewal". Among them, the regional environment, cultural characteristics, and architectural legacy are the focuses in Shanghai's urban governance, which is also reflected in the projects related to cultural and architectural heritage.

As an important project of Yangpu District's waterfront revitalization project, Green Hill reflects multi-faceted thinking on the reuse of industrial buildings and urban renewal. To rescue this former tobacco warehouse that once appeared on the demolition list, the design team came up with the idea of a "hilly city". The team arranges layers of green pallets and colony-like units into the original frame, intertwining them with the existing volumes. The building not only connects the rest of urban areas with the waterfront, but also serves as a multifunctional urban complex at an important location of the Yangpu section of the Huangpu River waterfront; it serves the purposes of public transportation, green space and service station along the Yangpu waterfront. In addition, the Shanghai Modern Art Museum, 1862 Fashion Art Center, Tank Shanghai, and M50 Creative Park are designed new functions, cleverly realizing the organic renewal of urban spatial form and sustainable improvement of functions. They have realized function changes while preserving the architectural legacy of Shanghai.

Under the Shanghai Urban Master Plan (2017-2035), organic urban renewal by means of refined governance has become a trend for Shanghai's future urban development. Planners and architects have to undertake a long-term task to understand the city's history, preserve its architectural legacy, and give new momentum to the city in its new life cycle.

3.5 Design innovation based on human scale and experience

In 2015, the Shanghai Municipal Government issued the Implementation Measures for Urban Renewal in Shanghai. According to the document, the renewal model would no longer be construction driven by blind urban land redevelopment, but refined governance focusing on quality improvement and functional revitalization of small and micro spaces. Urban renewal mainly focuses on the improvement of the daily lives of the society. It encourages practitioners to pay adequate attention to human scale and experience, and improve the function and quality of small public spaces or landscapes by means of innovative design, repairs and partial remodeling in the transformation process.

One feature of this round of innovation is the pinpointed interventions deriving from the everyday nature. The West Guizhou Road Residential Quarter in the Huangpu District is plagued by serious aging, small living spaces of the residents, and worn-out facilities. Due to the serious aging of the residents and their relatively low economic capacity, it is difficult to rally consensus in the community and the community's ability of self-renewal is declining. Under these conditions, the design team reorganized the existing elements in a minimal intervention manner while paying attention to the residents' daily activities and experiences. Its members created four spaces at lane entrances, 3+1 basic environment facilities, three sanitation facilities and a 1,800-square-meter shared space and a "midair study". These facilities provide necessary living spaces for residents, enhance communication and interaction opportunities, and also improve the quality of public spiritual life and strengthen the feeling of belonging in the neighborhood.

In addition, the attention to public nature and active integration into urban life is another highlight of this round of innovation. The Jiuzi Park, located on the south bank of Suzhou Creek in Huangpu District, is a rare public space in a dense urban residential area. The design team started from the urban design scale. It proposed a general strategy of "building a wall-less urban park" based on its relationship with the nearby functional and traffic facilities, removing the walls of the park and integrating it with the riverfront space by setting up a lookout platform. The latter only enriches the visiting experience of the park but also connects it with the Suzhou Creek more closely. At the same time, the design uses the green topography to form a weak confine of the surrounding urban space, actively engaging itself into the city and making visits to the park an integral part of citizens' daily activities.

The design innovation based on human scale and experience focuses on improving the quality of public spaces to meet their needs for living. In terms of characteristics and trends, the innovation applies pinpointed intervention—in line with the everyday nature—and macroscopic thinking at the citywide level--from the public nature. It can greatly improve the living experience through meticulous, humanized precise intervention, realizing the joint construction and sharing on both material and spiritual aspects.

4 Conclusion

Shanghai has already achieved good results since it entered the urban renewal phase. The acreage of its construction land has been effectively put under control. All urban renewal plans are all being implemented: The quantity and density of public space is steadily increasing; the frequency and quality of public activities are both improving, and the enthusiasm of communities in all districts for exploration is gaining increasing momentum. Since coming into force on September 1, 2021, the Shanghai Urban Renewal Regulations have ushered in a new stage of systematic urban renewal in Shanghai, serving to open up access to the waterfront. This regulation provides a strong rule of law guarantee for urban renewal work. This will enable Shanghai to implement the concept of "People's City" better and enhance the city's soft power in living environment, culture and public art. At this time point, looking back at past construction issues and achievements while looking ahead to 2035 will help us better understand urban space and governance models. Shanghai was once hailed as the foremost metropolis in the Far East and has created the miraculous market economy development. I believe that in 2035 Shanghai will also be renewed and truly become a city of innovation, humanity and ecology; its citizens can relish fantastic urban spaces of all kinds.

附录
APPENDIXES

设计·师说作者介绍
THE AUTHORS OF MASTERS' WORDS

38 涤岸之兴，向水而生
章明
同济大学建筑设计研究院（集团）有限公司原作设计工作室主持建筑师
同济大学建筑与城市规划学院景观学系系主任、教授、博导

LIVE AND THRIVE BY THE WATER
ZHANG MING
PRINCIPAL ARCHITECT OF ORIGINAL DESIGN STUDIO OF TONGJI ARCHITECTURAL DESIGN (GROUP) CO., LTD.
PROFESSOR AND DOCTORAL SUPERVISOR, HEAD OF DEPARTMENT OF LANDSCAPE, COLLEGE OF ARCHITECTURE AND URBAN PLANNING, TONGJI UNIVERSITY

102 黄浦江畔 工业历史的当代风景
柳亦春
大舍建筑设计事务所主持建筑师、创始合伙人
同济大学建筑与城市规划学院和东南大学建筑学院客座教授

HUANGPU WATERFRONT:
A LIVING INCARNATION OF SHANGHAI'S INDUSTRIAL PAST
LIU YICHUN
PRINCIPAL ARCHITECT AND FOUNDING PARTNER OF ATELIER DESHAUS
VISITING PROFESSOR, SCHOOL OF ARCHITECTURE AND URBAN PLANNING, TONGJI UNIVERSITY
AND SCHOOL OF ARCHITECTURE, SOUTHEAST UNIVERSITY

166 人类与水的共存
隈研吾
隈研吾建筑都市设计事务所创始人
东京大学教授

COEXISTENCE OF HUMAN AND WATER
KENGO KUMA
FOUNDING PARTNER OF KENGO KUMA AND ASSOCIATES
PROFESSOR, UNIVERSITY OF TOKYO

170 新建筑开发赋能
沈迪
全国工程勘察设计大师
华建集团总建筑师

FORGING A NEW TRAIL IN ARCHITECTURAL DEVELOPMENT
SHEN DI
NATIONAL ENGINEERING SURVEY AND DESIGN MASTER
CHIEF ARCHITECT OF ARCPLUS GROUP PLC

254 传承 延续
让·努维尔
Ateliers Jean Nouvel 工作室创始人
阿卡汗奖、2005 年沃尔夫艺术奖和 2008 年普利兹克奖获得者

PERPETUATION AND ENHANCEMENT:
JEAN NOUVEL
FOUNDER OF ATELIERS JEAN NOUVEL STUDIO
THE WINNER OF THE AKHAN PRIZE, THE 2005 WOLF PRIZE FOR ART AND THE 2008 PRITZKER PRIZE

258 为公共性赋形：景观基础设施与浦江两岸贯通
张斌
致正建筑工作室主持建筑师
同济大学建筑与城市规划学院客座教授

BUILT FOR THE PUBLIC: THE LANDSCAPE INFRASTRUCTURE AND THE HUANGPU RIVER WATERFRONT REVITALIZATION
ZHANG BIN
PRINCIPAL ARCHITECT OF ATELIER Z+
VISITING PROFESSOR, SCHOOL OF ARCHITECTURE AND URBAN PLANNING, TONGJI UNIVERSITY

288 舒适人居，趣味城市
吴蔚
gmp 建筑师事务所中国区合伙人

MAKING CITIES FUN, GIVING PEOPLE COMFORT
WU WEI
PARTNER OF GMP ARCHITECTS IN CHINA

326 苏州河的新生
唐玉恩
全国工程勘察设计大师
华建集团上海建筑设计研究院有限公司资深总建筑师

THE REBIRTH OF THE SUZHOU CREEK
TANG YU'EN
NATIONAL ENGINEERING SURVEY AND DESIGN MASTER
SENIOR CHIEF ARCHITECT OF SHANGHAI ARCHITECTURAL DESIGN AND RESEARCH INSTITUTE CO., LTD

364 从标志性到公共性与日常性：上海城市空间转型的新维度
支文军
同济大学建筑与城市规划学院教授
《时代建筑》杂志主编

FROM ICONIC LANDMARKS TO PUBLIC SPACE FOR EVERYDAY USE:
A NEW DIMENSION OF SHANGHAI'S URBAN SPACE TRANSFORMATION
ZHI WENJUN
PROFESSOR, COLLEGE OF ARCHITECTURE AND URBAN PLANNING, TONGJI UNIVERSITY
EDITOR-IN-CHIEF OF TIME ARCHITECTURE MAGAZINE

参编单位
Institutions Involved

主编单位：	上海市规划和自然资源局
参编单位：	华建集团上海建筑科创中心
	上海市城市规划设计研究院
协编单位：	

上海市测绘院
上海市城建档案馆
上海城市公共空间设计促进中心
浦东新区规划和自然资源局
宝山区规划和自然资源局
杨浦区规划和自然资源局
虹口区规划和自然资源局
黄浦区规划和自然资源局
徐汇区规划和自然资源局
闵行区规划和自然资源局
静安区规划和自然资源局
普陀区规划和自然资源局
长宁区规划和自然资源局
上海地产（集团）有限公司
上海杨浦滨江投资开发有限公司
上海北外滩（集团）有限公司
上海外滩投资开发（集团）有限公司
上海陆家嘴（集团）有限公司
上海东岸投资（集团）有限公司
上海西岸开发（集团）有限公司
上海宝建（集团）有限公司
上海闵行城市建设投资开发有限公司
上海苏河湾（集团）有限公司
上海长风投资（集团）有限公司
上海新长宁（集团）有限公司

Organizer: Shanghai Municipal Bureau of Planning and Natural Resources
Executors: Shanghai Archi-scientific Creation Center, ARCPLUS GROUP PLC
Shanghai urban planning and Design Institute
Supporters:

Supporters: Shanghai Surveying and Mapping Institute
Shanghai Urban Construction Archives
Shanghai Design and Promotion Center for Urban Public Space
Pudong New Area Planning and Natural Resources Bureau
Baoshan District Planning and Natural Resources Bureau
Yangpu District Planning and Natural Resources Bureau
Hongkou District Planning and Natural Resources Bureau
Huangpu District Planning and Natural Resources Bureau
Xuhui District Planning and Natural Resources Bureau
Minhang District Planning and Natural Resources Bureau
Jing'an District Planning and Natural Resources Bureau
Putuo District Planning and Natural Resources Bureau
Changning District Planning and Natural Resources Bureau
Shanghai Land (Group) Co., Ltd.
Shanghai Yangpu Riverside Investment and Development Co., Ltd.
Shanghai North Bund (Group) Co., Ltd.
Shanghai Bund Investment (Group) Co., Ltd.
Shanghai Lujiazui (Group) Co., Ltd.
Shanghai East Bund Investment (Group) Co. Ltd.
Shanghai West Bund Development (Group) Co. Ltd.
Shanghai Baojian (Group) Co., Ltd.
Shanghai Minhang Urban Construction Investment Development Co., Ltd.
Shanghai Suhewan (Group) Co., Ltd.
Shanghai Changfeng Investment Development Co., Ltd.
Shanghai NCN (Group) Co., Ltd.

摄影师
Photographers

奚仁杰、章鱼见筑、战长恒、章勇、陈颢、田方方、钟雪峰、庄哲、刘文毅、克里斯蒂安·加尔、何威、陈晨、陈旸、清筑影像、马库斯·布雷特等。

Xi Renjie, ZY Architectural Photography, Zhan Changheng, Zhang yong, Chen Hao, Tian Fangfang, Zhong Xuefeng, Zhuang Zhe, Liu Wenyi, Christian Gahl, He Wei, Chen Chen, Chen Yang, CreatAR-Images, Marcus Bredt, etc.

附录
APPENDIXES

后记
POSTSCRIPT

　　城市建筑与空间跟人们的衣、食、住、行息息相关，承载人民对美好生活的向往；城市建筑与空间见证历史、现在与未来，承载岁月与文化的厚重。城市建筑与空间之美，是生活之美，文化之美，是城市精神和品格之美。

　　上海一江一河沿岸城市空间从"城市锈带"转向"城市客厅"，涌现了许多具有示范意义的建筑与空间实践案例。这些案例汇集海内外大师名家，引领前沿设计，致力于营造高质量的城市滨水空间，创造高品质的市民生活。上海"海纳百川、追求卓越、开明睿智、大气谦和"的城市精神和"开放、创新、包容"的城市品格，正蕴含在这些新旧融合、多元创新、活力共享的城市建筑与空间之中。

　　本书的编著从资料收集到最终定稿出版持续了一年多的时间，在此衷心感谢各位专家，感谢建设单位、设计单位提供的案例资料，感谢各参编单位的大力支持。希望本书能从更专业的视角向读者展示自新世纪以来一江一河沿岸建筑与空间，传达设计美感与创新，讲述建设者、设计师与社会公众的思考与追求。希望本书能推动更多理念探讨，引领专业实践。期待有更多优秀的探索，百花齐放，百家争鸣，共同塑造更有魅力、更有温度、更具影响力的卓越全球城市。

编者

Urban architecture and spaces are closely related to people's clothing, food, accommodation and transportation, bearing people's aspirations for a better life. As memorials of history, they witness the past, present, and future. Their aesthetics reflect the beauty of life and culture, echoing the spirit and character of the city.

Since the beginning of the 21st century, the urban space along Shanghai's riverside has shifted from an "urban rust belt" to an "urban living room". Many exemplary architectural and spatial practice cases have emerged. These cases bring together masters from home and abroad, showcasing cutting-edge design, dedicated to creating high-quality urban waterfront spaces and creating high-quality civic life. Shanghai's city spirit of "being inclusive, pursuing excellence, enlightenment, wisdom and modesty" coupled with the open, innovative, and inclusive quality of the city are embedded in these urban buildings and spaces. They mix diverse elements old and new, embracing the spirit of sharing.

The compilation of this book took more than a year from data collection to final publication. We would like to sincerely thank all the experts, construction, and design institutions for providing materials, including all the participants for their great support. We hope this book can illustrate the beauty and innovation of buildings and spaces along the Huangpu River and Suzhou Creek in a professional manner, revealing the insights and pursuits of designers, constructors, and the public. We hope that this book will spark conceptual discussions and provide reference for professional practices. We look forward to more excellent cases from practitioners of different schools, so that we can jointly build a charming, people-friendly, and influential global city of excellence.

Editorial Committee

图书在版编目（CIP）数据

　　一江一河：上海城市滨水空间与建筑：汉英对照 / 上海市规划和自然资源局编著 . -- 上海：上海文化出版社 , 2021.12（2023.6 重印）
　　ISBN 978-7-5535-2403-0

　　Ⅰ . ①一 ... Ⅱ . ①上 ... Ⅲ . ①城市景观－景观设计－研究－上海－汉、英 Ⅳ . ① TU-856

　　中国版本图书馆 CIP 数据核字 (2021) 第 197613 号

出 版 人　　姜逸青
责任编辑　　江　岱　郑　梅
装帧设计　　上海纸道

书　名　一江一河：上海城市滨水空间与建筑
　　　　　上海市规划和自然资源局　编著
出　版　上海世纪出版集团　上海文化出版社
地　址　上海市闵行区号景路 159 弄 A 座 2 楼　201101
发　行　上海文艺出版社发行中心
　　　　　上海市闵行区号景路 159 弄 A 座 2 楼 206 室　201101
印　刷　上海雅昌艺术印刷有限公司
开　本　787mm×1092mm　1/12
印　张　32
印　次　2021 年 12 月第 1 版　2023 年 6 月第 2 次印刷
书　号　ISBN 978-7-5535-2403-0/TU·013
定　价　360.00 元

告读者 如发现本书有质量问题请与印刷厂质量科联系 T：021-68798999